The Messina Strait Bridge

A challenge and a dream

Credits and contributors

Editor: Stretto di Messina S.p.A. – Rome, Italy

Authors:

Fabio Brancaleoni
Professor of Structural Engineering, *Roma Tre University* – Italy

Giorgio Diana
Professor of Applied Mechanics, Director of the Wind Tunnel, *Politecnico di Milano* – Italy

Ezio Faccioli
Professor of Earthquake Engineering, *Politecnico di Milano* – Italy

Giuseppe Fiammenghi
CTO *Stretto di Messina S.p.A.* – Italy

Ian P.T. Firth
Director *Flint & Neill Limited* – Great Britain

Niels J. Gimsing
Professor Emeritus – *DTU Technical University of Denmark* – Bridge Designer, Denmark

Michele Jamiolkowski
Professor of Geotechnical Engineering – *Politecnico di Torino* – Italy

Peter Sluszka
Vice President *Ammann & Whitney Consulting Engineers* – NY, USA

Giovanni Solari
Professor of Structural Engineering – *University of Genova* – Italy

Gianluca Valensise
Senior Scientist, Department of Seismology and Tectonophysics *INGV Istituto Nazionale di Geofisica e Vulcanologia* – Italy

Enzo Vullo
Chief of Suspension Bridge Engineering, *Stretto di Messina S.p.A.* – Italy

Additional contributions from: G. Lucangeli, M. Marconi, and E. Micalizio
Stretto di Messina S.p.A. – Rome, Italy

Final editing: I.P.T. Firth, E. Vullo
With contributions of: U. Carletti & Technical Department, Stretto di Messina S.p.A. – Italy

Co-ordination: G. Fiammenghi,
J. Blom, CRC Press/Balkema, Leiden, The Netherlands

Illustrations: Authors/Stretto di Messina S.p.A./sources mentioned in the text

Cover: Marco Kuveiller

The Messina Strait Bridge

A challenge and a dream

Stretto di Messina S.p.A., Rome, Italy

CRC Press is an imprint of the
Taylor & Francis Group, an **informa** business
A BALKEMA BOOK

Typesetting: Vikatan Publishing Solutions Pvt. Ltd. Chennai, India

First published 2010 by CRC Press/Balkema

Published 2019 by CRC Press
Taylor & Francis Group
6000 Broken Sound Parkway NW, Suite 300
Boca Raton, FL 33487-2742

ISBN: 9780415468145 (hbk)
ISBN: 9780367577254 (pbk)
ISBN: 9780429182914 (ebk)

Library of Congress Cataloging-in-Publication Data

The Messina Strait Bridge : A challenge and a dream / by Fabio Brancaleoni ... [et al.]; Stretto di Messina S.p.A.
p. cm.
ISBN 978-0-415-46814-5 (hardcover : alk. paper) 1. Suspension bridges – Italy – Messina, Strait of. 2. Messina, Strait of (Italy) – History. I. Brancaleoni, Fabio. II. Stretto di Messina S.p.A.

TG400.M467 2009
624.2'30945–dc22

2009022132

http://www.taylorandfrancis.com

and the CRC Press Web site at
http://www.crcpress.com

Table of contents

Foreword

A bridge connecting the Mediterranean to Europe

Recent public debate in Italy about large infrastructure projects, and in particular the Messina Strait Bridge, has adopted strong political undertones strangely inclined to underestimate their most evident technical, scientific, strategic and economic advantages. By contrast, no-one questions the strategic relevance of an efficient transportation network for the competitiveness and the social, economic and cultural integration of our country, or at least for its alignment with other important partners within the European Union. Similarly, there is general agreement about the need to improve the network of principal routes in Southern Italy and to achieve a re-balancing of the whole system in favour of rail transportation.

Against this background, the Messina Strait Bridge is expected to drive the modernisation and improvement of transport infrastructure in Southern Italy and to represent the completion of the North-South road and rail corridor that will bring Europe into the heart of the Mediterranean. The priority given by the European Union to the Palermo-Berlin rail line, with the Messina Strait Bridge as an essential hub, reflects the re-discovery of the central role of the Mediterranean as a wide trading area whose growth and development is important for the entire continent.

The Mediterranean is certainly a major resource, but a favourable geographic location is not enough for Southern Italy, particularly for Sicily. The core of the issue regarding infrastructure development and related economic growth lies in enabling our Southern regions to capture the many trade opportunities offered by the Mediterranean area. This is a concrete goal that should be accomplished as soon as possible.

Due to its geographic position in the middle of the Mediterranean, Southern Italy has traditionally played a natural role in the exchanges between Europe and Eastern, North American and North African countries. However, it is presently prevented from acting as a major logistic platform by the lack of adequate transport infrastructure.

The Messina Strait is also a hindrance to smooth circulation of people and goods. A permanent rail connection would be the best way to meet this need and satisfy the demand for a more efficient and modern connection between Sicily and the continent. Completion of the bridge would also facilitate the development of other infrastructure projects in Southern Italy within the framework of Government strategies in this regard. Only a permanent connection to the continent will allow Sicilian ports to become competitive with other Mediterranean ports. A stronger system would be obtained by strengthening and exploiting the capabilities of different transportation modes, air, sea and land, giving true value to the so often asserted concept of "intermodality".

Apart from its strategic value, the strong contribution that this project can give to reviving domestic economy should not be overlooked, with an estimated impact on Italy's GDP of above six billion Euro. Positive economic effects will be engendered immediately, establishing good conditions for overcoming the currently unfavourable situation. Accordingly, the Bridge is once again a high Government priority, allowing the realisation phase of this exciting project to start at last.

This book is also a witness to the contribution given to Italian and International research by the enormous in-depth studies and design development carried out by Stretto di Messina SpA with many consultants and experts worldwide to bring the project to this point.

Pietro Ciucci
Chief Executive Officer
Stretto di Messina S.p.A.

Preface

This book tells the story of a huge undertaking: the story of a project whose enormous scale, complexity and sheer daring commands attention by all those who have interest in major construction projects, particularly in long span bridges.

When completed, the Messina Strait Bridge will stand undisputed as the world's longest single span bridge. Stretching out a full 3300 metres, it will exceed the current record holder located in Japan by more than 50 percent. Furthermore, this huge accomplishment will be achieved by a bridge that not only carries both road and rail (which as we will see involves considerable extra complexity for long spans) but also occupies a site where environmental conditions are extremely harsh. The Messina Strait is in the extreme South of Italy, and creating a fixed road and rail link across the Strait at this strategic location in the centre of the Mediterranean provides an important infrastructure link between Europe and North Africa. The narrow strait of Messina is known for its powerful whirlpool, which in ancient times linked it to the legend of Scylla and Charybdis, two great monsters that crushed many a Greek vessel. Strong winds, earthquakes, deep water, and rapid currents are just a few of the hazards that must be faced by designers seeking to bridge the Strait. But although this leap into the hitherto unknown domain of man-made spans is a giant step, it is not in any way a leap in the dark.

The Messina story has unfolded over a long period leading to this present moment where the bridge is at last poised to become a reality. This book tells that story. Starting with the desire for a fixed crossing of the Messina Strait and dealing with the genesis of the project from ancient history to the present time, the authors describe the principal technical, environmental, and socio-economic factors that combine to make this one of the most challenging construction projects ever undertaken. As this book testifies, an enormous amount of work has been carried out by leading practitioners all over the world over many years, involving both theoretical and practical research and development. Through their efforts, the various challenges involved with designing and building at this scale have been properly identified and quantified, and the residual uncertainties have been reduced to acceptable levels. The book describes not only this particular project, but also worldwide developments in long span bridge technology. There are parallels here with other major infrastructure projects, and readers will find that many of the lessons learnt on Messina and reported here will have wide application – not just for suspension bridge projects.

Although most of the book is written for the reader without a specialist's background, the authors have elected to include detailed technical information on a number of topics in the appendices (thus explaining, for example, why aerodynamics of long spans is so significant to this project).

Various chapters address the main technical issues involved in the bridge's design, such as wind and aerodynamics, seismic and tectonic actions, geotechnical and foundation conditions, highway and railway loading, structural behaviour, steel box girder design, and so on, with very useful reference lists for further reading. Other chapters address some of the contractual, project management and logistical challenges involved in such a huge undertaking. In addition, the book touches on the main political, economic, social, and environmental factors that have strongly influenced the project from its beginning.

The different aspects of the project are handled by several different individuals, many who have been involved in the project for years. Each offers a unique narrative derived from his own particular expertise and experience.

The authors wish to acknowledge the enormous contribution made by those who worked on this project from its beginning but were unavailable to contribute to this book. The vision for a single span of 3300 metres originated among them, and it is noteworthy that all the extensive research and design development in recent years has proved the validity of that vision, turning all those early dreams into a real project at last.

For many bridge engineers the Messina project represents the goal or benchmark of human endeavour; the limit to what is currently considered feasible for a single span. Although longer spans have been contemplated and even studied, bridge designers around the world are waiting to see Messina built before daring to consider the next great leap. The construction team is waiting, and it is with eager anticipation that the project stands ready for take-off.

Giuseppe Fiammenghi
Chief Technical Officer
Stretto di Messina S.p.A.

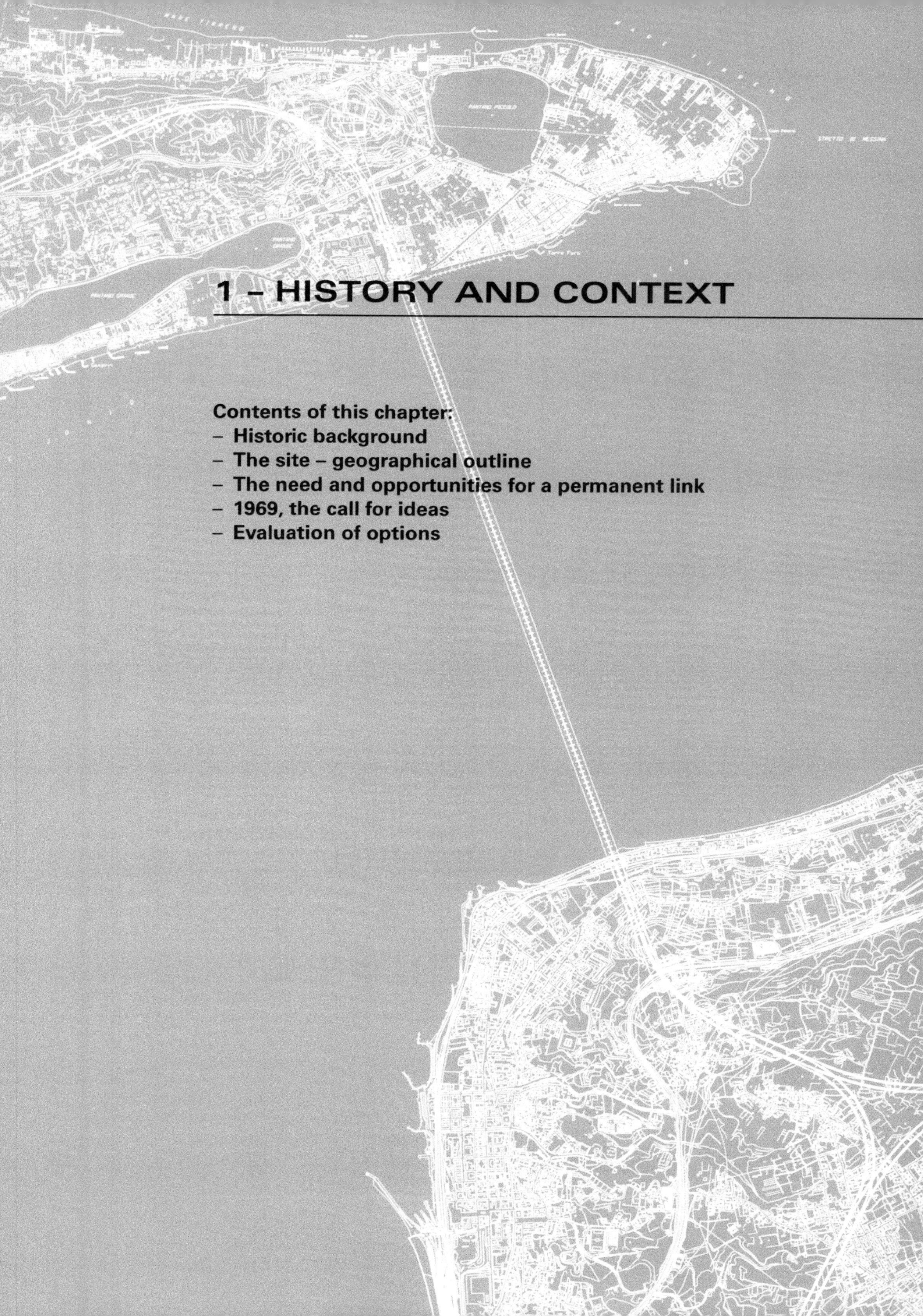

1 – HISTORY AND CONTEXT

Contents of this chapter:

- **Historic background**
- **The site – geographical outline**
- **The need and opportunities for a permanent link**
- **1969, the call for ideas**
- **Evaluation of options**

This chapter deals with the historical and geo-economic context of Messina Strait, starting with the centrality of the area in the Mediterranean, its historical relevance the myths and legends surrounding the Strait, the ancient inhabitants and their heritage. Peculiar geographic features of the region are then described, focusing on the problem of crossing the Strait. Sicily's desire to be connected with the Italian Mainland, which dates back to ancient times, is dealt with together with relevant opportunities offered by the link.

The chapter goes on to consider the needs and opportunities of a fixed link in relation to the present and predicted infrastructure and socio-economic framework of the area, and touches on the main pros/cons of the project as perceived by people living in the region.

Finally, the process that saw the project evolve from early naïve ideas to a technically sound and well developed proposal from the late sixties to the early nineties is described, focusing in the end on the final solution of a daring world record single span suspension bridge.

1.1 Historic background

Reference has often been made to the Mediterranean centrality of Sicily and of the area of the Messina Strait. As a matter of fact, Sicily is placed in a strategic position in the centre of Mediterranean, less than 200 km from North Africa, constituting a kind of bridge platform between Europe and Africa, but also between Eastern and Western countries of the Mediterranean, and has long been considered thus.

According to Strabo's[1] geography, the axis of the world inhabited 2,000 years ago, from west to east, was measured on a line drawn from the Pillars (Strait of Gibraltar), then through Messina Strait, Athens, Rhodes, and the Taurus, to the eastern sea. (Figure 1.1)

The geographic parallel passing through the Strait represented an axis of symmetry between inhabited countries of Europe and North Africa. At that time therefore the Strait of Messina was right in the middle of the civilized world, that was considered to be formed by Europe, North Africa and Middle East. Its central position along the ancient routes connecting the Aegean and the Tyrrhenian seas made the Strait an historic crossroad for ancient peoples and cultures.

The Strait was considered a fabled channel inhabited by "Scylla" and "Charybdis", mythological monsters who made passage of the Strait hazardous for ancient mariners, as described, among others, by Homer, Thucydides (6th century B.C), Eratosthenes (3rd century B.C.) and later by Dante during the Middle Ages.

Scylla can now be identified with the high, rocky promontory on the Calabrian shore (today named Scilla) and Charybdis with the strong, variable sea currents and whirlpools of the Strait.

[1] Greek geographer of I century B.C.

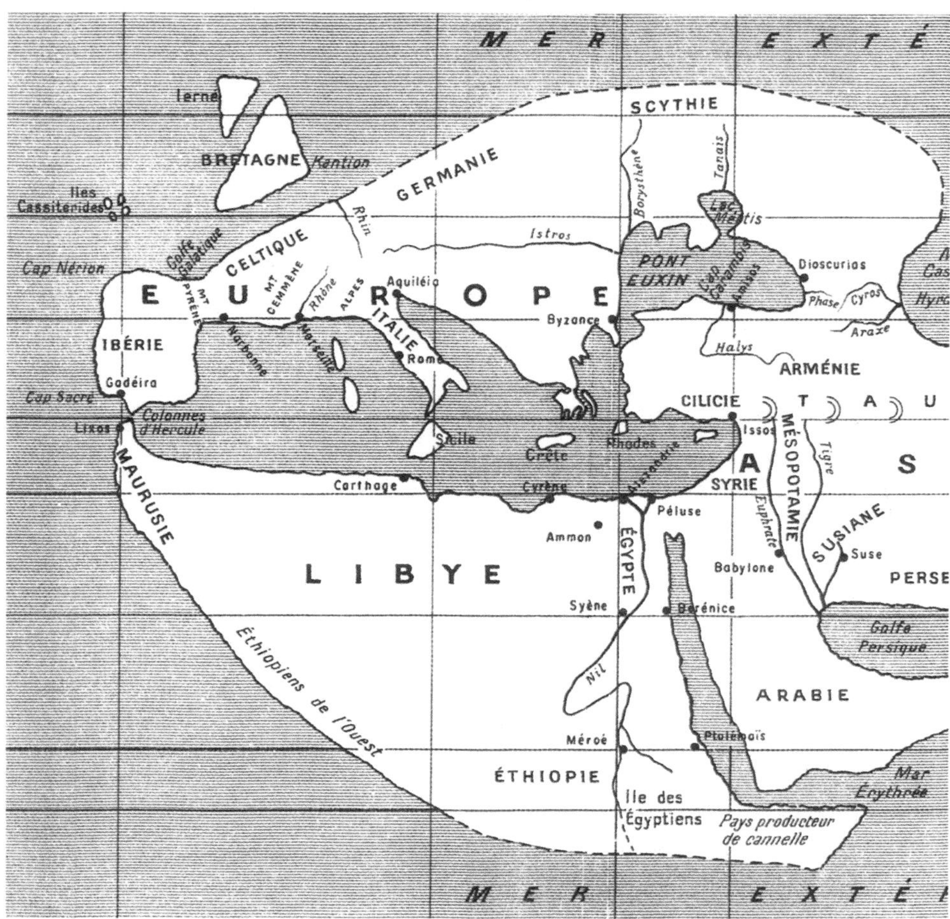

◄ Figure 1.1
The World and the Mediterranean as represented by Strabo (reconstruction by G. Aujac, Strabon Geographie).

From "Lo stretto di Messina nell'Antichità" – Stretto di Messina S.p.A. – 2005.

In the legend, Scylla was a half-woman and half-fish monster haunting sailors and Charybdis was a watery vortex on the Sicilian shore gobbling up ships and their crew. Homer describes these monsters in the XII canto of the Odyssey: *Scylla has six heads with tremendous mouths, each one catches and crushes a sailor ... three times per day. Charybdis causes a seaquake, as it sucks the sea water and throws it back creating currents and whirlpools ...*

Early Greek and Phoenician settlers established colonies in Sicily, and the island was later annexed by the Romans, invaded by the Arabs and dominated by the Normans, the French and the Spaniards. Finally, the liberation of Southern Italy by Garibaldi in 1860, which led to the United Kingdom of Italy, began in Western Sicily.

The area around the Strait played a particularly significant role in Greek, Roman and Byzantine civilizations. The Ionian and Tyrrhenian coasts of Sicily and Calabria housed, for instance, the first colonies of "Magna Graecia". In the Classic Age the whole region, and Sicily in particular, was an important centre of trading with the East. Later, during the age of the Roman Empire, Sicily also held great significance for the export of agricultural products (it was Rome's greatest supplier of wheat) as well as for the production of minerals, pottery, etc. Several times, Sicily played a strategic role in the Mediterranean politics, such as during the Punic wars between Rome and Carthage.

The whole area around the Strait, and Calabria and Sicily in general, has a unique historical heritage. Each ancient civilization that inhabited these regions left here traces of its culture and art, and as a result a significant part of the world's most precious archaeological and cultural heritage can be found in this area.

Most significant archaeological sites around the Strait area date back to the Greek and Greco-Roman era. They include, in particular, the first Greek settlements established in this area by the Euboeans of Chalcides in the 8th century BC., such as Naxos (the first Euboean settlement) just at the southern entrance to the Strait, Zancle (today Messina), Mylae (today Milazzo), Rhegion (today Reggio Calabria) and Metauros (today Gioia Tauro). To the west, in front of the Aeolian islands, Tyndaris was established in late 4th century BC. by Dionysius of Syracuse. Later Greco-Roman and Roman settlements, up to the 3rd and 2nd centuries BC., were added to the first colonies, mostly enlarging, rebuilding and repopulating former establishments.

Beautiful remains of these cultures, many of which are well preserved, include necropoles, cyclopic walls, achropoles, Greek theathers (such as in Taormina and Tindari) Roman thermae and residential villas (such as in Capo d'Orlando and in Patti), mosaics, wonderful sculptures (such as the Porticello and Riace Bronzes in Reggio Calabria) and plenty of artefacts preserved in several museums.

This important historic and cultural background adds a certain symbolic significance and value to the idea of building a permanent link between Sicily and the Mainland.

Moreover, it also suggests real prospects for the region. In fact, as will be made clear, if supported by efficient modern transport systems, Sicily has the potential of regaining its Mediterranean centrality and former significance within the present day political-economical geography.

1.2 The site – geographical outline

The Messina Strait, which has a minimum width of approximately 3 km at the north end, divides Calabria, the extreme southern region of the Italian Mainland, from Sicily, the major island of the Mediterranean and the second largest in southern central Europe.

Sicily is also the largest region of Italy (25,710 km^2), with a population of about 5.1 million inhabitants, (about 9% of the entire Italian population).

The Strait connects the Tyrrhenian and Ionian seas, and represents an important channel for international marine navigation, as one of the two possible waterways between the western and eastern parts of the Mediterranean.

The location presents many difficulties from the morphological, tectonic and marine point of view and for other related natural phenomena. These natural characteristics were certainly significant factors in making it difficult to cross the Strait by sea in ancient times. At the same time they have also strongly influenced the conception and design of a permanent crossing between the two shores. In fact, the feasibility of the different crossing solutions which have been considered was strongly dependent on the particular conditions encountered at the site. Comparing this site with that for other major sea crossings, it is worth noting that the design conditions in the Messina Strait are similar, for certain issues, to those at the Patras Strait in Greece (Rion-Antirion bridge) and the Akashi Strait in Japan (Akashi-Kaikyo Bridge).

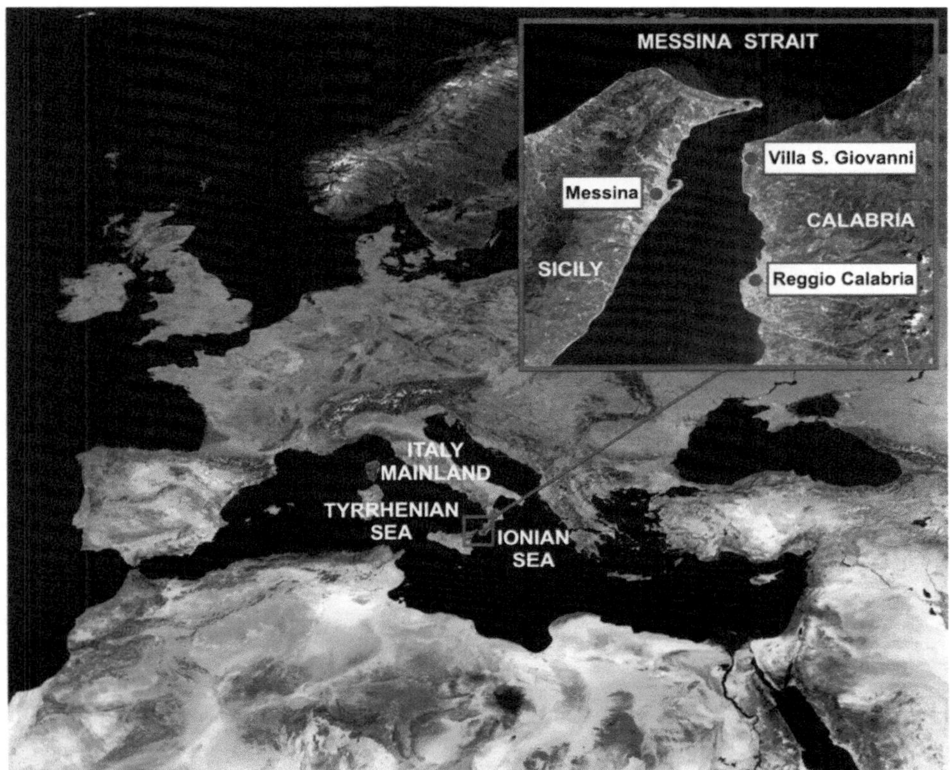

◄ Figure 1.2
The Messina Strait in the Mediterranean.

Background images from NASA (http://visibleearth.nasa.gov/).

◄ Figure 1.3
The narrowest section of the Messina Strait seen from Sicily. In foreground, Punta Pezzo, in Sicily, with the 242 m-high electrical tower and Ganzirri Lakes. In background, Villa S. Giovanni in Calabria.

The morphologic profile of the whole area is very complex, and the landscape reflects the action of natural forces and movements over many millennia.

The Messina Strait is an active tectonic area that is directly affected by the interaction between the African and the European continental plates. This interaction is highly complex and can be schematised with reference to two fundamental geodynamic occurrences: the north-westward motion of the African plate, and the subduction of the Ionian crust under the Calabrian Arc. The subduction process is believed to be the cause of the relatively rapid regional-scale uplift that has affected the entire region and most of the Apennines during the past million years at least. The Messina Strait also experiences significant upper crustal extension, which is responsible for the seismicity of the area and for the relative subsidence of the Strait with respect to adjacent landmasses (see Chapter 5.1).

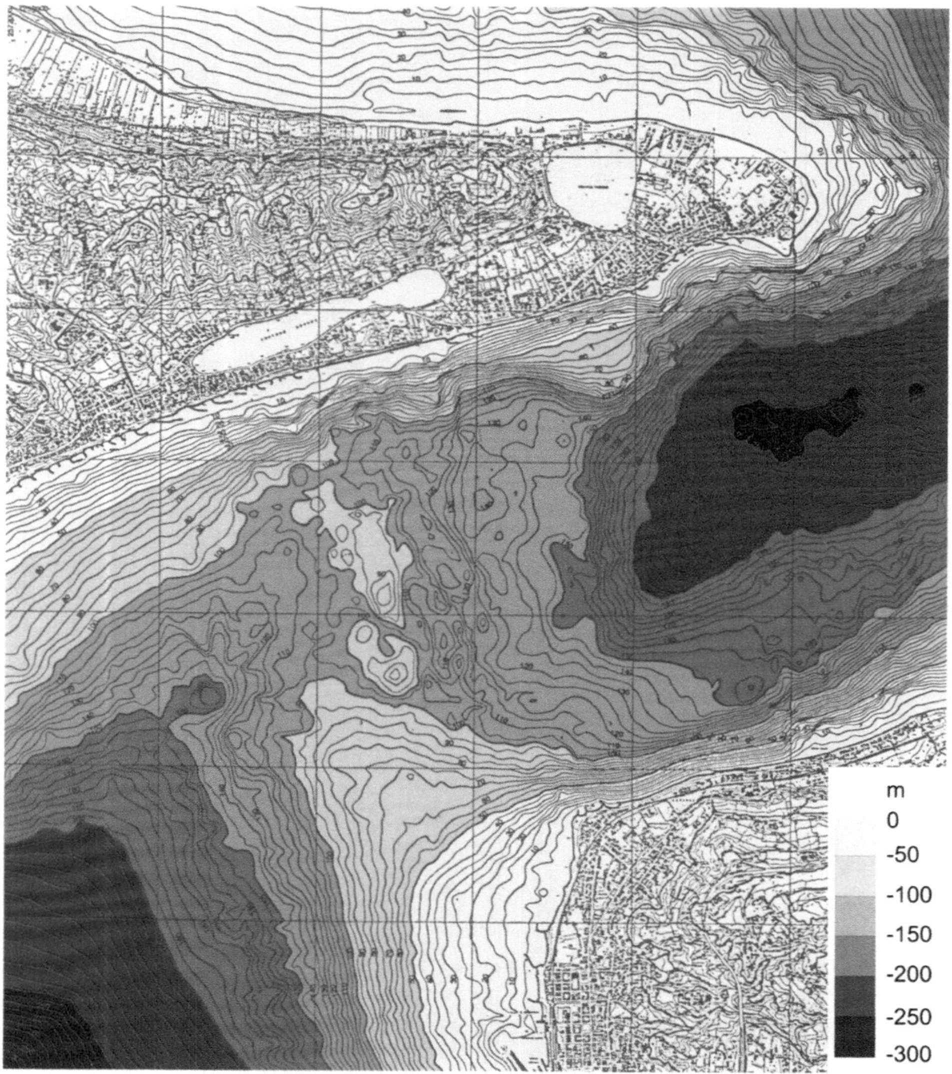

Figure 1.4 ▶ Bathymetry of the Messina Strait (depth in m).

The Strait itself is characterized by a deep seabed and strong currents. The coasts are bordered by high mountains. On the Sicilian side, the Peloritani Mountains gradually decline from 1,250 to 290 m above sea level, and on the other side, in Calabria, the steep and massif Aspromonte mountain reaches about 2,000 m above sea level and presents deep valleys perpendicular to the coast. These morphologic features strongly influence wind speeds and directions (see Chapter 5.4).

Three of the most important European emerged volcanoes (Etna, Stromboli and Vulcano) are located near the Strait. This area has one of the highest seismic levels of Italy and Europe, and in 1908 the Strait was hit by a magnitude 7.1 earthquake, which is the strongest ever recorded by instruments in Italy.

The marine environment of the Strait also exhibits very particular characteristics due the continuous alternating exchange of water flowing between the Ionian and Thyrrenian sea, which creates strong winding currents having speeds up to 5 m/s at surface level.

The strong sea currents in the Strait occur due to tidal oscillations of both sea basins that create significant sea level differences with an average slope of about 2 cm per km. The water flow can reach 750,000 m^3 per second, which is very high for such a relatively narrow channel, reversing direction roughly every 6 hours. Furthermore, complex phenomena are created by two peculiar situations: the particular shape of the sea bed and the difference in physical properties of the two seas. On one hand, the sea bed of the Strait has a particular saddle shape that abruptly rises from approx –300 m up to higher than –100 m. On the other hand, Ionian waters are colder and more dense than Tyrrhenian ones. These two factors cause a vertical mixing of the waters with a considerable vertical component of the current that creates a small-scale "upwelling" phenomenon. This upwelling of cold water, rich in oxygen and nourishments, creates very particular conditions for the marine ecosystem, especially for the sea fauna, and particular micro-climatic effects.

From the maritime navigation point of view, the Strait presents a complex morphology with poor visibility and little room for manoeuvre having a meandering profile and a width which reduces to less than two miles at the narrowest point. Together, the topographic, environmental and marine conditions can generate serious difficulties in steering large vessels through the Strait, particularly due to the strong currents and related phenomena. Two ports on each shore, with high ferry traffic, operate in these harsh conditions (as discussed further in Chapter 3).

As mentioned before, the marine ecosystem is particularly worth mentioning due to its high environmental value. This is one of the reasons why the environmental aspects played a significant role in the choice of solution type, including the bridge form, and in the design of the fixed link as a whole.

The region comprises one of the most densely populated areas of Southern Italy and includes the provinces of Messina (in Sicily) and Reggio Calabria (in Calabria), both on the shores of the Strait. The whole population of the Strait area amounts to about 1.5 million inhabitants.

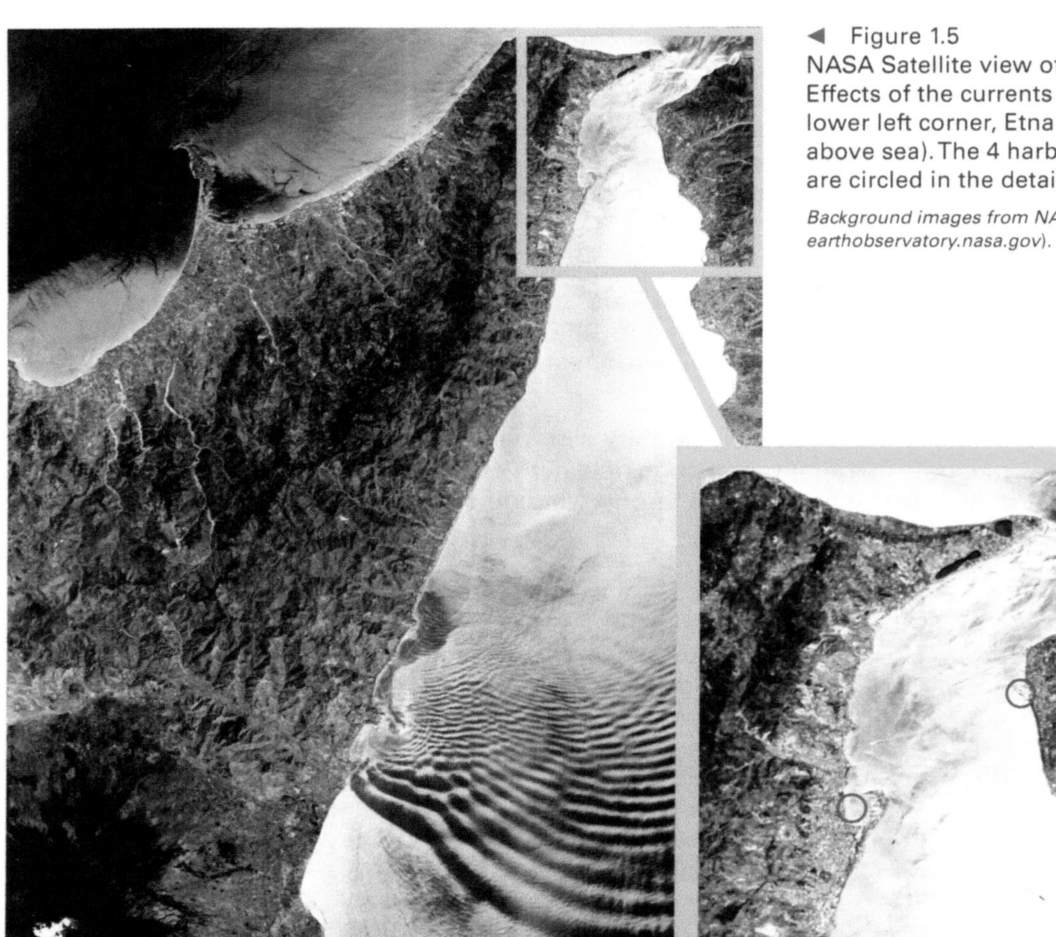

◄ Figure 1.5
NASA Satellite view of the Strait. Effects of the currents are visible. In the lower left corner, Etna volcano (3,360 m above sea). The 4 harbours of the Strait are circled in the detail above.

Background images from NASA (http://earthobservatory.nasa.gov).

The metropolitan area of Messina includes 51 municipalities constituting a continuous urban conurbation developing for 125 km along the Tyrrhenian and Ionian coasts. With its 480,000 inhabitants, this metropolitan area is the fifth largest in Southern Italy and the eleventh in whole country.

Reggio Calabria is located at the centre of a metropolitan area comprising 257,000 inhabitants and several municipalities, including Villa S. Giovanni which is located just opposite Messina, on the east side of the Strait. Villa S. Giovanni is the preferred departure and arrival point for the Strait crossing and has the most important ferry facilities in Calabria, for both road and rail transport. The other main Calabrian ports, at Reggio Calabria, are mainly used for pedestrian transport. In Sicily all the ferry crossing facilities are located in Messina.

50 km south of Messina, the metropolitan area of Catania begins, which includes 27 municipalities and about 750,000 inhabitants. Catania is the second most important industrial pole of Sicily and is probably the most dynamic Sicilian town thanks to the quantity and variety of its production and activities. Even though Catania has its own port and airport, it still depends on the Strait for road and railway connections

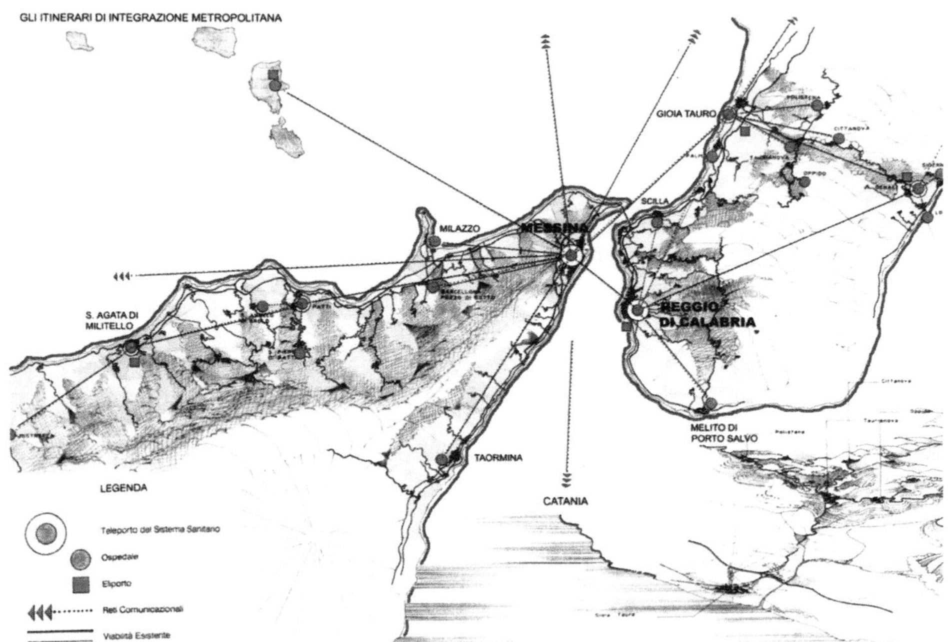

◄ Figure 1.6 Metropolitan integration scenarios: "The Health System".

with the Mainland. For this reason, and considering the proximity to the Strait, Catania substantially falls within the sphere of influence of the Messina Strait, from the territorial point of view.

The intensive relationships and exchanges between the settlements on the two shores, and the historical similarity of their populations, contribute to the creation of a "Region of the Strait". The idea of a metropolitan integration between the shores, creating relations and synergies to form a single metropolitan area, the "City of the Strait", having a population of about 740,000 inhabitants, is gaining more and more ground. The new access facilities provided by the Bridge would lead to the re-organization of settlement patterns, economic activities and general infrastructures as a basis for the formation of the "City of the Strait", giving a common plan to Messina and Reggio Calabria, whose communication is presently limited by the barrier of the Strait.

1.3 The need and opportunities for a permanent link

1.3.1 Historic demand of a link across the Messina Strait

The need and importance of a permanent link between Sicily and the Mainland was understood as early as 2300 years ago, as soon as ancient Rome started to extend its power and civilization in the Mediterranean. In fact, in the year 502 of Rome (251 B.C.), during the first Punic war, a great number of elephants were brought to Rome for the triumph of the Roman consul L. Caecilius Metellus. They had been taken in the victory gained by Metellus over the Carthaginians near Palermo (Western Sicily). Pliny and others report that the consul conveyed to Italy's shores about one hundred and forty elephants upon rafts supported by rows of barrels fastened together[2] (or upon a bridge of rafts, considering the alternative meaning of the Latin word "ratis"). This was, according to history, the first mass-bridging of Messina Strait with heavy loads.

By contrast, during the second Punic war, the Carthaginian Hannibal preferred to lead his army, also complete with many elephants, to Rome by walking through Spain and over the Alps, rather than crossing the Strait. This suggests that the Strait with its "monsters" was much feared by sailors in ancient times, as a kind of "Bermuda Triangle" of antiquity.

After the enterprise by L. Caecilius Metellus there was no link between the shores of the Strait until regular crossings by ferry-boats became established. Several ideas in principle for a permanent crossing have been put forward from time to time, with the first "engineered" ones emerging in the early 1900s. (The first one was actually a bored tunnel proposed by Carlo Navone in 1870.) However, any solution for bridging the Strait must deal with the unusual and hostile conditions related to the physical environment and seismicity of the area, and the final bridge solution would not emerge for nearly another hundred years.

On November 1st, 1899 the first ferry boats (predictably named "Scilla" and "Cariddi") started to provide regular shipping of trains between the ports of Reggio Calabria and Messina. They were operated by the regional railway company[3].

When in 1905 the Italian State Railway company (presently RFI – Rete Ferroviaria Italiana) started to operate the railway infrastructure, there were four ferries in total, as two additional ferries and connecting railways had been established for the second route between Villa San Giovanni and Messina.

Figure 1.7 ▶ A steam ferry, operated by the Italian State railway company, sailing from Villa S. Giovanni towards Messina in the early 1900s.

Typical features of first ferries

length	50.50 m
width	8.20 m
displacement	594 t
capacity	5 wagons
cruise speed	10.5 knots
steam engine – wheel propelled.	

"Scilla" and "Cariddi" ferries (1896–1923)

[2] "elephantos Italia primum vidit ... eadem plurimos anno DII victoria L. Metelli pontificis in Sicilia de Poenis captos. CXLII fuere, aut ut quidam CXX, travecti ratibus quas doliorum consertis ordinibus inposuerat" – Pliny the Elder, "Naturalis Hystoria" book 8, chapter 6. (also reported by T. Livy and Seneca).

[3] A Royal Decree issued on November 23rd, 1883 obliged the Sicily railway company to provide two daily ferry-boat runs between Messina and Reggio Calabria and later on two other runs between Messina and Villa San Giovanni.

In 1905 the number of railway vehicles (passenger coaches and freight wagons) shipped across the Strait was just 17,000 per year. Three years later, the whole Strait area suffered the effects of the disastrous Messina earthquake which occurred in 1908, significantly disrupting the transport infrastructure.

The traffic grew rapidly after the First World War reaching more than 140,000 train vehicles and 3,500 road vehicles in 1935, with a total of 1 million people. The traffic demand continued to grow in the following years, especially after the Second World War. In 1965, private companies joined the business of shipping road vehicles and pedestrians across the Strait and started a regular service using small bi-directional ferries which allowed more capacity and especially more frequent connections.

Traffic demand across the Strait increased strongly in the period 1950–1970, with high rates of growth of about 5%–6% per annum. The 1960s and 1970s were particularly important for the development of the area, due partly to the substantial completion of the highway system from Northern Italy to Reggio Calabria (completed in 1974) and to the construction of new highways in Sicily (started in 1962). It is worth mentioning in this context that during late 1960s and early 1970s the problems related to crossing the Strait were particularly evident at national level, thus attracting significant attention by the railway and highway authorities and by the Italian Parliament. (Figure 1.8)

In 1971 the Italian Parliament finally issued a State Law for the realization of a permanent road and rail link between Sicily and the Mainland. It was the law which established Stretto di Messina SpA (SdM), as discussed further in Chapter 2.1.

DOMENICA DEL CORRIERE

Anno 67 - N. 12 - L. 70 | Settimanale del CORRIERE DELLA SERA | 21 marzo 1965

La Sicilia diventa continente

Un ponte di quasi quattro chilometri unirà Punta Pezzo, in Calabria, a Ganzirri. Il progetto della grande opera che costerà circa novanta miliardi ricorda quello del celebre ponte di Nuova York, dedicato a Giovanni da Verrazzano. (Servizio di Vittorio Lojacono alle pagine 10-12).

Mille dollari facili per James Bond

L'inafferrabile agente 007 in una nuova puntata del romanzo «L'uomo dalla pistola d'oro». (Alle pagine 37-40)

42 anni di attesa per le pensioni di guerra

190 mila domande, 240 mila ricorsi: ecco le cifre da capogiro che illustrano drammaticamente la vicenda burocratica di chi attende dallo Stato ciò che gli spetta.

◀ Figure 1.8
In 1960s the interest stirred by the anticipated link across the Strait led to the publishing of several imaginative pictures of the future bridge: "*Domenica del Corriere*" national magazine, March 21st 1965.

The traffic across the Strait in the early 1970s can be estimated as approximately 1 million cars, 0.5 million trucks and 0.5 million railway vehicles, with a total of 6 million people and 6.5 million tons of freight.

In the subsequent years, road traffic demand continued to grow, although at lower rates. As far as the railway traffic is concerned, the number of passenger coaches remained approximately stable, while the number of freight wagons decreased. Although some contribution to this phenomenon can be related to an increased efficiency in loading freight, this reduction certainly indicates that the railways lost some share of freight transport. The phenomenon also occurred elsewhere and is obviously not limited to the Messina Strait. In this case, however, the trains were particularly penalized by the process of shipping and unshipping required by the ferry system. As a result, most freight traffic was progressively diverted to transport by road.

Furthermore, aircraft transport and Ro-Ro ships (the so called "*Motorways of the Sea*") developed substantially during the last 20 years, gaining important shares of the transport between Sicily and the Mainland. (See Chapter 4) Part of this growth is certainly because of the bottleneck due to the Strait being without an effective road and rail crossing.

1.3.2 Current transport and socio-economic development

The ferries crossing the Strait today carry approximately

- 2,500,000 cars per year;
- 1,000,000 trucks per year;
- 190,000 train vehicles per year.

Together these total about 11.4 million people and 13.9 million tons of freight per year (2006 data).

Ferries run a total of 123,000 yearly trips across the Strait, which is about 340 daily trips. In other words a ship departs every 4 minutes from either shore on average, 24 hours per day and 365 days per year.

The crossing by ferries still involves the same problems today as in the 1970s, but they are becoming more and more severe. Major problems include the following:

1 Time loss and cost of shipping trains
Shipping trains on ferries involves serious time losses due to waiting time, train disassembly, loading, unloading and reassembly, plus the navigation time. These times are longer than 90 minutes for Inter-city trains, and they can vary from several hours to more than one day for freight trains. Operation and maintenance costs of the system are very high, so that an economic contribution by the State is required. In the near future, shipping modern trains in this way will no longer be possible due to their locked assembly.

2 Time loss and cost of shipping vehicles
Similar problems exist for road vehicles, although the time loss and discomfort are less severe. Also the smaller vehicle-carrying ferries involve significant operating costs. In relation to the low efficiency and poor service level of the system, the fares are considerable: 25 € for a 4 metre car and 100 € for a 12 metre truck (one-way).

3 Traffic in urban centres
Shipping and unshipping of road vehicles requires cars and trucks to pass through the urban centres of Messina and Villa S. Giovanni. This causes further time loss, traffic jams, increase of accidents, noise and air pollution.

4 Interference between long-distance maritime navigation and cross-Strait traffic

The dense navigation of ferries across the Strait intersects and interferes with the longitudinal routes of long-distance ships sailing through the Strait from the Ionian to Tyrrhenian sea and vice-versa. Most of these are large ships, in particular container carriers coming from or heading to the Gioia Tauro container terminal (3,5 million TEU in 2007).

Risk of marine accidents:

- ships on longitudinal routes: ~20,000 per year, that is ~55 per day;
- ferries on transverse routes: ~123,000 per year, that is ~340 per day;
- 56 collisions occurred in last 50 years: 14 of them involved tankers, 7 caused casualties, 8 produced sinking, 3 caused oil spills into the sea. The most severe one, in March 1985, caused a tanker pouring into the sea 1,000 tonnes of crude oil. A recent one, on 15th January 2007, caused 4 casualties with about 100 injured.

The navigation in the Strait is subject to severe restrictions. Nevertheless the risk of marine accidents is still high resulting in hazards for human lives, goods and the environment. The ports of the Strait and the pertaining infrastructures are overloaded by the dense traffic of ferries crossing the Strait, as well as by the millions of vehicles headed for or coming from shipment.

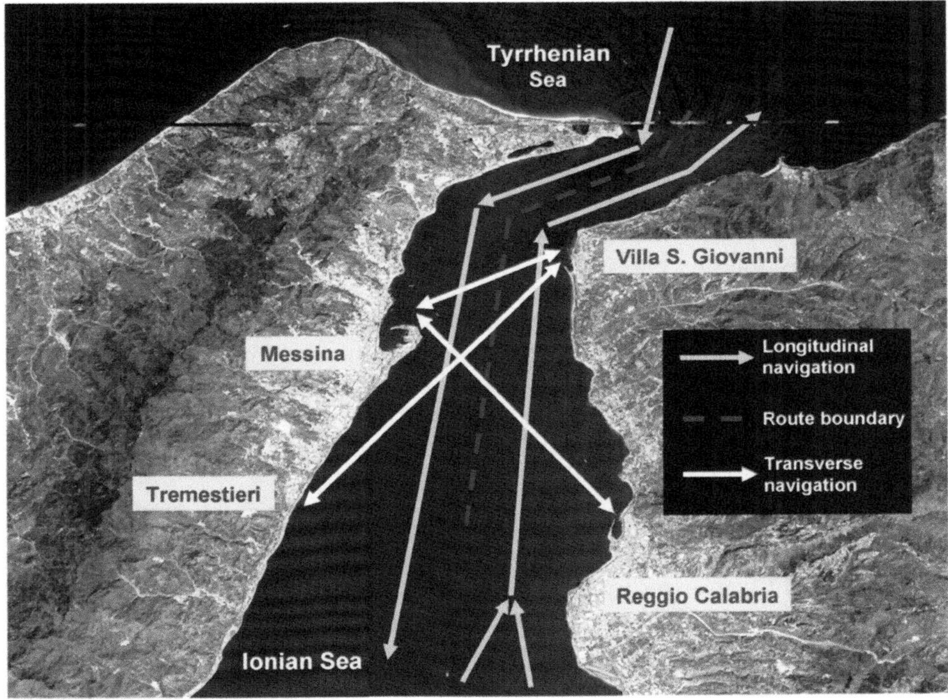

◄ Figure 1.9 Navigation in the Messina Strait.

Attempts to sort out these problems were recently made by the construction at Tremestieri, in the Southern outskirts of Messina, of a new port for the ferries carrying trucks. This solution has partially relieved the problem, but has also produced other difficulties, particularly to navigation.

5 Environmental Impact of ferries
Moreover, the ferries themselves, which run altogether more than 250 engine-hours per day, are highly impacting on the atmosphere (CO_2 emission) as well as on the sensitive marine ecosystem (noise, pollution and other damage).

1.3.3 Prospects of a fixed link to boost development

Spanning the Messina Strait means carrying the road and railway infrastructures from the Italian mainland to Sicily. In particular, for railways it would allow the extension into Sicily of the high speed, high-capacity railways that today end at Naples. This also reflects the planned extension of European transport infrastructures toward the centre of the Mediterranean and North Africa, which takes advantage of the natural central position and projection of Italy and Sicily in Southern Europe.

The area of the Strait is inhabited by about 1.5 million people, while the whole island of Sicily has a population comparable to some entire European countries, such as Denmark or Norway. It is well-known that these countries, among several others in the world, did not hesitate to build great highway and railway links to ensure continuity between different parts of their territory. Ensuring territorial continuity is in fact recognized all over the world as a fundamental requirement for boosting and sustaining socioeconomic development. Examples of great links similar to the case of Messina are, among others, the Great Bælt in Denmark, Öresund between Denmark and Sweden and the various bridges in Japan between Honshu and Shikoku islands. The investments made for such significant transport links in the world are in general comparable with the one foreseen for Messina, or even greater, while the population affected by the transport supply is in several cases less, as in the above mentioned examples.

In the 2000s, the Strait still constitutes a gap in Italy's (and Europe's) territorial continuity and a barrier to the flow of people and goods, generating a negative impact on social and economic development. Generally speaking, most of the so-called "*Mezzogiorno*"[4] suffers from a transport infrastructure that is inadequate and weak compared to the more highly developed and better served regions of northern and central Italy. The Mezzogiorno has in fact an index of potential accessibility scarcely higher than 50% of the average for the entire European Union due both to its peripheral location and to the poor quality levels of the infrastructure. This situation is also reflected in the weakness of local production and output, and of the South's economy in general.

The economy of Calabria and Sicily, similarly to what happens in the rest of the Mezzogiorno, has always been weaker and less developed than the Northern and Central regions of Italy. The average GDP of Mezzogiorno is presently approx 17,000 €[5] per capita, which is less than 60% of the average of North-Centre Italy.

---

[4] The term "*Mezzogiorno*" ("Midday" but also "South" in Italian) became popular in 1800 and is still used to identify the geo-economic area of all Italian territory located south of Rome. According to some (but not all) classifications, Mezzogiorno is intended to include both the major islands, Sardinia and Sicily, in addition to 6 entire regions of southern peninsular Italy.

[5] Economic data in this section are drawn from SVIMEZ 2007 report on the Mezzogiorno economy.

Before their incorporation into the Italian Kingdom (in 1861), these regions had originally a mostly agrarian economy, based on large estates, that was much weaker than the more modern and industrialized northern regions. Overcoming this socio-economic gap, that still persists, has long been the subject of studies and debates on the so called "Meridional Question".

The economy of the Mezzogiorno is presently growing (in 2006 GDP growth was equal to 1.5%), but with a lower rate than the Centre-North, (GDP growth in 2006 was equal to 2%). This means that it does advance but not fast enough to reduce the gap, that on the contrary continues to increase. The unemployment rate is particularly significant in causing a permanent or temporary transfer northward of many young people looking for jobs which fit their professional aspirations. This economic gap is evident even when compared with other economically weak areas of the European Union.

After the national stagnation during the period 2002–2005, the GDP of Sicily and Calabria recorded an upturn in 2006. In the period 2000–2006 the overall growth reached 4.8% in Sicily and 4.5% in Calabria, while the average rate in the Mezzogiorno stopped at 4.1% and the national average was equal to 5.4%. In the period 2000–2006 employment also increased, reducing the unemployment rate of 10% in Sicily and 6% in Calabria. Sicily alone accounted for one third of the total employment increase of the Mezzogiorno.

Sicily (≈5.1 million inhabitants) showed a GDP in 2006 equal to approximately 16,500 €/year per capita; that is about 66% of the national average. This value is primarily due to tertiary activities (77%), followed by industry (11.5%), construction (6.5%) and agriculture and fishing (4.8%). The chemical and petrochemical industry provides a significant contribution due to activities associated with oil extraction (90% of the whole Italian production) and refinery which are carried out in three main poles of petrochemical activity. The agriculture and food industries represent one of the natural inclinations and specialisms of Sicily. As a matter of fact there are several high-quality Sicilian food industries producing pasta, wine, olive oil, fish, etc. Tourism, which is the other natural inclination of the island, is rapidly developing, recording an annual growth rate of 5–6%.

The balance of payments shows a trade deficit on the whole, but a trade surplus for activities connected to refined oil products (51% of the export), vehicles, agriculture and food, non metal mineral products and electronic precision instruments.

Leading Italian and foreign Hi-Tech companies (such as StMicroElectronics, Nokia, etc.) recently invested in Sicily, opening new plants. The maintenance of appropriate tax and contribution policies could support the development of this beneficial trend.

Calabria (≈2.0 million inhabitants) showed a GDP in 2006 equal to approximately 16,100 €/year per capita; that is about 64% of the national average. The economic growth of Calabria is presently mainly due to the development of tertiary and construction activities, which constitute respectively 79% and 7.1% of the value added. On the other hand industry and agriculture, which represent respectively 9.1% and 4.8% of the value added, are facing a downturn. Tourism in Calabria is undergoing a continuous development, even if with lower rates than in Sicily. 2006 was marked by a revival of international tourism.

The Container Terminal of Gioia Tauro constitutes the "excellence case" in the development of Calabria. As a matter of fact it is today the most important logistic platform of Italy and the principal Container Terminal of the whole Mediterranean Area. The harbour was built in 1972 for other services and was re-structured in 1994 to become a transhipment hub, thanks to its central position between Suez and Gibraltar. Since 1995, in a very few years, Gioia Tauro has been attracting large shares of container transport in the Mediterranean, reaching 2.5 MTEU in 2002 and 3.2 MTEU in 2004 with the target of 4 MTEU in 2007–2008.

Nonetheless the capacity of distributing goods on railways and roadways from the terminal is presently limited by the lack of land infrastructures.

In general the problems hindering the economic development of the two regions (and of the whole Mezzogiorno) derive from an organizational and structural weakness, and in particular:

- a lack of transport infrastructure and services, despite the construction of important works in the last 50 years;
- the structure of the business system, with the majority of firms being small, traditional and not yet internationalised;
- the structure of the banking system which is not yet adequate;
- heavy bureaucracy;
- the presence of organized crime.

As far as the lack of infrastructure is concerned, several important projects are presently in progress, or have been planned. The most important of these, such as new highways and railways on the Ionian and Tyrrhenian coasts of Sicily and Calabria, converge towards the fundamental "hinge" represented by the crossing of the Messina Strait.

On the other hand Calabria and Sicily have great potential deriving from the possibility of establishing industrial settlements, developing agriculture and effectively exploiting their inestimable historical, artistic, cultural and natural heritage.

Furthermore, Sicily and Calabria could benefit from their central position in the Mediterranean by playing a leading role in international transport flows in the region and becoming the gateway into Europe for the increasing marine traffic flows in the Mediterranean, and particularly for:

- trade flows along transoceanic sea routes through Suez and Gibraltar which reach Europe from the Far East and America.
- Mediterranean flows of trade and know-how between European developed countries and the developing countries in North Africa and Middle East.

The bridge will obviously contribute to filling the existing infrastructure gap, creating favourable conditions to boost economic and social development. As it carries a railway as well as a highway, the bridge constitutes a fundamental step towards a modal rearrangement in favour of railway transport. Furthermore, its construction will spur the upgrading and development of road and railway infrastructure in the area.

The bridge will also produce other positive effects not strictly connected to transport. As a matter of fact it provides the opporturity to re-organize and re-qualify the towns of the Strait through innovative urban structures and functions, thus becoming the key element in powerfully boosting their integration into a metropolitan area.

In the European perspective, the bridge is the missing link in the 1st European Corridor (Palermo-Berlin axis) which is clearly incomplete without an efficient roadway and railway link between Sicily and the European Mainland. (Figure 1.10)

The permanent link across the Messina Strait could renew and reaffirm the central position of Southern Italy in the economic scenario of the Mediterranean. In fact, as already mentioned, the whole area and Sicily in particular has the potential to play a strategic role as a logistics platform in relation to the increasing flow of goods through the Mediterranean.

Recent studies (by Contship) show that the volume of freight exchanged between the Mediterranean and the Far East grew by about 19% in 2003, and by about 20% in 2004, and that this growth trend continues at comparable rates. A similar situation is reported for the exchanges between the Mediterranean and North America. Accordingly, in the central-western area of the Mediterranean a growth by 75% of container handling demand is expected by 2014. A competition is therefore expected, and is already occurring, between Atlantic and Mediterranean ports, as well as among the various ports within the Mediterranean. In this scenario, the availability of a fixed link across the Messina Strait could be very important. The actual opportunity for each port to share in these volumes of traffic is in fact strongly related to the efficiency of the connecting land infrastructure for distributing and dispersing the goods, especially if transport interchange and transformation activities can take place locally. In fact, the more efficient the synergy between different transport modes the more competitive the logistic system.

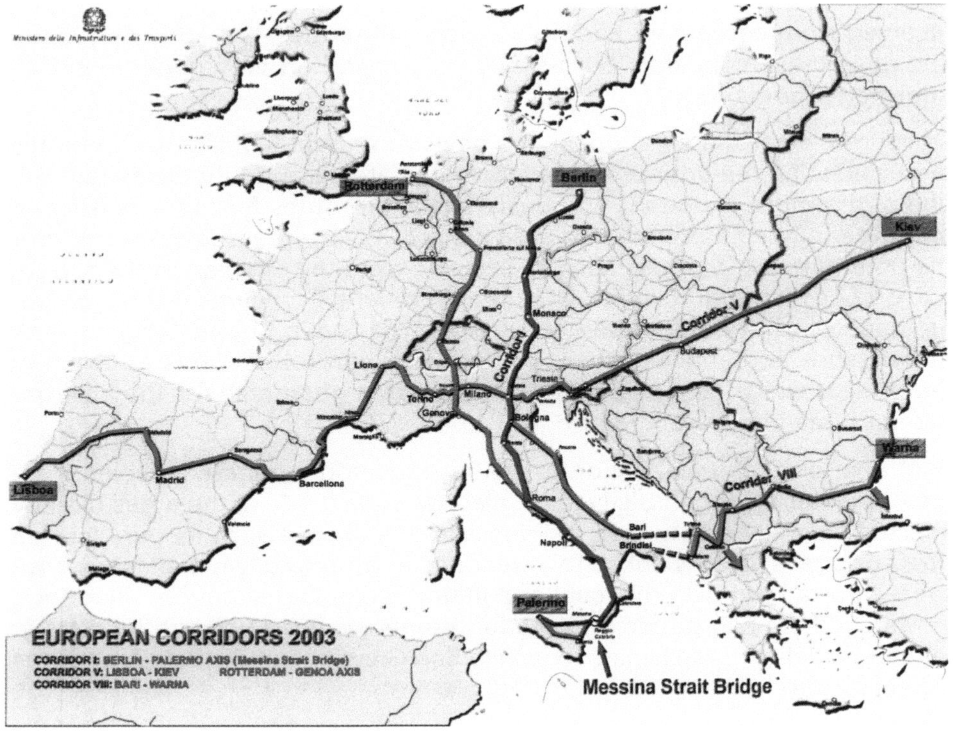

◄ Figure 1.10 Location of the Messina Strait in the European Transport System.

Redrawn with additions from www.infrastrutturetrasporti.it – Italian Ministry of Infrastructures and Transport website – 2004.

1.3.4 How the public feel about the project

Analysing public attitudes toward such an exceptional project is a very complex matter that would require a whole chapter. Therefore within the scope of this book it is only possible to present some short extracts.

There are, unavoidably, many conflicting opinions that are mostly based on ideological pre-conceptions among the population following more than 30 years of public and political debate about the bridge.

First of all, it should be noted that public discussions on the opportunity of bridging the Strait have always been subject to strong emotional factors related to the symbolic value and scale of the proposed structure and to the consequent changes to the region due to its construction. Favourable opinions of many who focus on the economic importance and positive benefits of a fixed link face opposition by some who perceive the insularity of Sicily as a precious heritage to be preserved.

As far as experts of transport and economy are concerned, those in favour of the Messina Strait Bridge consider it as one of the most effective and pressing strategies of inducing an overall up-grade and development of rail and road networks in Southern Italy and of promoting a revitalization of the economy of the area. In their view, the bridge would actually be a fundamental key in a modern infrastructure system which would be renewed through the logic of integration and interoperability. It appears feasible and effective considering that a significant traffic demand already exists and the bridge has the capability of being self-financed through the collection of tolls to a great extent. Conversely, some experts, associations and opinion groups who opposed to the project maintain that it is not a priority, preferring instead many other widespread improvements on the existing road and rail networks, or even further developments to maritime transport.

Unsurprisingly, objections over technical and financial feasibility were in the past the main arguments raised by opponents, particularly in relation to the unprecedented length of the main span and to the seismic hazard of the area. These concerns have been overcome by the large quantity of in-depth studies carried out by SdM and by the several approvals to the scheme received from competent and responsible authorities. Nevertheless, some still raise these concerns, but usually only based on very general and ill-informed arguments.

Of course, environmental questions have very great significance in public discussions on the project. For example, the area around the Strait borders one of the most important European migration routes for many species of migratory birds. Environmental associations have expressed concerns over the potential risk for these species deriving from the erection of the towers and other construction works. Other environmental concerns include potential impacts on marine mammals living in the Strait. Chapter 4 deals with these issues and analyses the environmental control methods foreseen in the project which help to provide an important guarantee for the environmental sustainability of the bridge.

SdM also pays close attention to the feelings and feedback received from the population living in Sicily and Calabria, particularly in the areas directly affected by the work. In fact, the success of a project such as the Messina Strait Bridge is directly linked to a clear and appropriate two-way flow of relevant information with the local population and external parties. Therefore, in March 2006, two public information points were established by SdM in the municipality of Messina and in Villa San Giovanni, and a Call Centre was also activated.

SdM voluntarily decided to invest in this consultation process in order to maximise the opportunities for ordinary citizens to be properly informed, and to have their concerns heard and taken into account, before the final design and construction phases commence.

Between March and May 2006[6] more than 14,000 questions were registered by the two information points and the Call Centre. In order to obtain a very preliminary picture of the concerns of the local population these questions have been divided into four main subjects: technical issues, environmental and socioeconomic impact, project finance and project organization.

Figures 1.11 and 1.12 summarise these preliminary results.

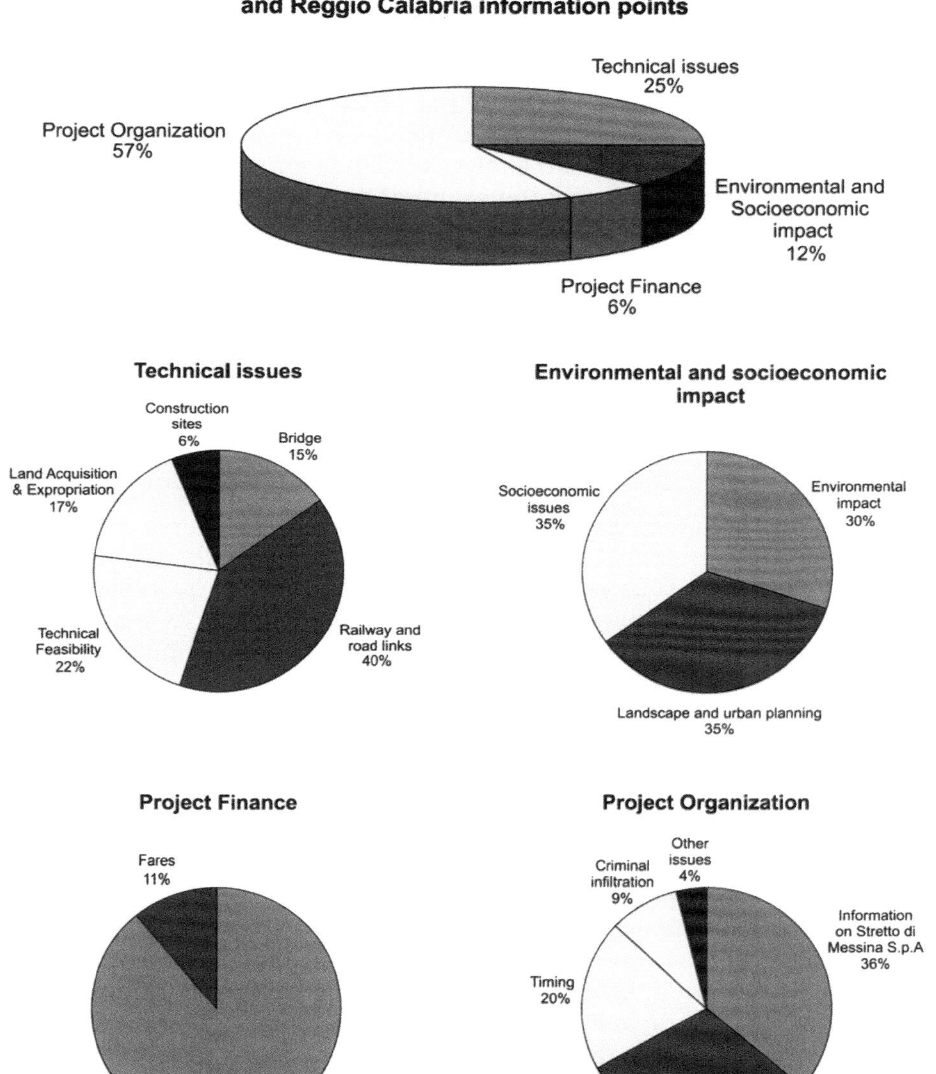

◄ Figure 1.11 Statistics on FAQs registered by SdM info points and call centre: main subjects.

◄ Figure 1.12 Statistics on FAQs – detail of each main subject.

[6] After May 2006 statistics are not representative as the new Government of Romano Prodi announced that the project was not a priority, thus leading to a drop of interest and to the focusing of questions on the future of the project.

About 75% of people registered at information points were in favour of the bridge. Most of their questions were focused on project organisation such as the organisation of the Concessionaire (SdM), contract arrangements and timing. Most of technical questions (25% of total sample) were focused on the feasibility of the bridge and the road and railway links that are included in the project. Only 12% of questions concerned the environmental and socioeconomic impacts, although these had been at the heart of most controversial issues under public debate in recent years regarding reasons in favour of or against the bridge.

The flow of people at the information points and their interest also showed that people living in Sicily are more interested in the project compared to those living in Calabria.

The image of the bridge in the minds of the population living within and outside the immediate area will also be studied during the final design phase. As a matter of fact, scientific, social and economic research will be critically important during the final design phase to ensure that the project delivers not only the construction of the bridge but also a modernization of the Strait area producing positive effects on population lifestyles. For this reason, further surveys and analyses will be carried out by SdM before and during construction within the scope of the "environmental and social monitoring". This work is described in Chapter 4.

1.4 1969, the call for ideas

This section and the two following are devoted to describing the process which, in a period ranging from the late sixties to the early nineties, saw the project evolving from early naïve ideas to a technically sound and well developed proposal, focusing in the end on a single span suspension bridge crossing.

Note that, while general drawings and illustrations of the alternatives considered are shown herein, no information is given in this chapter regarding details of the design such as the final deck cross section or quantitative data on the structural behaviour. This has been decided so as to avoid interrupting the historical narrative with an excess of technical elements. Such technical information is given in Chapter 6.

The first two milestones in the process were an international call for design ideas, issued by the Italian Ministry of Public Works in 1969, and the subsequent statutory law forming the concessionaire company Stretto di Messina SpA (SdM) in 1971. The intention of both was the same: a challenging road and rail crossing of the Messina Strait.

The call for ideas attracted enormous attention, with more than a hundred proposals from designers worldwide. Understandably, these produced a mixture of traditional, unconventional and altogether weird schemes. Six design outlines were awarded an ex-aequo first prize, comprising proposals ranging from single or multi-span suspension and cable-stayed bridges to three dimensional cable networks, underground tunnels and floating tunnels.

The six first prizes were:

Gruppo Ponte Di Messina (GPM) JV
This was the only group to submit alternatives, anticipating the evaluation process that was to be a focus of future SdM activities. Proposals comprised a twin suspension bridge with five offshore piers, a classic three span suspension bridge with two offshore piers on artificial islands, as well as for the first time a single span 3000 m span suspension bridge.

Technital JV
With main designer professor M.P. Petrangeli, this proposal featured a five span solution with hybrid suspension-stayed cable system, Dischinger style. Typical spans were 1000 m, with three offshore piers and foundations.

Calini-Montuori-Pavlo JV
The most traditional among the proposals awarded a prize, this featured a classical three span suspension bridge with two offshore piers.

Lambertini JV
With main designers F. Leonhardt and F. De Miranda, the proposal was based on a record-breaking 1400 m cable stayed bridge (not surprisingly for the famous German designer) with two deep water offshore piers.

Musmeci JV
As spectacular in shape as questionable in feasibility, the Musmeci 3000 m three dimensional cable network attracted attention from the media and the architectural establishment at the time. Even after forty years one can still find some of its faithful supporters today.

Grant JV
The only non-bridge solution considered worth a prize, this was a road and rail triple floating tunnel, located 30 m underwater and anchored to the seabed through steel cables, using ballasting and Archimedean thrust to minimise the stresses in the tunnel structure.

None of the proposals submitted demonstrated a definite feasibility at the time. Leaving aside the Grant floating tunnel and leaving also aside cost considerations, the bridge solutions lacked justification of feasibility on two main issues:

- For multi-span schemes, the possibility of building safe foundations and substructures on the Strait seabed, at a depth of more than 100 metres, in a highly seismic zone and with strong marine currents, when the available information was at a geological scale only, without any direct geotechnical data, was not certain.
- For single span schemes, the aerodynamic stability of a suspended span of 3000 m or more had not yet been proven, when the longest span at the time was the Verazzano Narrows Bridge with 1298 metres.

Nevertheless, in spite of its limited technical outcome, the call for ideas had the great merit of stirring attention in Italy and around the world on the possibility of the crossing, giving impetus to the process leading to the constitution in 1971 of the State owned company Stretto di Messina SpA (SdM) (see Chapter 2.), whose statutory task was to design, build and operate a permanent rail and road link across the Messina Strait.

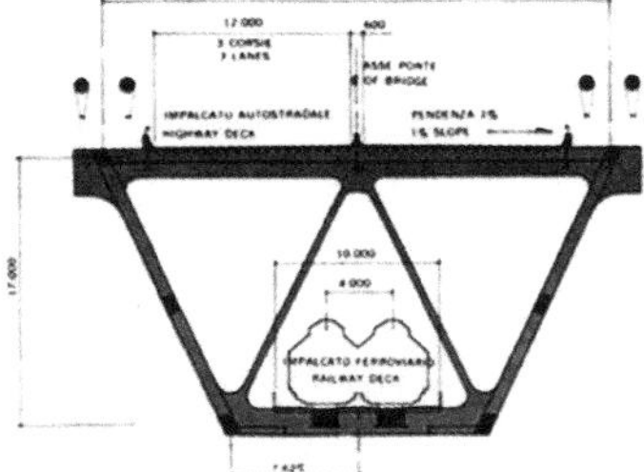

◄ Figure 1.13
Gruppo Ponte di Messina (GPM) JV.

Figures from 1.13 to 1.19 are drawn from "L'industria delle costruzioni" magazine March–April 1971 – courtesy of Edilstampa s.r.l. – Italy.

Figure 1.14 ▶ Technital JV.

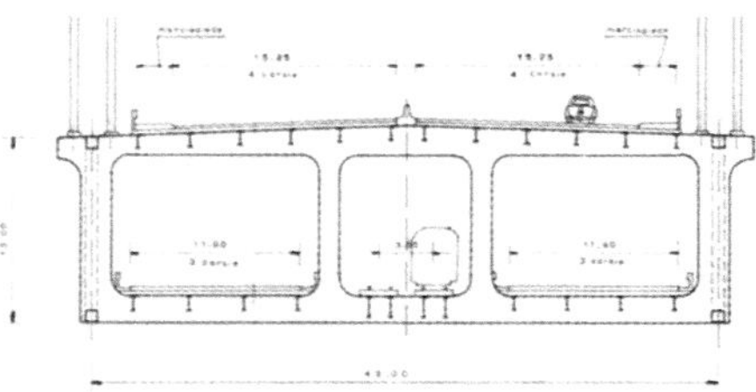

Figure 1.15 ▶ Calini-Montuori-Pavlo JV.

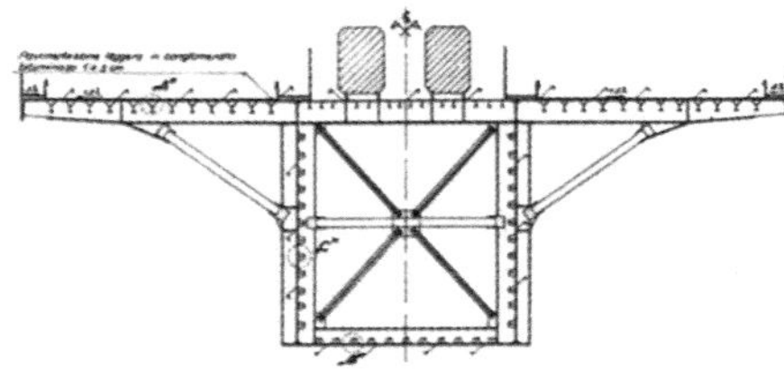

Figure 1.16 ▶ Lambertini JV.

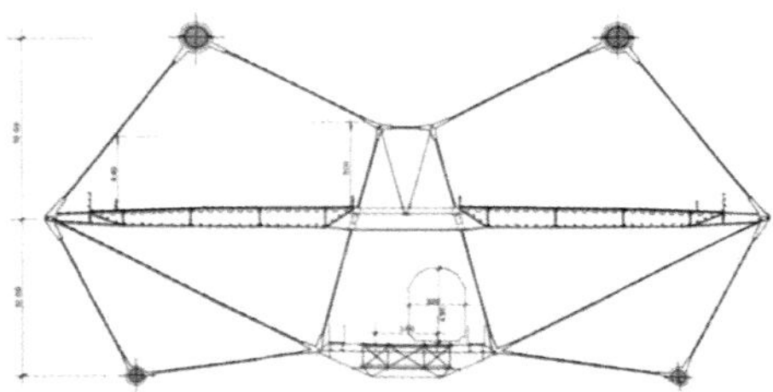

Figure 1.17 ▶ Musmeci JV.

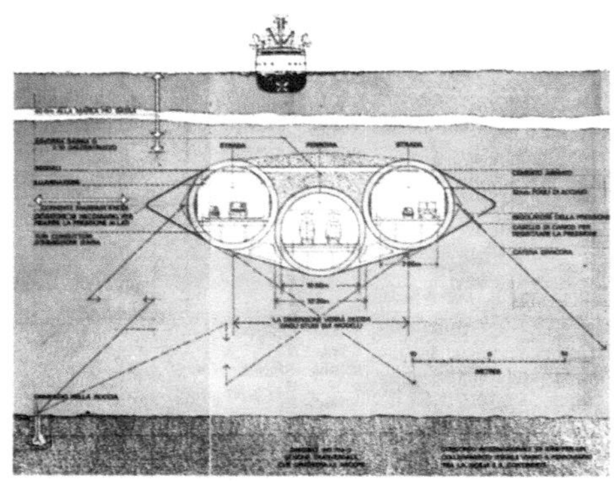

Figure 1.18 ▶ Grant JV.

◄ Figure 1.19
The six proposals awarded a first prize – general elevations (in the same sequence as above).

Gruppo Ponte di Messina (GMP) JV

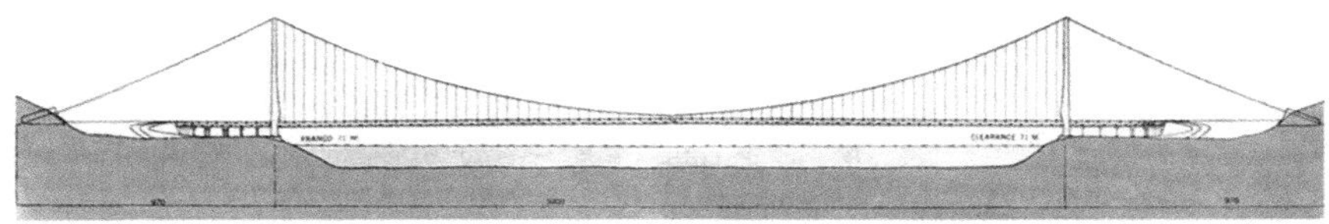

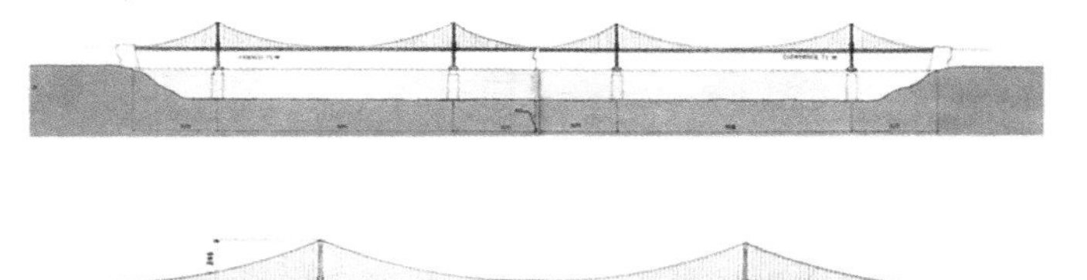

Technital JV

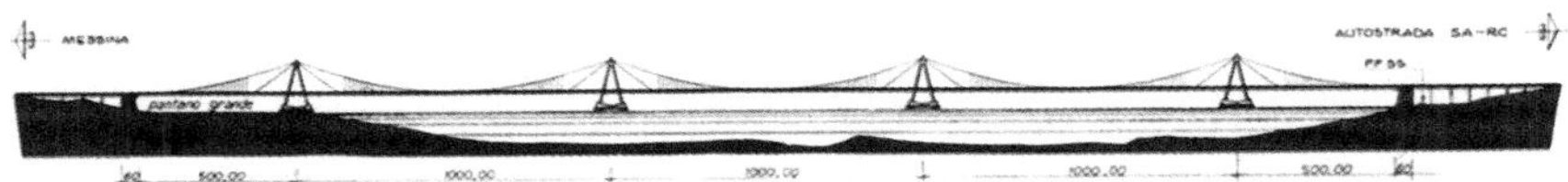

Calini-Montuori-Pavlo JV

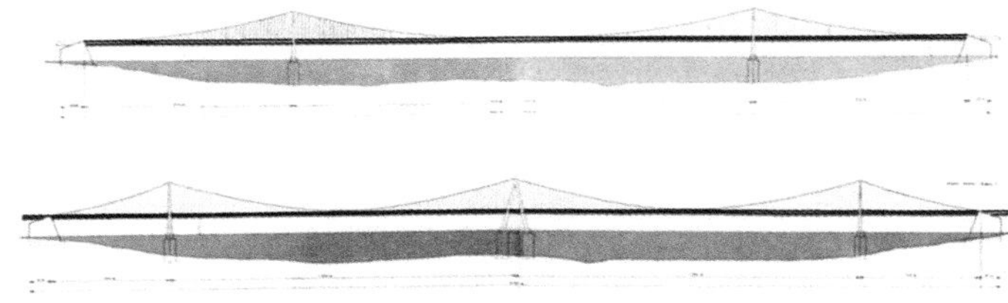

Lambertini JV

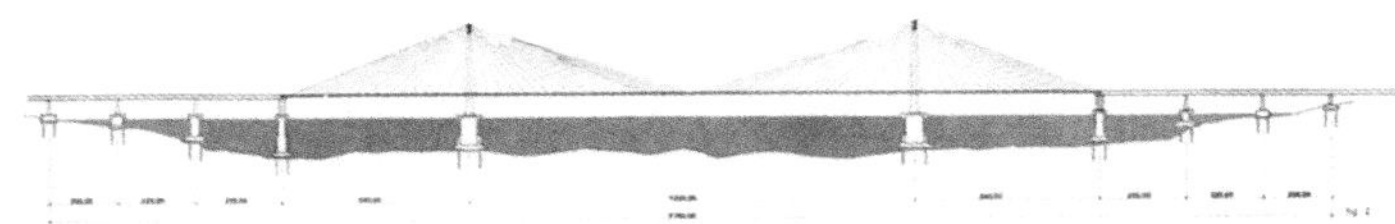

Musmeci JV

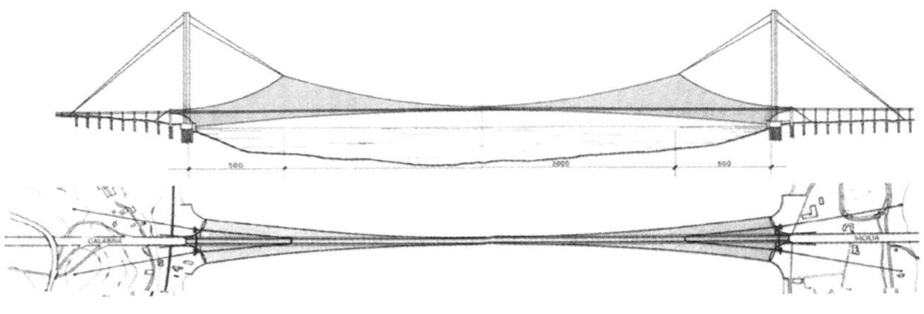

Grant JV

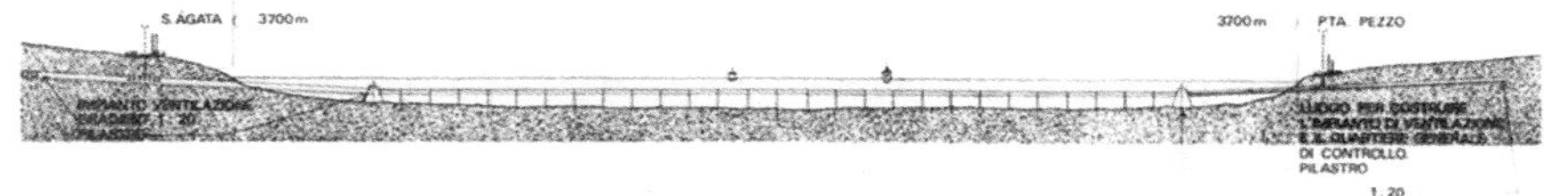

Curious proposals for the Messina Crossing

A witness to the peculiar interest in the project is found in the flourishing of proposals from all kind of sources, including non-specialist professionals and ordinary people, both during the call for ideas and afterwards. None of these provided any engineering contribution, but they showed how deeply and emotionally the problem was felt in the national awareness. Such ideas were often put forward to or through newspapers, politicians or authorities in general, reaching in some cases considerable media attention, more because of the stubbornness of their authors than any actual technical merit. Among the hundreds of these, some examples are shown below, including an enormous simply supported perfect circle, about which no specific comment on feasibility is needed!

Figure 1.20 ▶ Brasini proposal.

Figure 1.21 ▶ Rasconà proposal.

Figure 1.22 ▶ Sicily newspaper, November 1860.

Figure 1.23 ▶ Fichera proposal.

Figure 1.24 ▶ Perugini proposal.

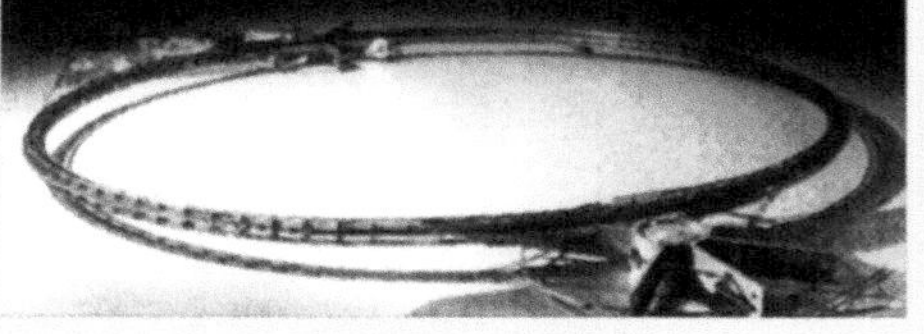

Something that attracted considerable attention was the idea of having something supported by floating elements. These included floating artificial islands, or a floating bridge supported by vessels (such as a collection of out-of-service aircraft carriers), or, at the most extreme, a proposal envisaging corkwood islands having the double benefit of solving both the Strait crossing problem and reviving cork tree plantations in Sardinia which were declining at the time. Then there was the idea of closing the Strait with a dam, carrying the road and rail on top. Conscious of the hydraulic problems involved, the author suggested allowing water to flow through a narrow, deep channel cut at the north-east point of Sicily, with a short span bridge to allow the crossing. Or, most intriguing, was a proposal based, according to the author, on the concept of "Cyclopean anti seismic half spheres" which comprised a large bridge supported by gigantic half spheres resting on the seabed, which in case of earthquake would roll and remain undamaged.

1.5 Evaluation of options

1.5.1 Gruppo Ponte di Messina (GPM) feasibility studies

SdM did not immediately become active after its constitution in 1971. On the contrary, one of the six groups awarded a prize in the call for ideas, the private company Gruppo Ponte di Messina (GPM), which was owned among others by important Italian companies such as Fiat and Impresit, decided to invest considerable efforts and resources in the project, aimed at carrying out a technically founded feasibility assessment. A team of experts and technical staff was established and started systematically collecting available data on the site and considering possible alternatives for the crossing.

One of the main assumptions made by GPM was that no type of tunnel solution, floating or underground, was of any possible interest, given the specific situation. Hence they focussed their attention on bridge schemes only.

◄ Figure 1.25 Submarine Investigations by J.Y. Cousteau.

Gathering of site data mainly involved the assembly, correlation and rationalisation of existing information, with less effort at this stage being devoted to conducting new site investigations. Some of the activities were also more media-centred than of any technical interest, such as the underwater images taken by the famous French oceanographer Prof. Jacques Yves Cousteau (Figure 1.25).

Some specific geotechnical campaigns were undertaken, but these were of limited extent, being either onshore or in shallow waters, and hence did not provide any substantial contribution in relation to deep water foundation design.

However, bridge design development and tests were on the contrary tackled with considerable attention and in some depth. Many possible crossing solutions were considered. Starting from the three solutions presented by GPM in 1969, the following considerations were soon developed.

For multi-span schemes, it was immediately understood how formidable, even if feasible, would be the construction of offshore deep waters piers and foundations, not only from a technical standpoint but also because of the extremely high costs. Therefore two multi-span schemes of 1969 which featured five or two offshore piers were abandoned; a step that put aside any "classical" scheme for the crossing. (Figure 1.26) Instead, an innovative two span suspension bridge scheme was developed, with a centre anchor tower and pier and two 1600 metre spans. The central anchor tower was needed to limit deflections under imposed loading conditions with one span loaded only. (Figure 1.27)

Only slightly larger than existing spans at the time, such a scheme was indeed certainly feasible as far as the suspension structure was concerned, and in the mid seventies became the favourite proposal within the GPM design team as the most promising solution. The single span alternative was not dismissed but was considered with caution, as being something too far from existing technical knowledge. Significant attention was therefore devoted to the study of solutions for the off-shore bridge tower foundation,

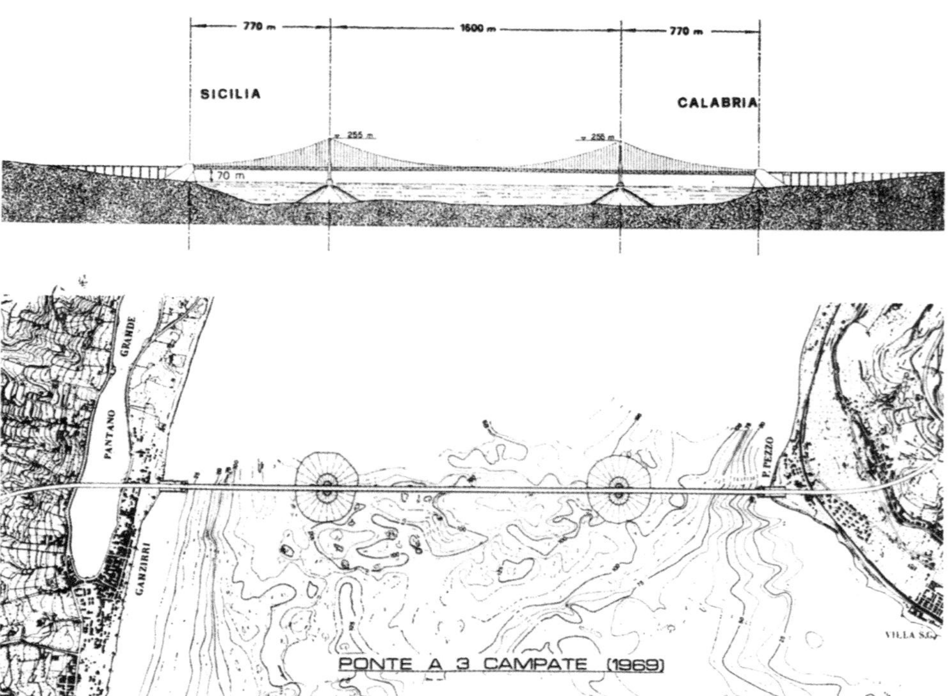

Figure 1.26 ▶ The 1979 GPM three span scheme with two offshore islands.

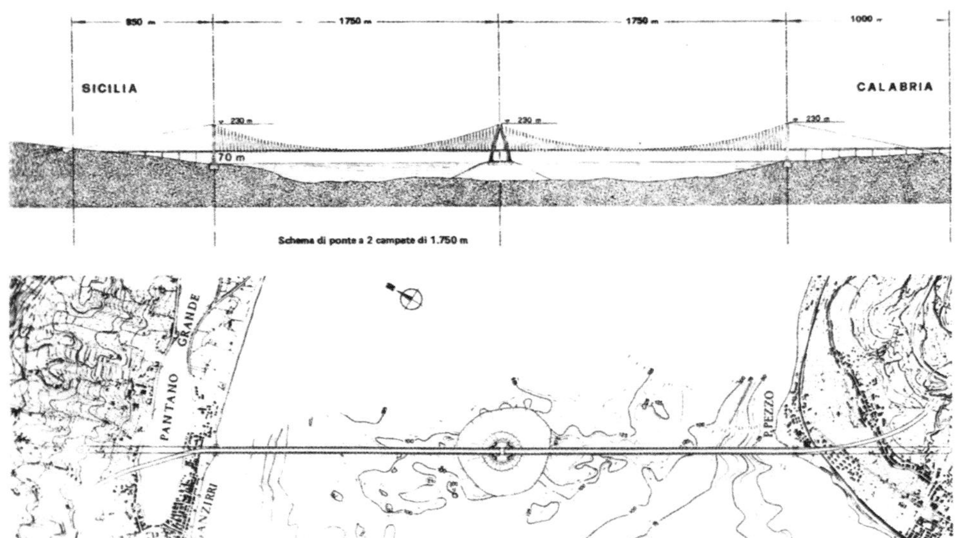

◀ Figure 1.27
The 1979 GPM two span scheme with single offshore island.

considering several different reinforced concrete and steel alternatives, as well as an artificial island. (Figure 1.28) The analyses carried out for the design saw the innovative use at the time of finite element sub-structure analysis techniques and modal response spectrum procedures.

Nevertheless, while work was progressing, the difficulties involved with constructing deep sea foundations in the Strait became more and more apparent. The artificial island was at the limits of feasibility and of a size such as to induce disputable consequences for the current flow. Concrete piers were considered feasible but with two substantial uncertainties, namely the actual geotechnical and fault conditions about which little was known, and the constructability in the complex marine environment of the Strait.

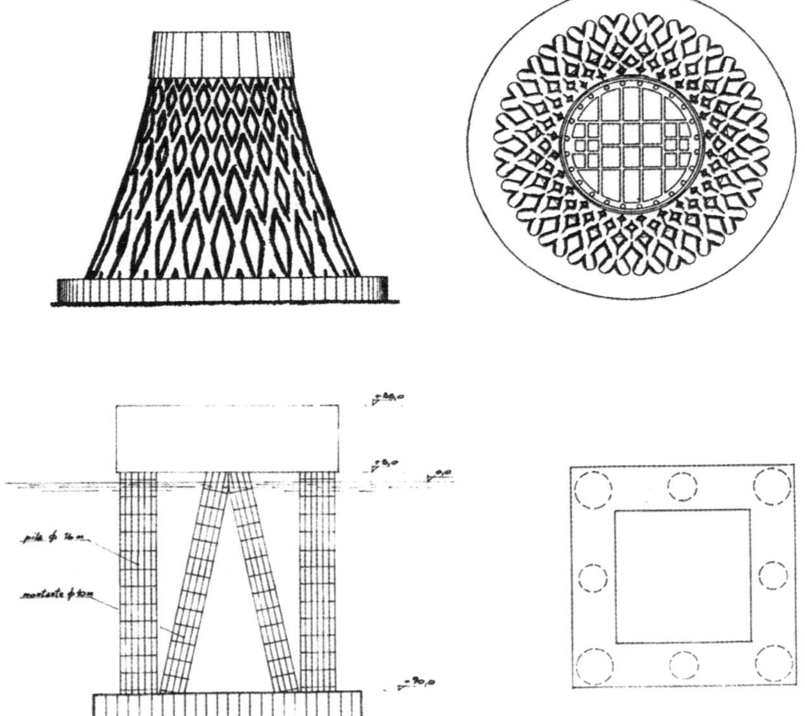

◀ Figure 1.28
1979 GPM hyperboloid concrete pier (left) and large struts concrete pier (right).

On the ot her hand, the design solutions considered and the detailed evaluations carried out showed many advantages for a single span solution in terms of improved reliability and performance. A significant contribution in such a direction was made at this time by the entrance into the picture of Dr. William C. Brown, the well known Scottish bridge designer. It is reported that, after due consideration, in a meeting with the then CEO of GPM, Gianfranco D. Gilardini, who was reluctant to leave the two span scheme to which so much effort had been devoted, Dr. Brown stated that "In Messina you have a splendid case for a single span suspension bridge". Two points were of paramount importance in moving in this direction:

- The in-depth understanding which had been reached regarding the behaviour of a suspension bridge under heavy railway loads. This had been made possible through the development of some of the first advanced large displacement software packages for the analysis of suspension bridges, not only in static conditions but also for determining railway runnability. This showed clearly that the stiffness of the large span was such as to make deflections and slopes under railway loads less critical than for shorter spans. In other words, there were no fundamental strength or serviceability problems for road or rail traffic.
- The birth in those early years of the embryonic concepts for aerodynamically designed highly stable bridge decks that were going to be fully developed ten years later. The design was supported for the first time by a systematic campaign of wind tunnel tests, enabling the comparison of alternatives and providing experimental proof that achieving sufficient stability at such spans was indeed possible.

It also emerged that while the construction cost of the two alternatives was close, the uncertainties and risks in the construction process were larger for the two span solution due to the need to engage with the deep sea environment.

GPM concluded its activities in 1979, issuing a feasibility report that stated as feasible both a two span solution with uncertainties surrounding the construction of the offshore foundation and the scarce knowledge of the geotechnical conditions of the sea bed, and a single span solution which was considered to be the preferred one in terms of robustness and performance. (Figure 1.29)

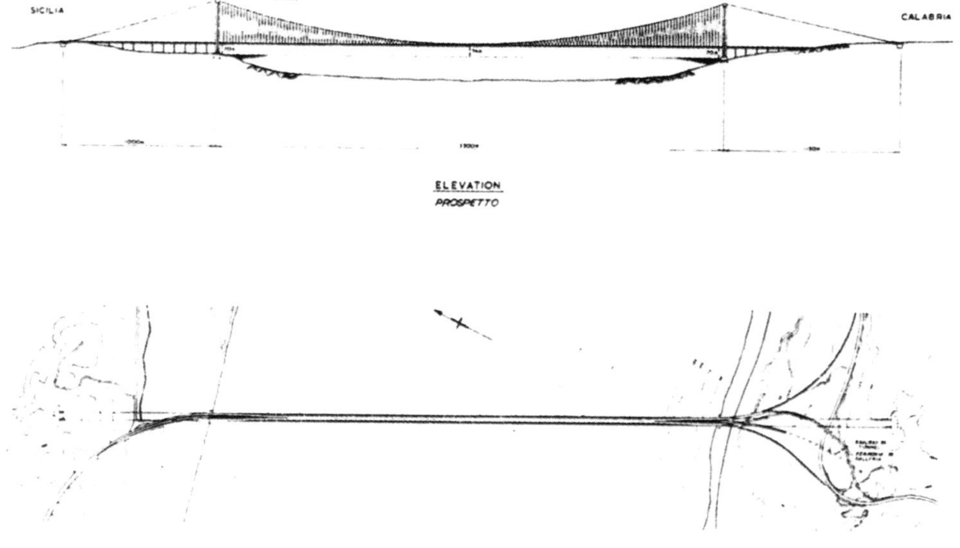

Figure 1.29 ▶ GPM single span scheme.

1.5.2 Stretto di Messina SpA feasibility studies

After a break of a few years that did not see any activity, SdM eventually came into the picture in the early eighties, tackling the challenges with unprecedented vigour and a global approach, thanks largely to the availability of adequate resources. The main steps forward accomplished in the following ten years of intense activity were concerned with:

- Campaigns and site survey activities, allowing the preparation of a database of information and an understanding on the Strait global environment, whose synthesis will be given in Chapter 5. Significant progress was achieved regarding:
  - The geology and tectonics of the Strait, through a vast onshore and offshore deep seismic survey, allowing the definition of the geological strata, together with their mechanical properties, and the mapping of the fault system in the area.
  - The wind environment, through the installation of a permanent wind and climate monitoring system located on the towers of the existing electrical crossing of the Strait, which has now collected more than 25 years of continuous information.
- Systematic wind tunnel experimental testing, starting from the assessment of the performance of the GPM solutions and extending to further schemes and their optimisation.
- Full scale experimental testing on large existing suspension bridges, aimed at improving and calibrating the numerical models and refining the concepts adopted for the Messina studies based on actual bridge behaviour. Such campaigns comprised:
  - Wind response, measured on the Humber Suspension Bridge;
  - Thermal, traffic and environmental response, measured on the Fatih Sultan Mehemet (Bosporus 2) Bridge;
  - Exchange of data with Honshu Shikoku Bridge Authority in Japan regarding the railway loading response of the three road and rail suspension bridges on the Kojima-Sakaide route (the Shimotsui Seto, Kita Bisan Seto and Minami Bisan Seto bridges).
- Systematic, in-depth analysis, comparison and development of all possible design alternatives for the crossing.
- An organised technical framework for feasibility studies and design development, equipped with state of the art computing technology and research level innovative software.

SdM classified the alternative crossing solutions under consideration into three categories:

- Underground tunnels;
- Floating tunnels, whether cable-anchored at the sea bed or supported by underwater piers;
- Bridges, both single and multi span.

Possibilities were systematically explored, reviewing the solutions of the 1969 call for ideas as well as those proposed by GPM and other groups, in addition to developing further schemes.

This huge work ended in 1986, with the issue of a feasibility report, whose main conclusions were:

- Underground or underwater tunnels are virtually unfeasible from a technical stand point, as well as being hugely expensive beyond any cost/benefit ratio of possible financial interest;
- Among bridge schemes, solutions with more than two spans are considered to be of lesser technical and financial performance, with genuine

interest being restricted to two-span and single span suspension bridge schemes;
- The robustness and performance of the single span solution is superior to the two span option.

Such conclusions may appear to be similar to those already reached by GPM, and in concept they were indeed, but they carried significantly more weight because they were supported by a level of environmental knowledge and an advanced innovative design solution which was of a different order of magnitude altogether.

The feasibility studies also involved the participation and final review of an International Consulting Board, comprising leading professionals and scientists in the field (F. Biesel, K. Borok, D.M. Brotton, A.D. Davenport, J.L. Durkee, N.J. Gimsing, C.C. Ladd, R.I. Madariaga, A. Prud'homme, A.F. Prunieras, E. Rosenblueth, J. Shlaich, P. Tapponnier, J. Tajima, A.F. Van Weele). More details on the reasoning behind the above conclusions are given in the boxes that follow.

Why the 1969 & other bridge alternatives were discarded

Arguments can be given for the broad classes of different options, selecting the dominant aspects for each.

Multi-span bridges. Several schemes, whether suspension, cable-stayed or hybrid schemes, did see a relevant number of deep water foundations, from one to several. It has already been outlined how, at the time, information on the local geotechnical conditions was very poor, and in this sense any such solution did lack proven feasibility, being inevitably based on general, or generic, geological data only.

Figure 1.30 ► Option proposed by Nervi.

From "L'industria delle costruzioni" magazine March–April 1971 – courtesy of Edilstampa s.r.l. – Italy.

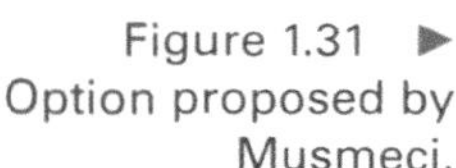

Figure 1.31 ► Option proposed by Musmeci.

Figures from 1.31 to 1.33 are drawn from "L'Ingegnere e l'Architetto", official magazine of the Italian Association of Engineers and Architects, no. 11 – 1971 – courtesy of ANIAI – Italy.

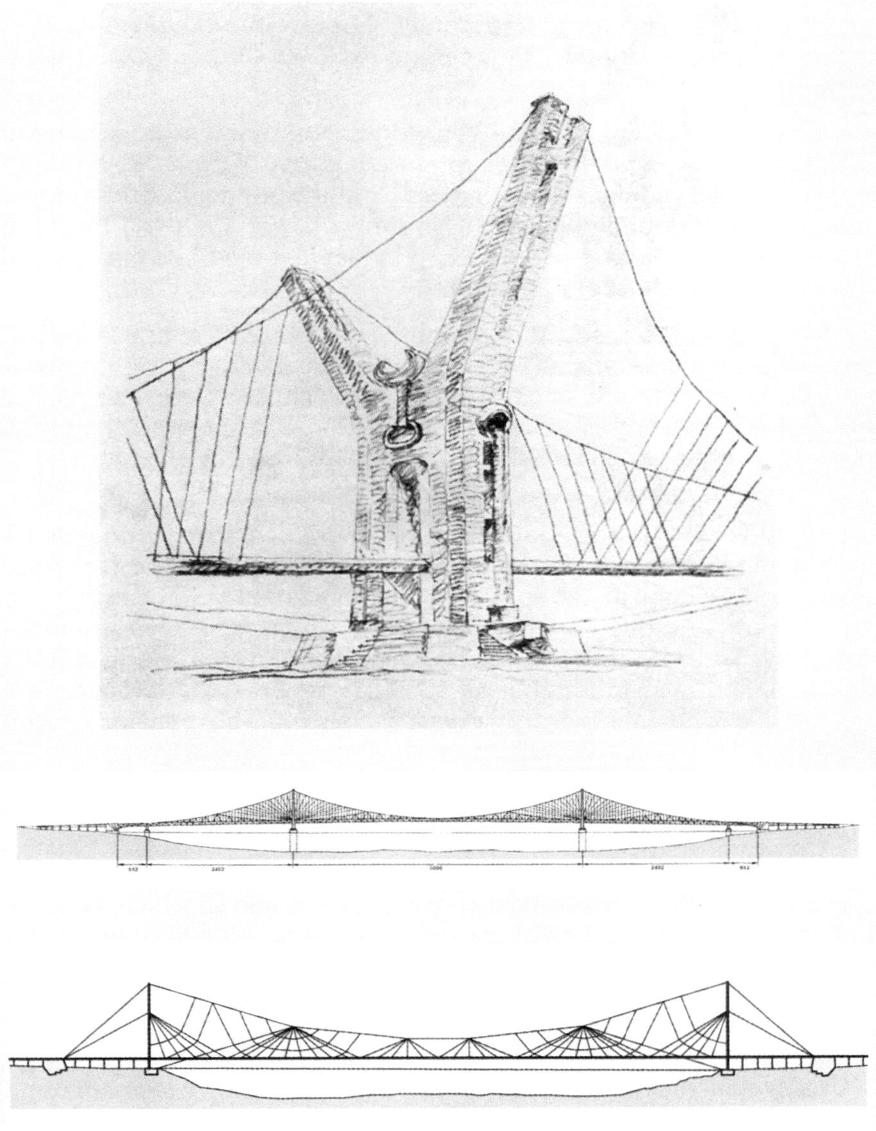

◄ Figure 1.32
Option proposed by Samonà.

◄ Figure 1.33
Option proposed by Maugeri.

◄ Figure 1.34
Option proposed by CMF.

This uncertainty was going to be resolved only in the late eighties, see paragraph 1.5.3, with conclusions not yet anticipated in 1969. Furthermore, and more important at the time, it was soon clear that the cost of such foundations was very significant and that envisaging more than one of them was definitely not cost effective for any bridge solution. This limited further considerations to two span bridges with one offshore foundation only.

1969 single span suspension bridge solutions. The key element of the feasibility of a single span suspension bridge (3000 m or more) is the aerodynamic stability. The deck configurations proposed in 1969 for the only such scheme, presented by GPM, were based on classic large truss girders, statically effective but definitely aerodynamically inadequate. Hence all the 1969 schemes as presented were without question unfeasible.

3D cable network single span bridges. The main example of these is the Musmeci scheme already presented and shown again in Figure 1.31, but other parties also presented similar, though simpler, concepts. Proposals by Nervi, Samonà, Maugeri and the CMF Company are shown in Figures form 1.30 to 1.34.

However, it is evident that the very idea behind any such scheme is basically ill-conceived. The motivation to seek such complex arrangements is usually concerned with increasing the stiffness of the bridge, often on the unfounded assumption that a larger span will be more deformable than a smaller one, when in fact just the opposite applies. This is discussed further in Chapter 6.

When the span increases, the geometric or "gravity" stiffness associated with the cable tension increases dramatically, and even if absolute displacements are large, relative displacements such as slopes and cross-falls definitely decrease. A very large suspended span thus has less need to be "stiffened" than a smaller one.

In other words, such complex cable arrangements are basically unnecessary. If existing rail or road suspension bridges have not needed a 3D stiffening cable system then indeed the same will stand for larger ones. Such systems are adopted only for very small or light structures, e.g. suspended pipes on small spans and suspension bridge temporary catwalks. Note that already on the Akashi Kaikyo bridge with its 1991 m span it was found that no 3D cable network ("storm system") was needed for the temporary catwalk and a simple suspension proved adequate.

Furthermore the complexity of such a 3D cable system implies a large number of drawbacks:

- Increased main cable stresses due to the opposite downwards and sideways pull of the network, with consequent size and cost increases.
- Considerable difficulties in construction: handling a 3D network for single small ropes is standard practice, but the technology needed for handling the main cables in a 3D network at this scale is currently not available.
- The need for further ground anchorages of considerable size.
- Unsolvable maintenance problems, with very many elements and connections which are difficult to reach and inspect as well as virtually irreplaceable. If the maintenance of simply suspended main cables is already difficult enough, a 3D network would be a nightmare.

In addition, some schemes envisage fully unfeasible details, such as the mid-air connection of a straight cable/stay with the main cable saddle point of the Musmeci scheme. This is easy on a small rope but impossible for a suspension bridge cable one metre in diameter. Such aspects show a deep lack of awareness of scale effects, so important for cable supported bridges. (See again Chapter 6.)

In conclusion, such schemes have no merit beyond the fascination of their appearance and were not seriously considered for further development. As a final remark on this matter we will refer to a statement by Jackson Leland Durkee, the recently deceased American cable bridge designer, also as a memory to his lucid engineering understanding. In his final report as a member of the International Board, Jackson wrote:

"Turning now to the aerial typologies, it is evident that some of the crossing concepts and proposals are altogether impractical and fall

into the category of 'structural acrobatics'. I would include in this category the Musmeci Group proposal .... The cable networks as sketched in those proposals portray structural technology that is not only speculative, but also quite unnecessary. There is no need to resort to such complicated, unproven and indeed bizarre supporting systems when practical, well established and straightforward structural configurations are available ...... The Report (the 1986 SdM feasibility report, n.a.) declares cable-network tension structures to lack feasibility, and I certainly concur."

Why underground tunnel alternatives were discarded

The reasons to discard bored tunnels in the Strait are straightforward. The only possible crossing area is the underwater ridge immediately south of the narrowest section, on a stretch of water of about 3500 m width, see figure below.

The water depth is 100–120 metres for most of the way, and the geology mostly comprises the Messina gravel formation, superficially cemented and loose in the lower strata. Thus, with such a permeable material and high water pressure, tunnel boring would not be easy, requiring special techniques to permit advancement.

In such a scenario, it was determined that in order to allow sufficient cover for a TBM to operate, the tunnel should be placed not less than

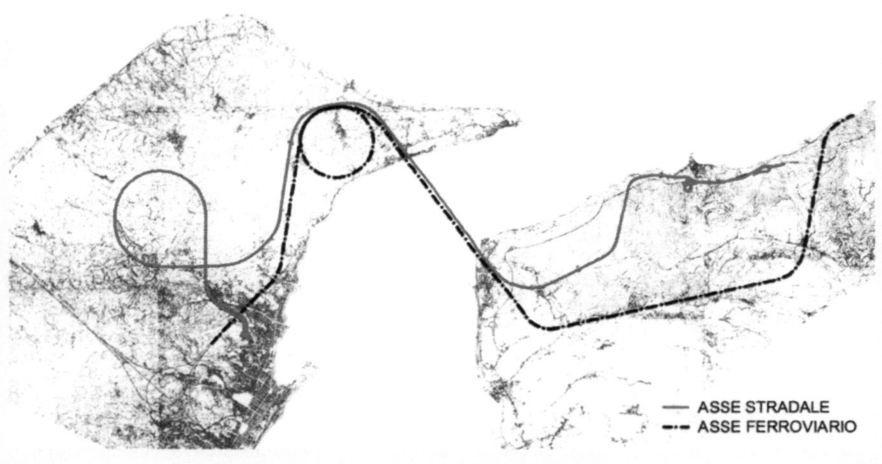

◄ Figure 1.35 Bored tunnel alignment.

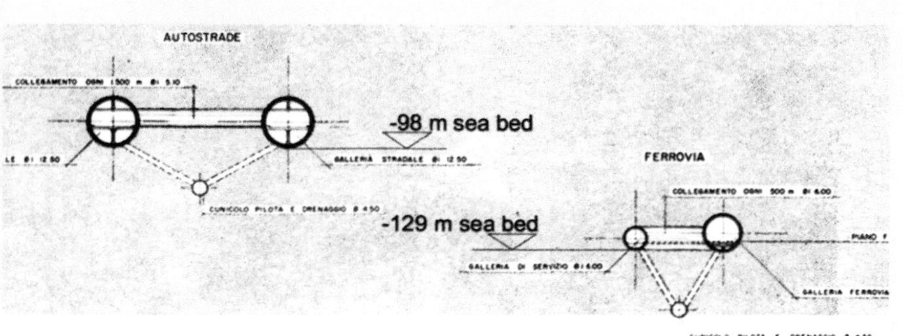

◄ Figure 1.36 Bored tunnel cross section.

50 metres underground, i.e. about 150–170 metres below sea level. As a result, the length of the access tunnels on land would be several dozen kilometres, in particular for the railway tunnels due to slope limits for train runnability.

The tunnel would also encounter several potentially active seismogenetic faults, and would therefore need to accommodate significant local displacements. Suitable measures are not unthinkable, e.g. based on double linings, but the complexities and risks involved are substantial.

In summary, while bored tunnels in the Strait are certainly feasible, their principal drawbacks are:

- Difficult boring conditions at the depths required in the specific geotechnical conditions;
- Uncertainties in costs and technical solutions for the advancement through the fault areas;
- Extremely high construction costs, evaluated to be several times higher than for any bridge scheme;
- All the negative functional aspects of very long tunnels with free road traffic connected with exhaust gas handling;
- The risks connected with accidents-terrorism-sabotage within the enclosed tunnel body, for a piece of infrastructure that would possibly become an internationally sensitive target.

Hence such schemes were not selected for further consideration.

It is worth noting that no party has ever seriously supported bored tunnel options in the recent history of the solutions proposed for the crossing.

Similar arguments, with certain due differences, also apply for immersed tube tunnels resting on the seabed and artificial dams supported on the seabed.

Why floating tunnel alternatives were discarded

The floating tunnel alternatives proposed at and since the 1969 call for ideas can be classified into two types:

- Tunnels supported by piers, with buoyancy forces nearly balanced by ballasting, hence in a nearly hydrostatically "neutral" condition.
- Tunnels anchored by cable systems to the sea bed, with buoyancy forces larger than the ballasting so as to keep the anchor cables tight in service. Besides the Grant proposal, a further example is the scheme developed by SdM illustrated below.

Several variants are possible for both types, such as for cross-section (e.g. with a single road and rail tunnel or with three separate ones), for materials (concrete or steel) and for other aspects.

Regarding the general configuration and structural behaviour, some considerations apply to both types, while others are specific for each. A basic common aspect stems from the particular configuration of

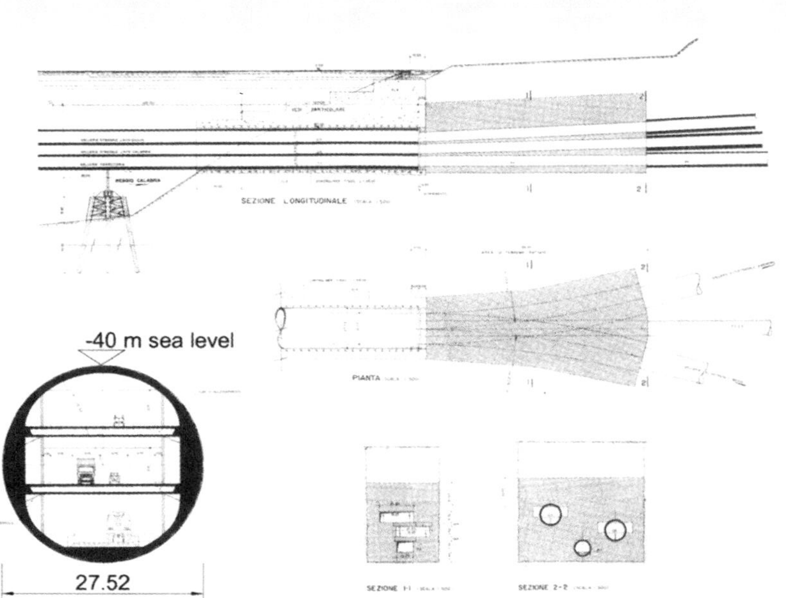

◄ Figure 1.37 SdM floating tunnel proposal.

the sea bed in the Strait (see also Chapter 5). The shallowest section (100–120 metres in depth) is also almost the narrowest one – about 3.5 km, compared to 3 km at the narrowest point). This fact, and the plan alignment of the Strait at this point, causes violent tidal currents, as described in Chapter 5, due to the flow of sea water from the Tyrrhenian to the Ionian sea and vice versa. Daily current speeds peak at 3–4 knots, with complex flow geometries, such as to strongly influence navigation, even for modern vessels.

This makes it simply unthinkable to propose a floating tunnel solution with the alignment located at the narrowest point, as the currents and sea conditions are such as to render the construction infeasible. This has been demonstrated by the far more modest recent experiences of placing power lines on the sea bed and of the marine geotechnical investigations carried out by SdM.

For a flow current speed low enough to permit construction, it is necessary to select an alignment with a much larger distance between the shores and also a much larger depth. Possible locations exist further south, with a shore to shore distance of about 6 km and a sea depth of about 300 metres. The larger length implies higher costs, although this is partly balanced by the more direct connection between the main towns of Messina in Sicily and Reggio Calabria on the mainland. (See Figure 1.38)

A further consideration is that to avoid excessive influence of marine waves at the surface, the depth of the floating tunnel must be not less than about 40–50 metres.

In terms of structural configuration and response, the two types show a number of differences. Floating tunnels supported on piers must have rather large spans of at least 500 metres, which would require the construction of perhaps 10 to 12 underwater piers whose height from the seabed would be about 100 to 250 metres. Smaller spans with more piers would imply added construction complexity and costs taking the scheme beyond any engineering good sense. Such large spans can easily handle the live loads, and the response under some accidental loads

Figure 1.38 ▶ Alignment of the SdM floating tunnel proposal.

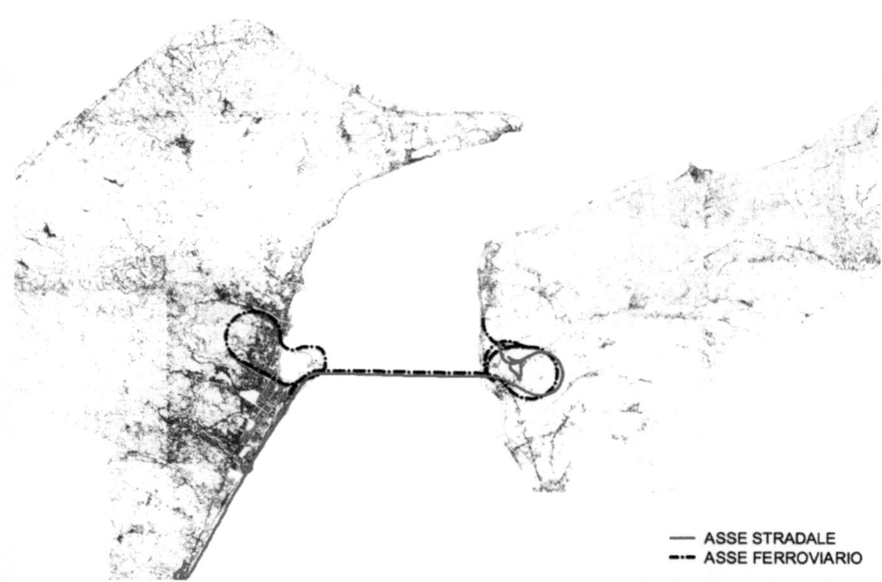

(e.g. sinking ships) can also be tackled. On the contrary, what has been found to be absolutely uncontrollable is the seismic response due both to the very large hydrodynamic inertial effects and the direct propagation of rarefaction-compression waves from the ground through the water, connected to the known seismic sources in the vicinity.

Cable anchored solutions would have the same problems, although mitigated by the shorter spans as the cable supports must be placed at about 50–100 metres, much closer together than the piers. Nevertheless, initial cable forces must be very high to compensate for the extremely large stress variations induced by the hydrodynamic response to earthquake events and avoid cable slackening that would cause uncontrollable flexure in the tunnel body. Such high tension imposes significant problems for the seabed anchor systems. It has been shown that to control the response it would be necessary to construct, at the said depths, a mixed mechanism anchor system which resists horizontal forces by gravity and vertical uplift using bored piles, since a gravity only system to handle both actions would reach excessive dimensions. This implies the construction of hundreds of anchor points in the sea bed using technologies which, if not unthinkable, are at the limits of present capabilities and would require long and difficult research and development.

An issue for either type of solution is also the provision of expansion joints or deformable sections capable of handling the relative movements between the shores which are predicted to occur mainly during earthquake events.

In addition to these specific points, and very many other lesser concerns, further weaknesses of such a solution are:

- Difficulties and risks concerned with long underwater construction in open sea waters, with technologies at or beyond present limits;
- Uncertainties in determining a robust estimate of costs, whose preliminary evaluations show values several times higher than for any bridge scheme;

- All the negative functional aspects of very long tunnels with free road traffic connected with exhaust gas handling;
- The risks associated with accidents-terrorism-sabotage within the enclosed tunnel body, for a piece of infrastructure that would possibly become an internationally sensitive target;
- For the cable anchored solutions, the risk of damage to the cables themselves, accidental or not, implying the possible loss of the whole structure if half the cables at any section are severed.

In such a context, it is considered unthinkable to propose a floating tunnel solution in any tender procedure requiring a contractor to take responsibility for the outcome of the process, or to find any reliable company that would provide an insurance guarantee. To quote from the report of the International Board: *"The underwater tunnel has many unexplored and unproven features and introduces unforeseeable environmental difficulties, making it speculative with respect to cost and construction time. Accordingly, the feasibility may be questioned and we are unable to endorse this alternative."*

In essence, although to say that a floating tunnel solution is absolutely unfeasible may be too strong a statement, an accurate conclusion on the matter can be found in the words of professor Jörg Schlaich: *"If the destiny of mankind depended on the construction of this tunnel and we had unlimited time and resources available, then I believe we could make it".* The authors of this book concur.

These solutions were thus discarded after careful consideration, and it is indeed surprising to still find some supporters for such an application in the Messina Strait.

Why a single span bridge was the top choice already in 1986

The feasibility studies carried out by SdM clearly showed that:

- The added cost due to the large 3 km span, arising mainly in the suspension system, towers and substructures, was well balanced by the high cost of a deep water foundation and pier for a two-span scheme which also showed greater construction uncertainties and risks.
- Due to the much higher geometric stiffness, the service performance of the single span scheme was superior in general to the twin span solution and excellent in terms of maximum gradients for railway runnability.
- Avoiding an off-shore pier in the middle of the Strait was a definite benefit for navigation, in waters with high international longitudinal traffic and difficult conditions due to strong and complex currents.
- Avoiding altogether any marine construction added considerable certainty in the time and cost evaluation in general.

The big question for the single span was the ability to prove adequate wind stability, and the concepts leading to highly stable deck configurations had already been laid in the late seventies and by 1986 had been developed far enough to provide the basis for a confident feasibility assessment.

1.5.3 Stretto di Messina preliminary and tender design

The following two years, 1986-1988, were devoted, besides further development of concepts and schemes, to the fundamental task of comparing the two span and single span alternatives, with the second preferred but with the former closer to existing experience. Preliminary designs were prepared for both, looking in depth at construction and cost issues. The general arrangements are shown below. (Figures 1.39 to 1.41) Note how both alternatives are characterised by stiff, slender triangle towers. For the single span this was influenced by the not yet completely overcome thinking that extra stiffness was needed with respect to the pure suspension system.

While the conclusions of the earlier feasibility analyses were confirmed and strengthened by this work, indicating several reasons for preferring the single span and abandoning the two span scheme, such a move away from existing or planned experience was too daring a decision to be taken lightly, so further geotechnical studies were undertaken. (Figure 1.42)

The uncertainty on the multi-span bridge options had always been the scarce knowledge of actual detailed geotechnical information for the offshore pier foundation. Hence it was decided in 1988 to embark upon the large effort involved in undertaking a deep water geotechnical offshore campaign. This allowed for the first time, and not without considerable difficulty, ground samples and strata definition to be obtained from borings in the seabed, via the use of a large multi engine dynamic positioning vessel.

The soil conditions encountered (see next box and Chapter 5) were such as to forbid, at the current state of technology, any feasible solution for a substructure foundation in the middle of the Strait. The same statement it is believed would apply today, even after twenty more years of technology advance.

Figure 1.39 ▶ SdM two span solution.

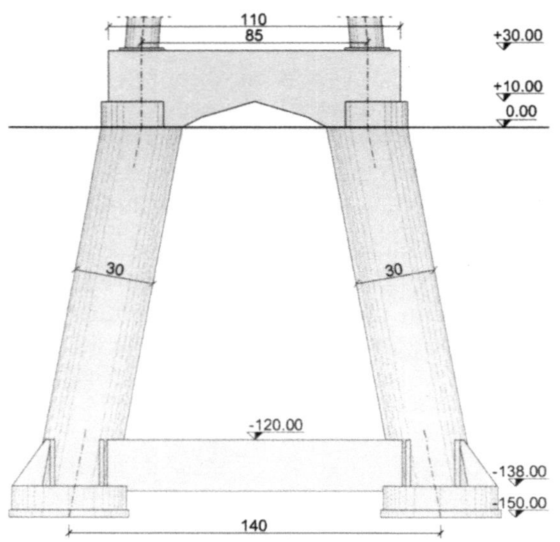

Figure 1.40 ▶ SdM two span solution, centre pier.

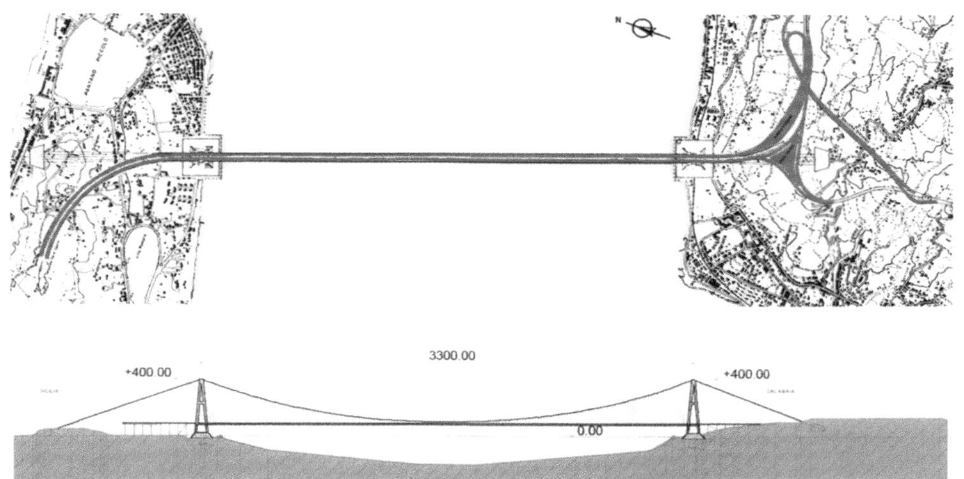

◄ Figure 1.41
SdM 1986 single span solution.

◄ Figure 1.42
The 1988 offshore geotechnical campaign.

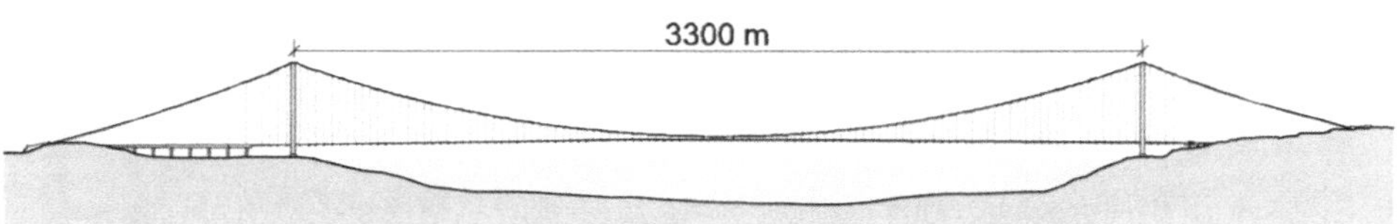

◄ Figure 1.43
The 1992 "Progetto di Massima" design.

With such a strong statement, the way forward was clear: a record breaking 3300 single span suspension bridge from shore to shore. This decision was taken when the longest bridge in the world was the Humber Bridge with a 1410 metre span, and construction of the Akashi Kaikyo bridge with a planned 1990 metre span had just been initiated.

The challenges involved and the solutions found in connection with the particular evolution of suspension bridge characteristics and responses with such a long span will be discussed in Chapter 6. This present chapter closes with the remark that such a target, while at the frontier of the state of the art, was no longer such a distant, awesome possibility but something approaching reality, thanks to the understanding formed during more than ten years of effort leading to the first completed tender design *("Progetto di Massima")* of the bridge in 1992. The chronicle of the path leading from early concepts to the final design for lightweight, highly stable bridge decks is given also in Chapter 6.

The elevation of the 1992 bridge design is given. (Figure 1.43) Note how the towers have changed into simple, longitudinally flexible frames, having fully appreciated that the intrinsic stiffness of the cables is more than sufficient to achieve adequate tower performance.

Why an offshore foundation in the Strait is unfeasible

At the intermediate pier location for the two span suspension bridge the water depth ranges between 146 and 157 metres. The subsoil at the site can be summarized as follows: the sea bottom is covered by a few metres of loose sand, overlaying a layer of very hard conglomerate having a thickness 2 and 3 metres, laying in turn on slightly cemented, very dense sand and gravel formations of great thickness, locally named Messina Gravels.

In such conditions it is not possible to have a superficial foundation resting directly on the sea bed, as the thickness of the conglomerate layer is far from sufficient. It would on the contrary be necessary to break through the conglomerate, remove part of the underlying sands and gravels and prepare an even platform of about 34000 square metres, possibly treating the soil locally, and forming a bed of regular stone/rock elements. All this would have to be done at the said water depth, combined with strong currents which change direction every six hours.

The equipment and technologies existing in the early nineties were absolutely inadequate for such a task (and with reference to the design alignment, the same perplexities exist today). The development of special remotely controlled equipment able to prepare the foundation platform is not unthinkable, but it would be a tremendous task in itself, with radical uncertainties in time, cost and outcome effectiveness, and this would need to be carried out before defining a tender design and cost. Similar challenges would apply to the problems of constructing the pier itself. All the several solutions studied, mainly based on prefabrication procedures drawn from North Sea experience with oil platforms, showed a number of problems, not least the touch down accuracy.

Even if all the erection difficulties were overcome, the structural robustness of the foundation under the required seismic conditions would be questionable, due to the direct and soil-water-structure interaction effects.

Under such circumstances a reliable preliminary design for the intermediate off-shore pier was considered not feasible, with the issue to be re-considered only if the single span solution proved to be a dead end.

References and Further Readings

[1] Stretto di Messina S.p.A. (2005). *Lo Stretto di Messina nell'antichità* (history and archaeology of Messina Strait) – Edizioni Quasar, Rome.

[2] Giuseppe Campione (1976). *I Trasporti e gli Effetti Indotti nell'area metropolitana dello Stretto di Messina* (transports in Sicily and effects induced by a fixed link in the metropolitan area of the Strait) – edizioni La Loggia dei Mercanti (Camera di Commercio di Messina).

[3] Giuseppe Campione (1998). *Italy's sea, problems and perspectives* – section: *The multinodal role of the Strait system* – Società Geografica Italiana, Rome.

[4] L. Jannattoni (1975). *Il treno in Italia* (history of railways in Italy) – Editalia, Edizioni d'Italia, Rome.

[5] SVIMEZ (Association for Industrial Development of Mezzogiorno) – *Rapporto SVIMEZ 2007 sull'Economia del Mezzogiorno* (2007 report on the economy of Mezzogiorno) – www.svimez.it/.

[6] Official websites of Sicily and Calabria Regions: www.regione.sicilia.it/; www.regione.calabria.it/.

[7] CENSIS (Centro Studi Investimenti Sociali) for Stretto di Messina S.p.A. (1992). *Scenari per il terzo Millennio: Europa-Ponte-Mediterraneo* (scenarios of economic development for the 3rd millennium with the Bridge).

[8] Pietro Busetta et al. for Fondazione Curella (2005). *Un collegamento per lo sviluppo, le ragioni del si per il ponte sullo stretto* (economic reasons for the Messina Strait bridge) – Liguori Editore, Napoli.

[9] Accademia Nazionale dei Lincei (1979). *L'attraversamento dello Stretto di Messina e la sua fattibilità* – (proceedings of the symposium on the crossing of Messina Strait held in Rome on July 4th-6th 1978).

[10] ANCE (Italian Assosciation of Building Contractors) official magazine *L'industria delle Costruzioni – numero speciale sull'Attraversamento dello Stretto di Messina* – March-April 1971 (special edition on the call for ideas for the crossing of Messina Strait) – Edilstampa s.r.l. – Italy.

[11] ANIAI (Italian Association of Engineers and Architects) official magazine "*L'Ingegnere Italiano*" n° 11 – November 1971 on the call for ideas for the crossing of the Messina Strait – "*Il progetto del Gruppo Lambertini, Il Progetto Maugeri, Il progetto del Gruppo Nervi, Il Progetto del Gruppo Samonà*."

[12] Gruppo Ponte di Messina S.p.A. (1981). *Rapporto di fattibilità tecnica, imprenditoriale, ambientale, economica e finanziaria per un collegamento viario e ferroviario fra la Sicilia e il Continente* volumes I to VII (overall feasibility studies for the fixed link).

[13] Stretto di Messina S.p.A. (1985). *Rapporto di sintesi degli studi di fattibilità* volumes I to VI (executive summary of the feasibility studies for the fixed link).

[14] Stretto di Messina S.p.A. (1986). *Rapporto di fattibilità* volumes I to X and related "GV" basic studies (overall feasibility studies for the fixed link).

[15] Consulta Estera di Q.A for Stretto di Messina S.p.A. (1986). *Esame qualitativo degli studi di fattibilità* (International Consulting Board – quality evaluation on the feasibility studies for the fixed link).

[16] Stretto di Messina S.p.A. (1987). *Riscontro allo studio di fattibilità per un attraversamento alveo dello Stretto di Messina proposto da Saipem-Snamprogetti-Spea-Tecnomare* volumes I to X (examination of a semi-floating tunnel solution proposed by S.S.S.T. joint venture).

2 – PROJECT MANAGEMENT AND FRAMEWORK

Contents of this chapter:

- **The administrative framework**
- **Project organization architecture and principal players**
- **The method**
- **Project cost and project financing**
- **Tendering and contracting**

This chapter deals with the overall framework of the project in terms of institutionary, legal and administrative aspects, organization problems and project finance.

The realization of the bridge across the Messina Strait with related highway and roadway connections presents a particularly complex and demanding challenge, due to:

- Its technological features: it will be the bridge with the longest span ever built, carrying both highway and railway traffic, and it will be located in a highly-seismic area.
- The high value of the contracts: as a matter of fact it is the largest infrastructure project ever contracted in Italy.
- Its location in a densely populated area with particularly significant environmental, natural and sociological factors.

Furthermore the project, which was defined by the Italian Government as *"a strategic project of main national interest to update and develop the Country"* represents one of the first applications of the procurement procedure introduced by the innovative Italian Law on public works-Legge Obiettivo.

The realization of such a project demanded extraordinary organisational and management procedures to ensure success. These included the management of severe time and cost constraints and the achievement of very high standards of quality, environmental sustainability, safety and security against organized crime.

In 2003, the concessionaire company Stretto di Messina S.p.A, attained the final technical and administrative Governmental approvals after a very long and complex procedure, due not only to the scale and complexity of the project but also to the variety of public entities involved.

The company made important strategic decisions to enable the project to proceed in relation to:

- The organization structure of the project.
- The management methods and procedures.
- The arrangement and structure of the necessary contracts.

At the same time it was essential to develop a financial strategy, based on project financing, able to ensure the viability of the project.

Finally, contract documents had to be prepared and tenders invited from international consortia to select the contractors and consultants capable of carrying out the works and providing the necessary services.

2.1 The administrative framework

Stretto di Messina S.p.A. (SdM) is a state owned company established by law[1] in 1981 as the concessionaire for the financing, design, construction and operation of the fixed link between Sicily and the Italian mainland. It is

[1]The special law for the permanent roadway and railway link between Sicily and the Italian mainland: Law n° 1158 dated 17th December 1971 modified and integrated by Legislative Decree dated 24th April 2003. This law declared the permanent link as a *"work of prevailing national interest"* and established Stretto di Messina S.p.A. as concessionaire for its study, design and construction, as well as operation of the roadway section. The concession agreement was granted in 1985 by ANAS and RFI and confirmed in 2003 by the Ministry of Infrastructure and Transport that assumed the role of grantor.

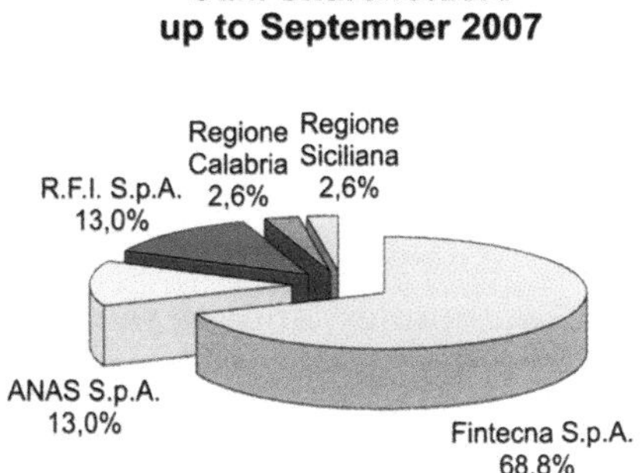

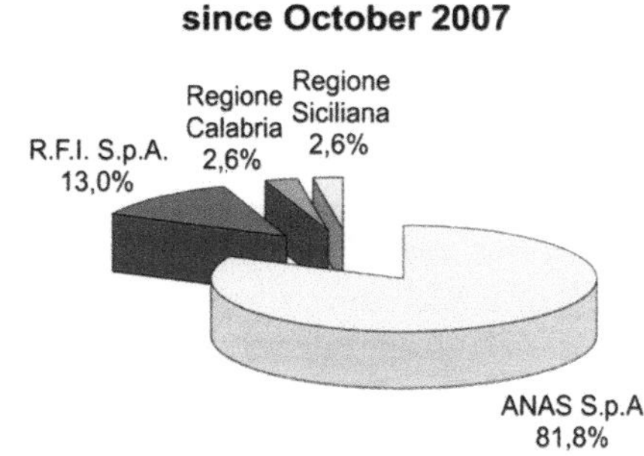

◄ Figure 2.1 Stretto di Messina shareholders.

thus responsible for the entire process, from a technical and operational as well as financial point of view.

SdM has an equity capital equal to approximately €383 million (at 31st December 2006). SdM's shareholders include the main Italian public companies for transport infrastructures and the two regions directly connected by the link. More specifically, SdM's shareholders are:

- ANAS S.p.A. now[2] the main shareholder. ANAS is the state authority for roads and motorways that directly operates the entire major roadway and motorway network (6,500 km of motorways and 20,000 km of national roads) on behalf of the Italian Government.
- RFI – RETE FERROVIARIA ITALIANA S.p.A., the state authority for railways that operates the whole national railway network (16,000 km railways) including "High Speed" railways on behalf of the Italian Government.
- REGIONE CALABRIA – the region of Calabria.
- REGIONE SICILIANA – the region of Sicily.

Since its establishment SdM needed to adapt its organization from time to time to suit the different project stages. Four main stages can be identified, each having different production and organization requirements, depending on the different targets:

- Feasibility Studies, 1985–1989. (SdM also incorporated all studies carried out from 1960 to 1985 by the private group "Gruppo Ponte di Messina").
- Outline Design ("Progetto di Massima"), 1990–2001.
- Preliminary Design ("Progetto Preliminare"), 2001–2002.
- Preparation of Tender Documents, 2003–2004.

The project was developed, from the earliest studies, with continuous interaction between the following organisations who each assumed various roles and responsibilities:

- **Stretto di Messina SpA**, as Concessionaire was responsible for all the studies and the design. SdM availed of its own staff and the work of a large number of experts, Italian and international engineering companies, research institutes and universities. The SdM staff played a co-ordination and guiding role, consulting specifically established permanent committees composed of notable experts in various disciplines.
- **ANAS** (national road/highway authority) and **RFI** (national railway authority, previously known as FS), as grantors of the concession and customers

[2] Before October 2007 (see fig. 2.1), most of equity shares presently owned by ANAS were owned by FINTECNA S.p.A. (formerly IRI), that is the holding for industry and service sectors, owned by the Italian Ministry of Treasury.

of the project. They played an active role in the design, as foreseen by the special law for the project, by giving guidelines and controlling the design through purposely-established commissions.

- **Consiglio Superiore dei Lavori Pubblici** (the National Council for Public Works and the State's most important consulting board) which played a control function during the design and provided advice on the design from the technical-scientific point of view.
- **Regione Siciliana, Regione Calabria** (regional authorities) and the **Ministry of the Environment** took an active part in the design for the assessment of the territorial[3] and environmental aspects, particularly after completion of the Outline Design.
- **The Ministry of Infrastructure and Transport** took part in the Preliminary Design, through a Technical and Scientific Committee, established in 2002 to provide a guiding role.

Two independent advisors were selected through public tender in 2000 to evaluate certain key issues related to the suspension bridge and its possible alternatives. They were appointed by the Ministries of Public Works and of Economy & Finance, upon request of CIPE (the Inter-ministry Committee for Economic Planning). These were:

- A joint venture led by **PriceWaterhouseCoopers**, appointed as advisor for studies to undertake a general review and further analysis of aspects related to territorial impact, transportation, environmental impact, cost-benefits and financial feasibility. The scope included, in particular, a comparison of the bridge with other transportation systems (i.e. maritime connections) in terms of efficiency and socio-economic effects.
- The **Steinman International – Parsons Transportation Group** (USA), appointed as engineering advisor to carry out technical investigations and in-depth studies on certain engineering aspects identified by the National Council for Public Works as needing further examination. The scope included, in particular, verification of satisfactory bridge performance, operation continuity and efficiency in severe conditions.

The PriceWaterhouseCoopers study concluded that both the construction of the bridge and the expansion of maritime connections were feasible and suitable even if only a modest future traffic demand is assumed. Both solutions obviously present different costs and benefits: the bridge is certainly more costly, but it constitutes a unitary, final and more efficient solution. In particular the bridge offers the most satisfactory solution for railway transport. The bridge was also found to present a stronger institutional reliability and allow important results of "territorial marketing" and of urban re-qualification. That is, due to its nature as a single exceptional work decided by law and committed to a State company, the bridge solution is more reliable than most in terms of funding, administrative matters, completion, performance etc. and it also serves to produce a "marketing" promotion of the region, for example by attracting further investment to the area. Furthermore the bridge can attract private funds, allowing partial self-financing of the project. The environmental impact is not prejudicial

[3] Throughout this book, the words "territorial" or "territory" are generally used to refer to a geo-economic entity that includes not only its physical environment and natural resources but also the land use, settlements, transport, services, administration and the socio-economy aspects.

for the bridge, neither in absolute terms nor in comparison with the alternative solutions. In fact the expansion of maritime connection was found to involve strong negative impacts due to jams of urban coastal areas, degradation of coasts and marine environments, and low energy efficiency.

Whichever solution is selected, the advisor recommended that a permanent link across the Strait is combined with the renewal and improvement of road networks in Sicily and in Calabria and with the expansion of long distance air and marine transport systems.

The Steinman – Parsons study examined the Outline Design or Progetto di Massima of the bridge, which had in fact already been taken to a relatively advanced level of detail. Using their own independent mathematical models, the consultants confirmed the technical soundness of the proposals. The study concluded that the bridge, as designed, was feasible and efficient, and that it met the highest international standards. The consultants also formulated some suggestions that will be important references for the detailed design.

In December 2001 the Messina Strait Bridge was included by CIPE in the Strategic Infrastructure Programme as part of the Government's priority "FastTrack" Strategic Infrastructures.

In 2002 Stretto di Messina prepared the Preliminary Design that derived from the one first proposed in 1992, updated to include the recent resolutions of the Technical and Scientific Committee, but also to comply with all new standards and regulations (safety, road design, environment, etc.), and to be more coherent with the new legislative framework for public works. The Preliminary Design documents also incorporated all recommendations issued in 1997 by the "Consiglio Superiore dei Lavori Pubblici" and in 2001 by the Government's technical advisor Steinman – Parsons.

In the meantime, the financial feasibility of the project was being studied and the financial plan established. (See section 2.4).

In 2003 the project obtained the overall approval of the Government (CIPE resolution no. 66 dated August 1st 2003). This represented the final approval in terms of technical, environmental, financial, territorial and transport conditions. At the same time, following a Government re-organisation, the Ministry of Infrastructure and Transport, became the grantor of the project.

Thus, on the basis of the approved design and Financial Plan, the **Concession Agreement** was signed between the Ministry of Infrastructure and Transport and Stretto di Messina defining the concession period (38 years), the project features, the construction plan, the toll determination and adjustment criteria, the financial details, and the methods and terms for the operation, maintenance and final handover of the project.

The Ministry of Infrastructure and Transport and Stretto di Messina also agreed with the Ministry of Economy, the Regions of Calabria and Sicily, RFI and ANAS a **"Framework Agreement"** defining the technical and financial commitments of the signatories involved in the Project. (See also Chapter 7.10)

On the basis of these important proceedings, the realization phase of the Project started. The first step was to define the overall organization and carry out four international tenders (2004–2006) (see section 2.5) to award the following contracts:

- Project Management Consultant.
- General Contractor.
- Environmental, Territorial and Social Monitoring
- Insurance Broker.

These contracts break significant new grounds, particularly in Italy, in view of:

- the very large project-financing operation;
- the very high contract values for the works (€4,400 million), the engineering services (€150 million) and the environmental monitoring (€37 million);
- the exceptional participation of large international tendering groups.

2.2 Project organization architecture and principal players

While the administrative procedures to obtain the necessary design approvals were in progress, SdM considered the best organization structure to successfully carry out the project.

The Company first concentrated on the experience of other large scale public works projects both in Italy and abroad concerning design and construction complexities similar to the bridge over the Messina Strait.

However, the information thus obtained had to be assessed in the light of the new Italian legislation for the construction of public works which would apply to the Messina Bridge. As a matter of fact the bridge is one of the "strategic projects" subject to the 2001 innovative special law ("Legge Obiettivo")[4] issued by the Italian Government to speed up the construction of infrastructures and strategic production settlements. Among other benefits, the law introduced:

- The possibility of entrusting all contract activities to a General Contractor who takes the full responsibility and carries out the project "by whatever means". This type of contract, an innovation in Italy, can be compared to international Design and Build Turnkey contracts.
- Simpler and faster procedures for technical and administrative approvals, including special Environmental Impact Assessment procedures as described in Chapter 4.
- New regulations regarding project-financing.

The sharing of tasks and responsibilities between General Contractor and Owner, as described in the following, was new in Italy. A fundamental difference from the former regulation for public works was the possibility of awarding the contract on the basis of the preliminary design, with the detailed design[5] and works supervision[6], traditionally carried out by the Owner, becoming the General Contractor's responsibility.

---

[4] Law 443/2001 and accomplishing Decree 190/2002, Law 166/2002 (Regulations on Infrastructures and Public Works facilitating Government plans on infrastructures).

[5] "Definitivo" design according to Italian law, representing the development of the Tender Design. It precisely defines all technical and economic aspects of the project and contains all the elements needed for the final approval for construction.

[6] Supervision, technical accounting, and administrative controls of the realisation of the project.

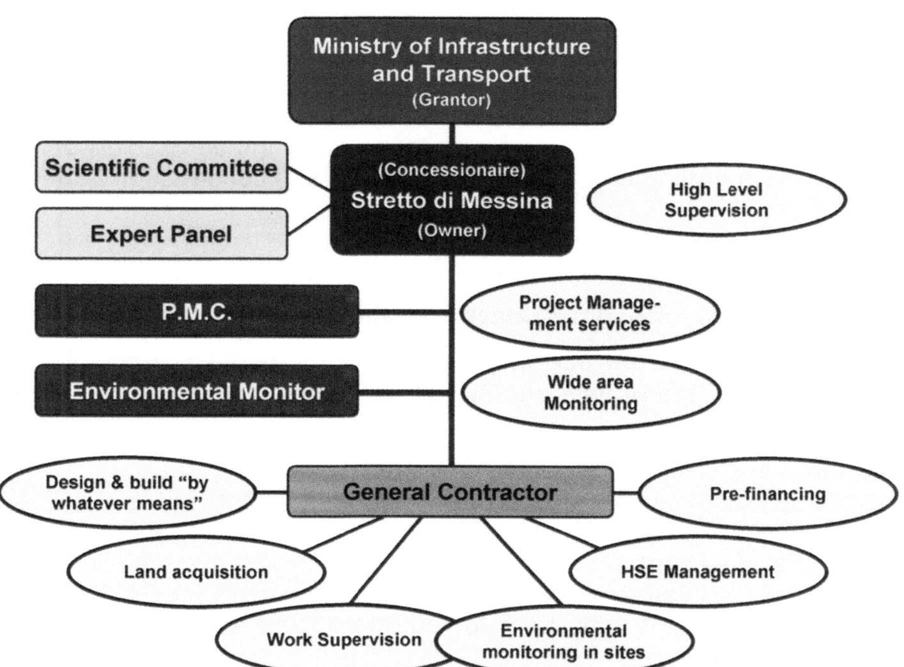

◄ Figure 2.2 Organizational architecture.

Once approved by CIPE, the Messina Bridge project became the first project to be approved within the terms of the Legge Obiettivo. Thus it was not possible to make reference to relevant previous experiences in Italian public works.

SdM decided to adopt an innovative organization, with the Owner carrying out High Level Supervision Activities, supported by a *Project Management Consultant (PMC)* with responsibility for closely controlling all the design and construction activities carried out by the General Contractor.

The PMC activities include validation of the final design through a qualified Control Board, as required by the Italian regulations for public works, as well as the Independent Check of the Final Design of the suspension bridge in accordance with best international practice.

The PMC role obviously requires particular technical and scientific experience, especially in the design and construction of long span bridges, as well as suitable project management and financial capacities. The idea of creating within SdM a technical staff that could carry out this role was examined and rejected. To form a staff comprising so many high-level and suitably experienced professionals in such a short time would have been neither easy nor cost-efficient for a single project whose technical and organizational needs would change from one stage to another.

Considering the importance and sensitivity of the territory affected by the project (ie. dense population, historical relevance, value of natural habitats[7], etc.), SdM decided to monitor an area much wider than that directly affected by the works to identify and evaluate any possible environmental, territorial and social changes being produced by the project. This innovative "wide area monitoring", with a much wider scope than usually required in Italy, will be carried out by a separate consultant, independent from the General Contractor, the *Environmental Monitor (EM)*.

[7] Ecosystem-sensitive zones are present in the area, such as "Special Protection Zones", "Sites of Community Interest" and "Important Bird Areas".

The adopted organization model, illustrated in Figure 2.2, provides for the Owner to be supported by the PMC controlling the activities carried out by the General Contractor and by the EM. The Company is also supported by a *Scientific Committee* set up by law[8], and by an *Expert Panel* formed by experts in structural, aerodynamics, seismic, geotechnical engineering. This panel is consulted from time to time on particular design and construction issues.

2.2.1 The Concessionaire

2.2.1.1 Tasks and roles of the Concessionaire

As Concessionaire, SdM performs High Level Supervision including the overall control of the project, monitoring all project activities from the following points of view:

- technical and scientific aspects;
- economic and financial aspects;
- organization and timing;
- fulfilment of all contract requirements;
- compliance with all relevant laws and regulations;
- proper relationships with the involved Boards and Authorities as well as with third parties.

In this context the following roles are particularly important:

a. to guide and direct the project, by promptly detecting possible deviations or problems and taking effective correction measures;
b. to check and approve for acceptance the main technical aspects of the design and the construction, even though the General Contractor remains entirely responsible for these activities.

2.2.2 The General Contractor (GC)

2.2.2.1 One or more General Contractors

The organization model was based on the decision to entrust the whole work to a single General Contractor. This choice required the evaluation of different scenarios and of their possible advantages and disadvantages.

Three possible types of contract schemes were compared (see Figure 2.3):

- A single General Contractor (Contract a + b + c).
- Two different General Contractors, one for the suspension bridge (Contract a) and one for the connected works on both shores (Contracts b + c).
- Three General Contractors, one for the bridge and two different contractors for the works on each shore (separate Contracts a, b and c).

The idea of granting the construction to several contractors had some contractual, management and operational advantages, such as:

- Lower technical and financial capacity required for the companies tendering for the works on the two shores, and higher and more specific requirements for those tendering for the suspension bridge;
- More competition among the tenderers, especially for road and railway works, with the possibility of lower bid prices;
- Entrusting homogeneous activities to specialized groups.

[8] According to article. 2 of Legislative Decree 114 of 24 April 2003, the Scientific Committee set up by the Ministry of Infrastructure and SdM *"gives its advice to the Board of Directors of SdM on the detailed and construction design and on its changes".*

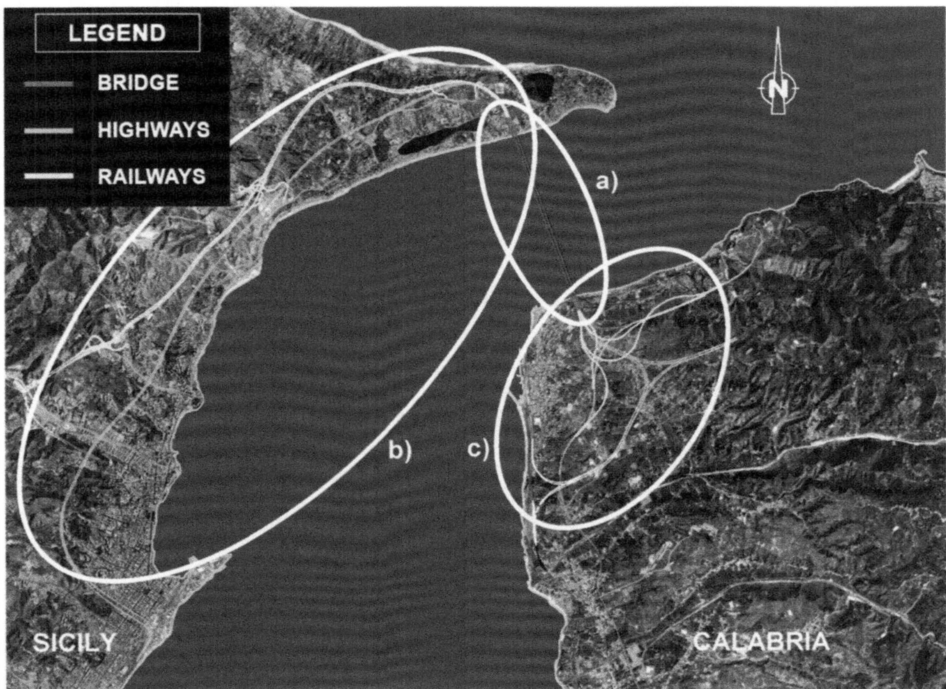

◀ Figure 2.3
Possible sub-division of the work into 3 lots according to their type and location. Overlapping of the three lots mostly concerns areas, equipment and activities of the construction sites.

On the other hand placing the contract with a single contractor presented many important advantages:

- Avoiding the difficulty of preparing and carrying out three important tenders at the same time or within a short time period.
- Overcoming the problems arising from the need to simultaneously complete and obtain approval for the detailed design of the different lots: as a matter of fact each possible delay of a contractor could affect the others creating significant contract problems.
- Best organization and planning of preliminary activities (land acquisition, construction design, site establishment, etc.).
- Best management of work areas, transport and common activities (such as earth works) considering the restricted available site dimensions on both sides of the Strait, and control of the possible time and space conflicts between the activities of the different lots.
- Overcoming any conflict between the different contractors realizing adjacent and continuous works, each one affecting the functionality of the others.

The analysis of these pros and cons led to the decision to appoint a single General Contractor, thus resolving the critical problems of interface coordination and responsibility, and reducing contract risks.

2.2.2.2 Tasks and role of the General Contractor

The contract was conceived according to the requirements of the Legge Obiettivo which precisely defines the tasks and duties of the General Contractor:

- Develop the detailed design[9] and undertake the technical and administrative activities necessary for the Owner to obtain the CIPE approval.
- Pre-financing of the project.

[9]The contract expressly states that the General Contractor undertakes the full responsibility of the design from the preliminary level by sharing and endorsing in his bid the tender design with any necessary changes.

- Land acquisition including stocking and spoil areas.
- Construction design.
- Execution of the works "by whatever means".
- Supervision of the works.
- Co-operation with the Owner for all activities and information control to avoid the infiltration of organised crime (ie. careful planning of sub-contracting, land acquisition, procurement, etc.) in agreement with the relevant Authorities.

2.2.3 The Project Management Consultant (PMC)

2.2.3.1 Tasks and role of the PMC

The PMC will assist the Owner in checking that design and construction activities performed by the General Contractor and by the Environmental Monitor are timely and correctly carried out. Thus PMC activities include, among others:

- Project Control: control of planning, organization and progress, including the economic points of view.
- Monitor and review the engineering, including the detailed design, environmental design, construction design and methods, as-built documentation, and operation and maintenance manuals.
- Independent check of the detailed design for the suspension bridge, according to internationally recognized standard and methods.
- Validation of the detailed design through an accredited quality control body.
- Assessment and control of all design and construction changes.
- Monitor and review all construction activities including land acquisition, commitment, prefabrication, assembly in remote yards, transport, site works and erection.
- Control of environmental monitoring activities.
- Supervision of all systems for quality management, quality assurance, health and safety and environmental management.
- Assistance with commissioning, testing and pre-operation activities.

2.2.4 The Environmental Monitor

2.2.4.1 Tasks and role of the Environmental Monitor

As already mentioned, SdM voluntarily decided to carry out a wider scale independent monitoring than is normally required by any law or regulation or is usually foreseen. The Environmental Monitor, directly appointed by SdM, will perform environmental, social and territorial monitoring over a wide area before, during and after construction. (See Chapter 4)

The Environmental Monitor will therefore act on a larger number of elements than the General Contractor. He will assess the possible direct and indirect impacts produced by the works and the efficacy of any adopted mitigation measures, thus checking the General Contractor's activity, monitoring the environment in areas affected by construction activities.

The data thus acquired will constitute a geo-referenced database that will be available to the regional authorities and the wider, scientific community.

2.3 The method

2.3.1 The organization model

Most public or state-owned concessionaire companies follow the traditional "tree organization model", with structures subdivided in directions and/or units. Each direction or unit tends to concentrate activities in its own hands and allocate roles and tasks through strict "job descriptions". As a result, many companies have a rigid overall organization and are not inclined to innovation.

Companies with a higher level of business diversification, specialization and flexibility, particularly carrying out high-tech activities, use a modern "mixed" type of organization model. These models foresee that the general functions follow the company formal organization, while the different production activities are ruled by procedures involving specific competences of the different company units, promoting "horizontal" co-operations.

This organization model based on "process-based management" through an integrated project team was considered by SdM to be the best reference.

2.3.2 Process-based management

To be consistent with the adopted organization model the company mission was subdivided into single operative activities. Each activity was analyzed, identifying macro-processes and processes to be attributed to the responsibility of a "process owner".

The analysis of the processes supplied a detailed Process Breakdown Structure (PBS).

Figure 2.4 represents the two macro processes: "Preparation and Managing of Tenders for Public Works" and "High Level Supervision", introduced in the System Improvement Model as per ISO regulations.

The process definition also enabled the identification of the structures and procedures to be implemented in the Company Quality Assurance System. The first macro-process was the preparation of the tenders that were in fact developed according to specific Quality Plans, one for each tender.

Figure 2.5 illustrates the second macro-process with its main activities and involved parties.

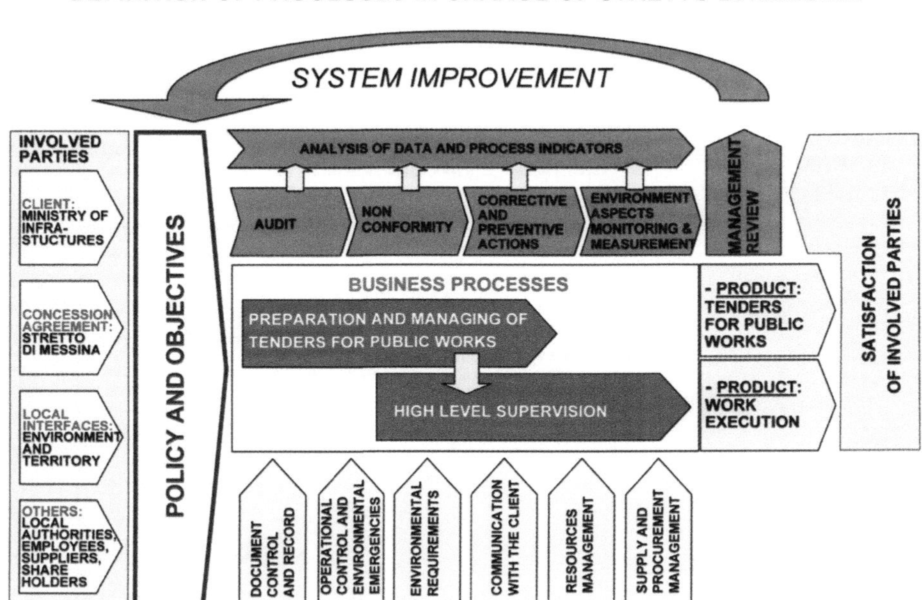

Figure 2.4
The macro-processes.

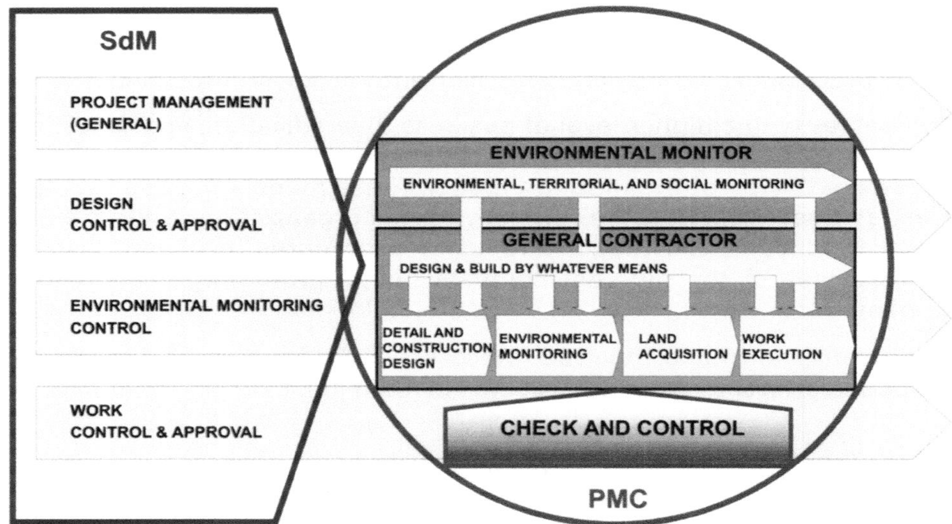

Figure 2.5 ▶ Activities and parties involved in the "High Level Supervision Process".

Figure 2.6 details how this macro-process can be subdivided into three vertical processes or phases (supervision of detail design, construction design and construction works) and furthermore into horizontal sub-processes.

The sub-division in single sub-processes that transversally invest the operative structure of the company allows, among other things:

- The coherent definition of responsibilities, tasks and roles as well as of interrelations.
- The best efficiency and optimum employment of specialized competences.
- Organic and integrated overall management through shared methods and tools.

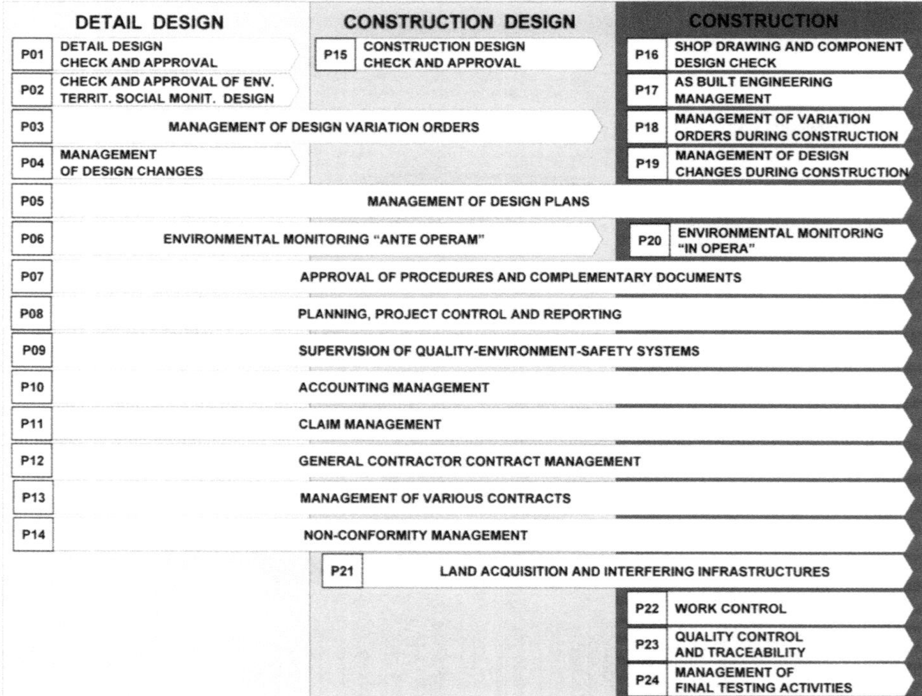

Figure 2.6 ▶ Processes and "horizontal" sub-processes ("P").

The correct development of the processes will be monitored through key performance indicators purposely defined for each process. These indicators will be constantly updated thus creating a system of Managerial Monitoring that, through periodical reports and a "Management Dashboard", will supply the necessary information on the progress of the project and on relevant performance levels.

2.3.3 Structure of the activities and of the organization

Planning, organizing and controlling such a complex project required, first of all, that all activities are identified, rationalized, split into elementary items and organized according to their logical sequence.

A detailed Work Breakdown Structure (WBS), structured into eight levels in the tender documents and to be developed more in detail by the General Contractor, was thus defined and adopted. (Figure 2.7)

0 SCOPE	1 MACRO ACTIVITY	2 UNIT	3 SYSTEM	4 LOT	5 COMPON.	6 SUBCOMP. OR ACTIVITY	7 PART	8 PAID ITEM	VALUE as % of total
GENERAL CONTRACTOR CONTRACT									100.00
	DETAILED DESIGN - H. & S. COORDINATION								(omissis)
						(sub-items are omitted)			
	LAND ACQUISITION								(omissis)
						(sub-items are omitted)			
	CONSTRUCTION DESIGN								(omissis)
						(sub-items are omitted)			
	WORK SUPERVISION - H. & S. COORDINATION								(omissis)
						(sub-items are omitted)			
	ENVIRONMENTAL MONITORING								(omissis)
						(sub-items are omitted)			
	CONSTRUCTION								
		MOBILIZATION & OPERATION OF SITES - DUMPING							(omissis)
						(sub-items are omitted)			
		SUSPENSION BRIDGE							57.91
			PERMANENT WORKS - STRUCTURES						56.33
				SUBSTRUCTURES					8.99
						(sub-items are omitted)			
				SUPERSTRUCTURES					47.34
					CABLE SYSTEM				20.42
						SHOP DRAWINGS & ERECTION DESIGN			n.a.
						MAIN CABLES			16.14
							Temporary Systems		
								Supply & Delivery	(omissis)
								Erection	(omissis)
							Spinning Plant		
								Supply & Delivery	(omissis)
								Erection	(omissis)
							Cable construction		
								Supply & Delivery	(omissis)
								Spinning	(omissis)
								Compaction	(omissis)
								Wrapping	(omissis)
								Surface Treatm.	(omissis)
								Walkways, etc.	(omissis)
						(other items are omitted)			

◄ Figure 2.7 Example of Work Breakdown Structure for the General Contractor contract, focussing on the main cables of the bridge. Progress payments will be made on the basis of level 8 percentages (here omitted), on completion of "paid items" and upon achievement of contractual milestones.

The WBS system constitutes the "backbone" of the organization and control of the work, as well as the common reference for the operational and contract management of the project. The WBS system is thus applied to all the activities performed by the General Contractor and the Environmental Monitor and also, as a consequence, to all the activities performed by the PMC.

As specified in the contracts, each party must define and apply a corresponding Organization Breakdown Structure (OBS), so as to associate each activity with the organization unit in charge of it. This is a fundamental element in the application of the Quality Assurance System and in the control of the organization by the PMC and the Owner.

2.3.4 Stretto di Messina's Corporate Policy and Social Responsibility

The success of such an important project does not depend only on its economic value, i.e. the production of goods and services, but also on its social and environmental value.

This is the reason why the Concessionaire decided to undertake a strategy to properly take account of public interest. By taking responsibility for the impact of project activities on local communities, shareholders, workers, users, and all stakeholders in general, as well as on the territory and environment, the Concessionnaire was thus operating in line with best Corporate Social Responsibility (CSR) international standards and practice.

This decision obviously implies wider responsibilities than those deriving from statutory obligations and commitments, and largely exceeds the standards set by normal good practice.

Particularly, the CSR aims at:

- Achieving the best environmental sustainability and the optimum territorial and social outcomes.
- Assuring high standards of Health & Safety in all workplaces and the best application of suitable competences.
- Assuring that all the activities are carried out in compliance with relevant national and European laws, without infiltration of organized crime in the project.

According to these CSR principles, SdM is aware that particular attention is due to the issue of possible criminal infiltration or influence on the project. Specific countermeasures have been foreseen to prevent this risk together with a specific operative plan, based on a strict monitoring of all the project activities. This monitoring will particularly focus on site activities, such as contracts, supplies, payments, but also financial transactions for the project financing and for the subsequent pay back.

Apart from the cautions required by law, specific memoranda of understanding have been executed with the investigation and financial authorities, the intelligence services and the police. These memoranda involve all the parties taking part in the project, and assure a continuous flow of information to the authorities, allowing the traceability of all activities carried out by the Concessionaire, the General Contractor and by all contractors and subcontractors.

A similar memorandum of understanding was executed by the national and regional secretarial offices of the trade unions to avoid illegal working practices and to ensure very high levels of health and safety, application of competences and development of corporate participation relationships.

An advanced Fleet Management system for the operation and security of the sites, based on tracking by GPS of all the vehicles and ships operating in the working area, has also been planned. All the modes of transport and the handling of goods are thus monitored in real-time from departure to arrival together with the identity of the carrier. Similarly the identity of all site workers will be controlled by an electronic identification device.

Furthermore, when moving towards the realization phase, SdM decided to uphold the highest principles of quality and environmental sustainability by the application of a corporate Quality and Environmental Policy within the overall framework of the Company's Corporate Social Responsibility policies. It particularly aims to achieving "top" targets in terms of:

- attention to human resources,
- application of most advanced methods and technologies,
- stakeholder satisfaction,
- monitoring and constant reduction of environmental impacts assuring a proper relationship between the project and the physical and social environment during the design and even more during the construction and operation phases.

Both the voluntary management and control methodologies included in the SdM environmental and territorial approach (described in Chapter 4) and the Health & Safety management system (described below), are strictly connected to this CSR strategy.

2.3.5 Quality, Environment and Safety Management Systems

In the above context SdM developed its Quality and Environmental Management System not only following relevant international standards (ISO 9001 and ISO 14001), but also allowing the inclusion of requirements prescribed by other regulations (e.g. safety regulations) or by voluntary or compulsory management models deriving from regulations or national laws such as Legislative Decree no. 231/2001.

According to this CSR approach SdM decided to apply its specifications and requirements regarding the Quality and Environmental Management System not only to the General Contractor but also to all the contractors and subcontractors.

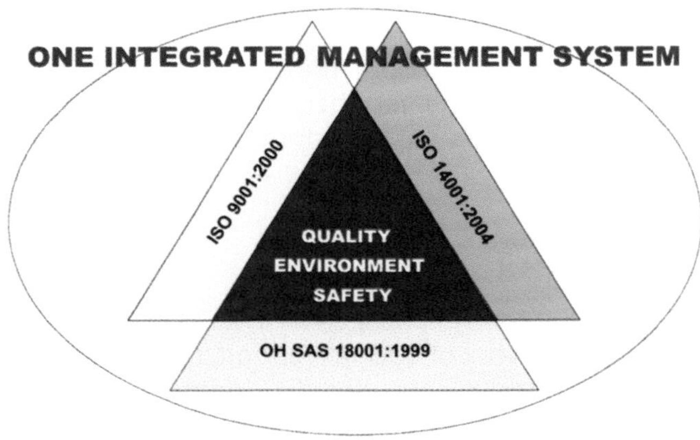

◄ Figure 2.8 QES Management System.

Furthermore SdM was the first public company in Italy to apply such a quality management system to public tenders and to obtain the ISO 9001 and ISO 14001 certifications.

Integrated Quality and Environment Policy

This represents the strategy adopted by the SdM management:

To carry out the Company mission consisting of *"the design, construction, operation and maintenance of a permanent link across the Messina Strait, thus playing an active role in the renewal of the Country and in the development of local areas, always paying attention to minimize the impacts on the environment" – "playing a leading role for the adopted solutions (approaches, methods, models) in the management of public works and in infrastructure operation".*

2.3.5.1 Contractors' Quality, Environmental, Health & Safety Systems

Targets defined by SdM's Quality Policy can be achieved only if the management system is extended in a co-ordinated and integrated way to all the parties involved the project.

The final target is to reach a level where "everybody speaks the same language". This is the reason why, right from the preparation of contract documents for the General Contractor, particular attention has been paid to definition of requirements implying the organization of an integrated quality, health, safety & environmental management system complying with ISO 9001, ISO 14001 and OHSAS 18001 regulations, while simultaneously meeting the contract technical requirements.

SdM's contract documents also require that the PMC and EM apply the same integrated management system of the GC which should be coherent and consistent with the system applied by SdM itself.

For this purpose the PMC was charged with monitoring and checking the whole system through the auditing of activities, the control of documents and the application of managerial monitoring based on several performance indicators. Ever since the tender phase, the contract documents included requirements concerning the Quality Assurance System with special attention to quality control and traceability of each element of the work. The priority of the control will be based on the potential impacts of production and assembling processes on the final quality of the work and on timing.

For this purpose a joint activity of SdM, the General Contractor, the PMC and the EM will be carried out at the very beginning of the work to implement the adopted information systems and adopt the relevant procedures.

2.3.6 Risk management

2.3.6.1 General

The planning of a public work of such exceptional dimensions with extraordinary economic and financial characteristics, demands an approach aimed at identifying, analyzing and mitigating the risks associated with the design, construction and operation of the asset. Technical literature abounds with examples of "extreme engineering" works whose cost and/or completion time significantly exceeded initial estimates.

An Enterprise Risk Management (ERM) approach was thus applied in this case. The importance and efficiency of this practice is well known for projects presenting substantial technical difficulties and complex contract conditions. The ERM approach is internationally acknowledged to be the best since it takes into account not only the technical uncertainties but also all the possible risks deriving from management systems, organizational and external factors, and particularly from:

- strategies,
- market evolution,
- changes of the political, legislative and macro-economic framework,
- working processes,
- adequate financial resources,
- adequate human resources,
- technological evolution.

This approach is based on the identification, analysis and mitigation (by the application of appropriate measures) of the risks that may occur during the different phases of the project, from design to completion and during operation and maintenance.

This activity is carried out in a dynamic time context, in order to monitor the evolution and changes in the type and number of risks, and to evaluate the proportion of overall risk transferred from the Owner to the other parties involved in the project.

To achieve the established goals the following guidelines have been determined:

- to apply a code-based, sensible and sustainable risk model,
- to not overlook critical risk factors,
- to share with the stakeholders the list of risks and critical conditions[10],
- to identify and share the mitigation measures to be taken,
- to identify and share the priorities of the measures to be undertaken on the basis of the effects on safety, environment and cost-benefit ratings.

The ERM approach is directly connected to the strategy of the Company and to its mission, goals, development plans and initiatives. A successful application of ERM must avoid possible mistakes such as failing to assume an in-depth and detailed approach, or inadequately defining responsibilities and risk treatments.

The adopted ERM process follows the various different operational phases, alternating with consulting phases during which the results are shared among all company units involved, as illustrated in Figure 2.9.

The following Australian/New Zealand Standards were found to best suit the Company's needs, and have been applied for these procedures:

- AS/NZS 4360:2004 "Risk Management";
- HB 436:2004 "Risk Management Guidelines".

[10] Communication and consultation are fundamental to achieve the established goals as reported in clause 3.1 of AS/NZS 4360:2004 Standard:
"...communication and consultations considerations should involve a dialogue with stakeholders with efforts focused on consultation rather than a one way flow of information from the decision maker to other stakeholders...".

Figure 2.9 ► Application phases of the Risk Management System.

A brief summary of the results obtained by the application of the Phase 1 Risk Analysis, undertaken in 2005 and 2006, is reported in the following.

2.3.6.2 Phase 1 Risk Analysis

This step involved the identification and analysis of all possible risks, with the support of the competent company units, and the identification of possible mitigation strategies.

Obviously the uncertainties (and thus the risks) are higher at the outset of a project, and they reduce as the different project phases are completed. (Figure 2.10)

Figure 2.10 ► Typical risk development curve.

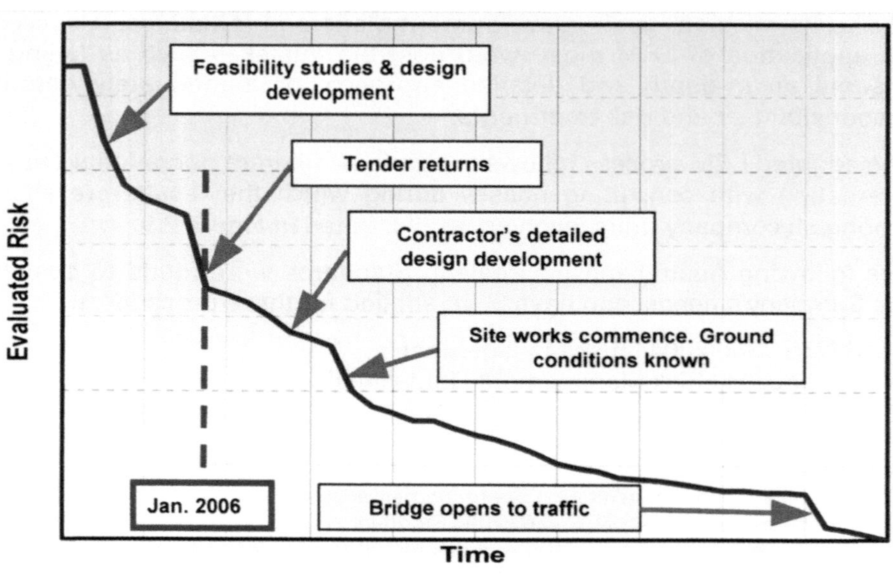

Identifying risks and risk categories

The risks were identified, listed and classified according to a Risk Breakdown Structure (RBS) that groups the risks into categories down to the lowest level consisting in the elementary uncertainties. The project RBS consists of three levels (macro-categories, categories and sub-categories), the highest of which includes macro-categories such as:

- **Financial Risks** (availability of financial resources).
- **Insurance Risks** (ineffective risk transfer).
- **Normative Risks** (existing and new laws and regulations).
- **Contractual Risks** (non-compliance with the contract by any party).
- **Management Risks** (strategies, organization, information management).
- **Political and External Risks** (political and territorial decisions).
- **High Level Supervision Risks** (ineffective control by Owner, PMC, EM).
- **Technical Risks** (design, construction, Quality Assurance System).
- **Environmental Risks** (effects of the environment on the works and vice-versa).

The diagram in Figure 2.11 shows the identified risks, divided into macro-categories and associated to the six macro-phases composing the Company's mission.

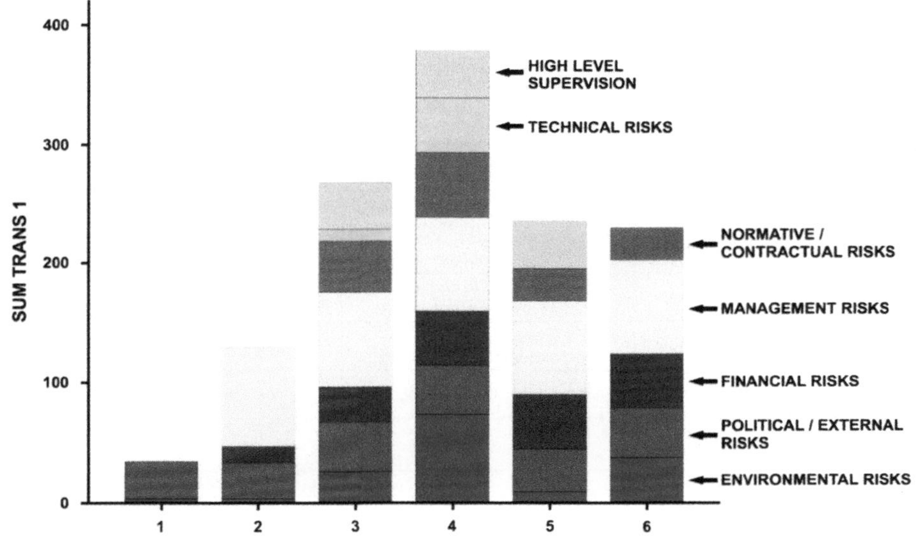

◄ Figure 2.11
2005 risk analysis (before contracts and relevant mitigation measures): risk count by macro-categories (segment colours) and by mission macro-phases (X axis – "Index1").
Macro-phases:
1. Preliminary design.
2. Tender.
3. Detail & construct. design.
4. Construction.
5. Test.
6. Operation.

Defining measurement metrics and evaluating risk criticality

The risk assessment was carried out applying the FMECA (Failure Mode Effect and Criticality Analysis) method; a tool which identifies and quantifies the risks and their consequences.

The following risk assessment methods are usually applied:

QUALITATIVE ANALYSIS

In this analysis the magnitude and the likelihood of potential consequences are presented and described in a qualitative way. Different descriptions may be used for different risks.

SEMI QUANTITATIVE ANALYSIS

In the semi-quantitative analysis some quantitative values are assigned to the descriptions used in the qualitative assessment. These values do not however realistically define risk consequences and likelihoods, which is the prerogative of the quantitative approach.

QUANTITATIVE ANALYSIS

In the quantitative analysis numerical values are assigned to both impacts and likelihoods. These values are derived from a variety of sources. Before the analysis can be carried out, drivers and homogeneous measurement metrics must be defined in order to avoid an assessment based on a single value or on not commonly agreed values. Once the general parameters have been determined, each party involved in the process will assign, in case no reliable database is available, numerical values to each elementary risk.

Clear measurement metrics have been determined to obtain homogenous assessments from the various company units according to their mission. Metric scales have been defined for probability levels (Figure 2.12) for impacts on time and costs (Figure 2.13) and for risk criticality deriving from the combination of likelihood and impact (Figure 2.14).

Figure 2.12 ▶ Metric scale for risk probability levels.

Value	Category	Definition	Probability value
5	High - Frequent	Likely to occur frequently	> 0.80
4	Medium / High - Likely	Likely to occur often (many times)	from 0.50 to 0.80
3	Medium - Occasional	Expected to occur occasionally (several times)	from 0.25 to 0.50
2	Medium / Low - Seldom	Expected to occur on a rare basis (some times)	from 0.125 to 0.25
1	Low - Unlikely	Unexpected but might occur	< 0.125

Figure 2.13 ▶ Metric scales for risk impact on time and cost.

Value	Category	Impact on project time
5	High	more than 1 year or unfeasibility
4	Medium / High	from 6 months to 12 months
3	Medium	from 3 months to 6 months
2	Medium / Low	from 1 months to 3 months
1	Low	less or equal to 1 month

Value	Category	Impact on project cost
5	High	more than € 500,000,000 or unfeasibility
4	Medium / High	from € 200,000,000 to € 500,000,000
3	Medium	from € 50,000,000 to € 200,000,000
2	Medium / Low	from € 5,000,000 to € 50,000,000
1	Low	less or equal to € 5,000

Figure 2.14 ▶ Risk criticality matrix (based on CEI EN 50126 Standard).

RISK CRITICALITY		RISK IMPACT				
		Low	Medium / Low	Medium	Medium / High	High
RISK LIKELIHOOD	High	bearable	undesirable	unbearable	unbearable	unbearable
	Medium / High	bearable	bearable	undesirable	unbearable	unbearable
	Medium	negligible	bearable	undesirable	undesirable	undesirable
	Medium / Low	negligible	negligible	bearable	undesirable	undesirable
	Low	negligible	negligible	negligible	bearable	bearable

The 15 most critical risks which were identified from this risk analysis are summarised below:

- PS 05 Change of political will on the bridge construction;
- NA 07 Difficulties/retards in managing the interferences;
- CR 05 Late or missing preliminary activities;
- CR 06 Late or missing functional works to be carried out by other parties (the Framework Agreement);
- PS 26 Opposition to land acquisition by the owners of the areas involved in the works;
- AM 01 Environmental Impact Assessment and/or the Detail Design have to be updated;
- CR 64 Economic claims laid by the General Contractor;
- NA 03 Litigations;
- PS 64 Contrasts, protests, obstructionism;
- CR 45 Late, missing or non-compliant completion of changes to the preliminary design;
- TE 05 Cost and availability of the steel for steelworks;
- TE 70 Lack of spoil dumping areas;
- TE 12 New requirements or needs identified by the General Contractor;
- TE 13 New design requirements imposed by local authorities;
- PS 18 Proposals, requirements or changes imposed by the local authorities.

The method identifies several overlapping and inter-connected areas of risk, but the following key overall aspects emerged from the Phase 1 assessment:

- The environmental risks are the most numerous.
- The normative, contractual and management risks outnumber the technical risks.
- The majority of risks occur during detail and construction designs and during construction.
- Many risks present medium criticality.
- The starting situation (i.e. before the application of mitigating measures) identifies several critical risks requiring management.

Timely treatments must be defined for high criticality risks in order to obtain conditions at least equivalent to those of similar public works. The strategy for implementation focuses on the well-known *"4T strategy"*:

- *Terminate* the activity that is likely to trigger the risk,
- *Treat* the risk to reduce its impact and/or frequency,
- *Transfer* the risk (through contracts, insurances, partnerships or joint ventures),
- *Tolerate* the risk since it is acceptably low.

Sensitivity analysis

The dynamic correlation between the different risks and the different activities was obtained by supplementing the design PERT (developed in terms of WBS) with the Risk Breakdown Structure, so as to associate the appropriate project tasks with the relevant risks. The WBS includes the macro activities concerning the tenders, design, approvals and land acquisition.

On this basis probabilistic distributions of time and cost variations of each task have been identified, and probabilistic simulations using the Monte Carlo method on the time/cost PERT and related risks have been carried out.

Figure 2.15 shows two examples of the confidence curves obtained for cost and time from the Monte Carlo simulations (1000 scenarios).

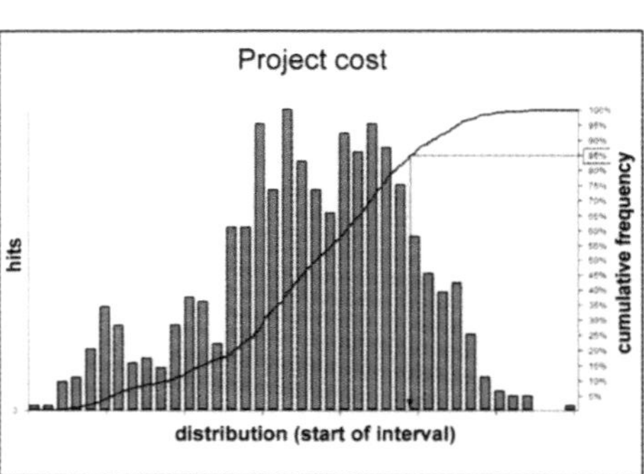

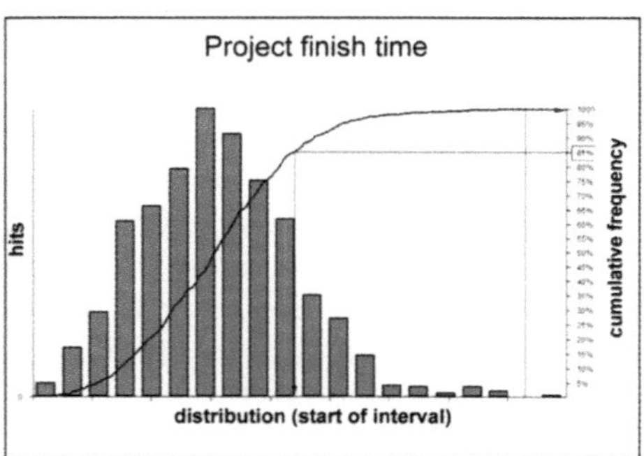

Figure 2.15 ▶ Confidence curves for project cost and finish time.

The first results of the sensitivity analysis identified the probabilities of various predicted time delays and cost overruns. Thus, for example, it was found to be highly unlikely that the project completion time and cost would overrun acceptable levels. However such a probability is low but not negligible, and thus it is necessary to manage all the possible risks to keep them "as low as reasonably possible".

Conclusions

The first results of the Phase 1 risk analysis supplied some interesting guidelines on the priority actions to be undertaken, such as:

- To define the shape and target limits of confidence curves.
- To detect new mitigating actions producing a demonstrable reduction in the residual risk.
- To record the stakeholder opinions and claims, to reach the best consent.
- To prepare a long-term communication plan.
- To be prepared to face a longer construction period, considering the impending and binding presence of the political risk.
- To analyze more thoroughly the planned, designed and applied procedures in order to equally share the risks among the parties, to reduce the risk of litigation and to transfer as much as possible the responsibilities onto the competent public authorities.

2.3.7 Information & Communication Technologies

The implementation of an automated management system for all project information was a strategic decision for this project. The Concessionaire thus developed a rigorous Information and Communication System constituting an interface between all the parties involved in the project that could manage all the information and use them as integrated inputs for all project processes.

For this purpose, a web-based portal (Enterprise Portal) was established constituting a single point of entry for all project information originating from the various parties involved. As established in the contracts, all communications and exchange of data and documents between the parties will have to occur through the Enterprise Portal.

The Information Technology system allowed the integration of advanced applications such as:

- Document Management System (DMS) that manages and traces all documents and communications in real time.
- Project Control Management System.
- Quality and Traceability Systems and a Product Life Cycle Management System that collects and manages all the information on products and

processes. This system will constitute the base of the management system of the project during operation.
- Geographic Information System (GIS) that facilitates the management of environmental and territorial data.

2.4 Project cost and project financing

2.4.1 General

Road and railway transport infrastructures are commonly acknowledged to be usually unable to be self-financed due to the high construction costs on one side, and the limited revenues, by tolls or otherwise, that comply with socio-economic constraints. Construction costs of infrastructures are, in Italy, particularly high for various reasons that include the difficult morphology of the territory, the high density of man-made and natural pre-existing constraints, and the important mitigation and compensation works required by local communities, often to overcome the "Not-in-my-backyard" (NIMBY) syndrome.

The project financing of a permanent link across the Messina Strait can however take advantage of certain particular situations including the following:

- a substantial transport demand has already become established over a long period,
- hundreds of millions of euro per year are presently being spent in shipping trains and vehicles across the Strait,
- a permanent link would not suffer effective competition by the maritime transport systems that would, at best, share a limited amount of local road traffic but no railway traffic at all.

Nevertheless, previous financial analyses (including those carried out in 2001 by PWC as advisor to the Ministry) found that project financing for both the bridge and the road & railway links would have required State grants-in-aid for about 50% of the total investment.

A thorough "*financial engineering*" was undertaken in 2003 to verify the full feasibility of the project without any State aids, aiming at the following main objectives:

a. Ensuring the coverage of the entire financial requirement for the completion of the project before opening the worksites, in order to minimize the risk that any lack of financial resources could postpone the timely completion of the works.
b. Eliminating the need for capital resources from the State budget (which could thus be used for other public works), as well as the distribution of grants-in-aid and the release of guarantees by the State.
c. Ensuring a substantial involvement of the financial markets in the project, following typical project finance models.
d. Ensuring shareholders will receive an acceptable return on investment under all considered scenarios. Although the shareholders are all Government controlled companies, they commit capital to the Project based on market oriented principles and on investment return analyses: these financial injections shall in no way represent a grant.

As a result, the Financial Plan (that formed an important basis for the Concession Agreement) ensures the coverage of the required amount (approximately 6 billion euros) as follows: (Figure 2.16)

- 40% of the overall amount required during the construction period through capital increase of €2.5 billion – to be injected progressively.

- Remaining 60% through project finance loans from the international debt markets. These loans would be guaranteed solely by the cash flow generated by operating the crossing during the concession period.

Figure 2.16 ▶ Project financing.

The financial plan proved the feasibility of the project without any State aid grants and its capability to pay back the entire capital invested as well as to pay interests and dividends.

The crossing has been inserted in the TEN-T (Trans Europe Network) priority axes, as part of the primary railway axis Berlin–Verona/Milan–Bologna–Naples–Messina–Palermo. The project is thus eligible for EU contributions. Nevertheless, for sake of caution, such contributions and/or EIB loans have not been taken into account.

The Financial Plan was developed by a thorough financial analysis with the support by an international financing advisor (PWC), based on the data and assumptions summarized in the following paragraphs.

2.4.2 Project cost

Following thorough cost estimate analyses, based also on bench-marking of the international market for long span bridges, the total project cost was determined to be approximately **€4.7 billion** (2002 value) broken down as follows:

Table 2.1 ▶ Project cost.

Suspension bridge	€	2.720	billion
Road and Railway links in Sicily and Calabria	€	1.270	billion
Land acquisition, interferences, design, testing, surveys, works supervision, Project Management etc.	€	0.450	billion
Mitigating & compensating works (as required by CIPE)	€	0.130	billion
Other costs – overhead expenses	€	0.115	billion
Total Project Cost	€	4.685	billion

The overall project financing requirement amounts, accordingly, to approximately **€6 billion**, allowing for inflation and financial charges over the construction period.

2.4.3 Traffic scenarios

The new data collected was used to update (to year 2000) the figures for traffic level estimates.

As usually adopted for traffic reports in economic and financial feasibility studies of infrastructures, a number of possible transport scenarios was taken into consideration, including those having negative impacts. All scenarios were based on year 2000 existing transport demand to and from Sicily.

Four main scenarios were developed for the evolution of passenger and freight traffic as a combination of:

- Two macro-economic scenarios driven by different GDP growth rates in Southern Italy (plausible high and low growth), in line with the objectives of the 2003 DPEF (National Economic & Financial Planning Document);
- Two transport scenarios driven by different hypotheses for the evolution of air and ferry transport (favourable and unfavourable evolution).

The four scenarios are shown in the following table:

Scenario 1	Scenario 3
High Economic Growth Favourable Transport Scenario	Low Economic Growth Favourable Transport Scenario

Scenario 2	Scenario 4
High Economic Growth Unfavourable Transport Scenario	Low Economic Growth Unfavourable Transport Scenario

◄ Table 2.2 Economic and transport scenarios.

2.4.4 Toll levels

For road vehicles (motorbikes, cars, trucks, buses), the tolls were assumed to remain in line with those currently applied by the ferry services crossing the Strait. Unlike similar projects, the set tolls will not be subject to increases as a result of the benefits granted to users by the bridge in terms of better service levels and a shorter crossing time (roughly one hour).

Average one-way 2003 toll levels are shown in the following table:

Type of vehicle	Tariff (€)
Motorcycles	5
Cars returning within 3 days	9.50
Cars returning after 3 days	16
Trucks returning within 6 days	50
Trucks returning after 6 days	63
Full buses	80

◄ Table 2.3 Toll rates.

2.4.5 Other fees

Concerning trains, the management of the railway on the bridge is entrusted to Rete Ferroviaria Italiana (RFI) under law. Consequently, a yearly lump-sum fee has been set to be paid by RFI to SdM for the use of the railway on the bridge, starting from the opening date of the bridge to traffic. This fee relates to the most prudent traffic forecast for the first year of operation and is based on the fee currently in force for shipping trains across the Strait. This fee was furthermore increased by a premium, given the huge time saving in crossing the Strait with the bridge (more than two hours) which will lead to higher efficiency with consequent savings in train running costs.

Therefore, the set fee implies no overcharges for end users and does not affect the railway investment policy in Southern Italy.

2.4.6 Depreciation and redemption value

The duration of the concession was defined in the Concession Agreement as 30 years of operation plus the design and construction phase, although the design life of the bridge (200 years) would allow a much longer concession.

As extending the first concession operating period to more than 30 years appeared financially ineffective, among the strategies to facilitate project financing a "flexible" depreciation of the investment was considered, and the following measures were agreed with the Grantor:

- Depreciation of at least 50% of the investment during the operational period (the first 30 year concession);
- A cash value ("redemption value"), equal to no more than 50% of the investment, to be paid to the Concessionaire at the end of the 30-year operation period.

Providing a "flexible" redemption value on the one hand recognises the long useful life of the infrastructure (more than 100 years). On the other hand, it allows the depreciation of the asset to be based on the actual operating results of the infrastructure, avoiding the problem of being excessively conservative in traffic forecasts and cost estimates, which are expected in such a long term forecasts. The concept here is that by having a kind of "financial buffer", the usual over-conservatism of very long-term forecasts of traffic revenues and costs can be avoided.

The redemption value may be reduced by strong operating results. In any case, even a 50% value will easily be provided by the State using the revenues from the awarding of a new concession. In fact, the "residual value" that could be achieved by the Grantor by tendering out the infrastructure operations for an additional period of 30 years, has been estimated to be in the range of €6.2–€12.8 billion for the "low growth" and "high growth" scenarios respectively, assuming 1% annual traffic growth rate after the first concession.

2.4.7 Conclusions

The results obtained from the financial analyses show the ability of the financial plan to reimburse and remunerate the capital to shareholders and financiers, in every scenario, without any contribution by the State.

The financial structure includes a plan to eventually privatise the Concessionaire. This may even be feasible shortly before the completion of the works, concluding the process during the operation phase.

Finally, further to the tender for the selection of the General Contractor, the successful bid provided for a ~12% reduction in the construction cost, that is about €0.5 billion. This allowed the Concessionaire to increase the provisions for contingencies during the construction phase.

2.5 Tendering and contracting

In accordance with all criteria and choices so far described, four international tenders have been prepared, invited and the corresponding contracts awarded. Relevant main data are reported in the following tables.

◄ Table 2.4
Tender 1.

TENDER 1 – GENERAL CONTRACTOR			
Restricted procedure 3 short-listed candidates	call	awarding	commitment
	April 2004	Nov. 2005	March 2006

Contract scope and amount

1. Bridge steel superstructures	€2,300 million
2. Bridge substructures and technical plants Road/Rail Links	€1,695 million
3. Engineering (design and works supervision, "post-operam" environment monitoring)	€235 million
4. Land acquisition	€65 million
5. Environmental impact mitigation & compensating works	€130 million
TotalTender Sum	€4,425 million
Total Contract Sum after bid reduction	**€3,880** million

Note: tems 1–3 are fixed lump sums, other items are estimates based on contract unit prices and shall be fixed further to the detailed design.

Award criteria

The most economically advantageous bid based on following points: price (max 45/100) – organization (max 15/100) – technical value of variants (max 15/100) – utilization & maintenance cost (max 10/100) – execution time (max 5/100) – number of nominated subcontractors (max 5/100) – value of pre-financing (max 5/100).

Winning bidder

Temporary association formed by **Impregilo** S.p.A.(Italy), the principal company, and **Sacyr** S.A. (Spain), **Società Italiana Per Condotte D'Acqua** S.p.A. (Italy), Cooperativa Muratori&Cementisti-**C.M.C.** (Italy), **Ishikawajima-Harima Heavy Industries** CO Ltd. (Japan), and **A.C.I.** S.c.p.a – Consorzio Stabile (Italy), the agent companies – Nominated designer: **COWI** A/S (Denmark).

First non winning bidder

Temporary association formed by **Astaldi** S.p.A. (Italy), the principal company and **Ferrovial Agroman** S.A. (Spain), **Maire Engineering** S.p.A. (Italy), **Ghella** S.p.A. (Italy), **Vianini Lavori** S.p.A. (Italy) and **Grandi Lavori Fincosit** S.p.A. (Italy), the agent companies – Nominated designer: **Chodai** Co. Ltd. (Japan).

◄ Table 2.5
Tender 2.

TENDER 2 – PROJECT MANAGEMENT CONSULTANT			
Restricted procedure 7 short-listed candidates	call	awarding	commitment
	Jan. 2005	Nov. 2005	Jan. 2006

Contract scope and amount

Project management services (see chapter 2.2.3)	€150 million
Total Contract Sum after bid reduction	**€120** million

Award criteria

The most economically advantageous bid based on following points: price (max 30/100) – organisational, methodological, technical and qualitative characteristics of the proposal (max 70/100).

(*Continued*)

Table 2.5 ▶ (*Continued*)

TENDER 2 – PROJECT MANAGEMENT CONSULTANT			
	call	**awarding**	**commitment**
Restricted procedure 7 short-listed candidates	**Jan. 2005**	**Nov. 2005**	**Jan. 2006**

Winning bidder

Parsons Transportation Group inc. (USA)

First non winning bidder

Consortium led by **Systra** S.A. (France) working in association with **Technital** S.p.A. (Italy).

Table 2.6 ▶ Tender 3.

TENDER 3 – ENVIRONMENTAL, TERRITORIAL AND SOCIAL MONITOR			
	call	**awarding**	**commitment**
Restricted procedure 2 short-listed candidates	**June 2005**	**Jan. 2006**	**Apr. 2006**

Contract scope and amount

Equipment, surveys, measurements, samplings, controls and services for the environmental, territorial and social monitoring over a wide area.	€37 million
Total Contract Sum after bid reduction	€**29** million

Award criteria

The most economically advantageous bid based on following points: price (max 35/100) – organisational, methodological, technical and qualitative characteristics of the proposal (max 65/100).

Winning bidder

Temporary association formed by **Fenice** S.p.A. the principal company and **Agriconsulting** S.p.A., **Eurisko** NOPWorld s.r.l., **Nautilus** S.c., **Theolab**, the agent companies (all Italian).

First non winning bidder

Temporary association formed by **Spea** Ingegneria Europea S.p.A. the principal company and **CESI** S.p.A., **Telespazio** S.p.A., **URS** Italia S.p.A., **Elsag** S.p.A., the agent companies (all Italian).

Table 2.7 ▶ Tender 4.

TENDER 4 – INSURANCE BROKER			
	call	**awarding**	**commitment**
Restricted procedure 6 short-listed candidates	**July 2005**	**Jan. 2006**	**Apr. 2006**

Contract scope and amount

Consulting and Brokerage Services for insurance policies and management, including assessment, management and monitoring of risks.	Fee on the insurance premiums (charged to insurance companies)

Award criteria

The most economically advantageous bid based on following points: price, as percent fee, (max 20/100) – organisational, methodological, technical and qualitative characteristics of the proposal (max 80/100).

Winning bidder

Marsh S.p.A. (Italy branch)

2.5.1 Preparation of the technical specifications and tender documents for the Bridge

The challenges of the project included developing and defining appropriate design and construction specifications to be included as part of the contract documents. Many aspects were relatively conventional and could rely on well established and generally acceptable criteria already defined in Italian and international codes and standards. However, other aspects required a more detailed project-specific approach, building on best international practice and extending it to encompass the scale and particular characteristics of this project, especially for the main bridge.

There were no pre-existing Italian, European or International codes or standards directly relevant to a bridge of this size and form, as no such structure had previously been built. However, there was considerable international experience in the design and construction of major suspension bridges, and a large body of research which had already been undertaken in connection with this bridge. Accordingly, SdM extensively involved the Scientific Committee (see end of section 2.2) for this task and also appointed international consultants, experienced in this field, to assist with the preparation of the necessary specifications.

The task included writing new design, materials and workmanship criteria, mostly adapted from existing specifications which were known from experience to have worked well on previous projects and which were suitably applicable in this case.

In order to support this work it was necessary to carry out further analyses of the bridge to verify satisfactory performance under the proposed design specifications both from a safety and a serviceability point of view. The work on the fundamental basis of design for the bridge continued in parallel with the development of the more detailed specifications, and the consultants helped to ensure that consistent performance levels were defined throughout.

Suitable proving tests and trials were defined in the documents which were required to be carried out by the General Contractor in order to verify satisfactory performance. These included tests and procedures to prove the adequacy of the design (such as further wind tunnel tests and geotechnical investigations for example) and also trials to be carried out to demonstrate the suitability of construction processes (such as main cable compaction and dehumidification for example). These tests and procedures were based on best international quality control practice.

The specifications also included the definition of aspects concerned with ongoing operation of the crossing, including specific requirements regarding access to all parts for inspection and maintenance. Requirements for the Life Cycle Costing (LCC) and Reliability Centred Maintenance (RCM) procedures were drafted in keeping with the philosophy for these already adopted by SdM. A sophisticated instrumentation system for the monitoring of structural health and performance was also included, together with requirements to demonstrate satisfactory railway runnability and expansion joint movement performance.

3 – TRANSPORTATION, ENVIRONMENT AND SOCIO-ECONOMIC BACKGROUND

Contents of this chapter:

- **Transportation studies**
- **Environmental background**
- **Socio-economic effects**

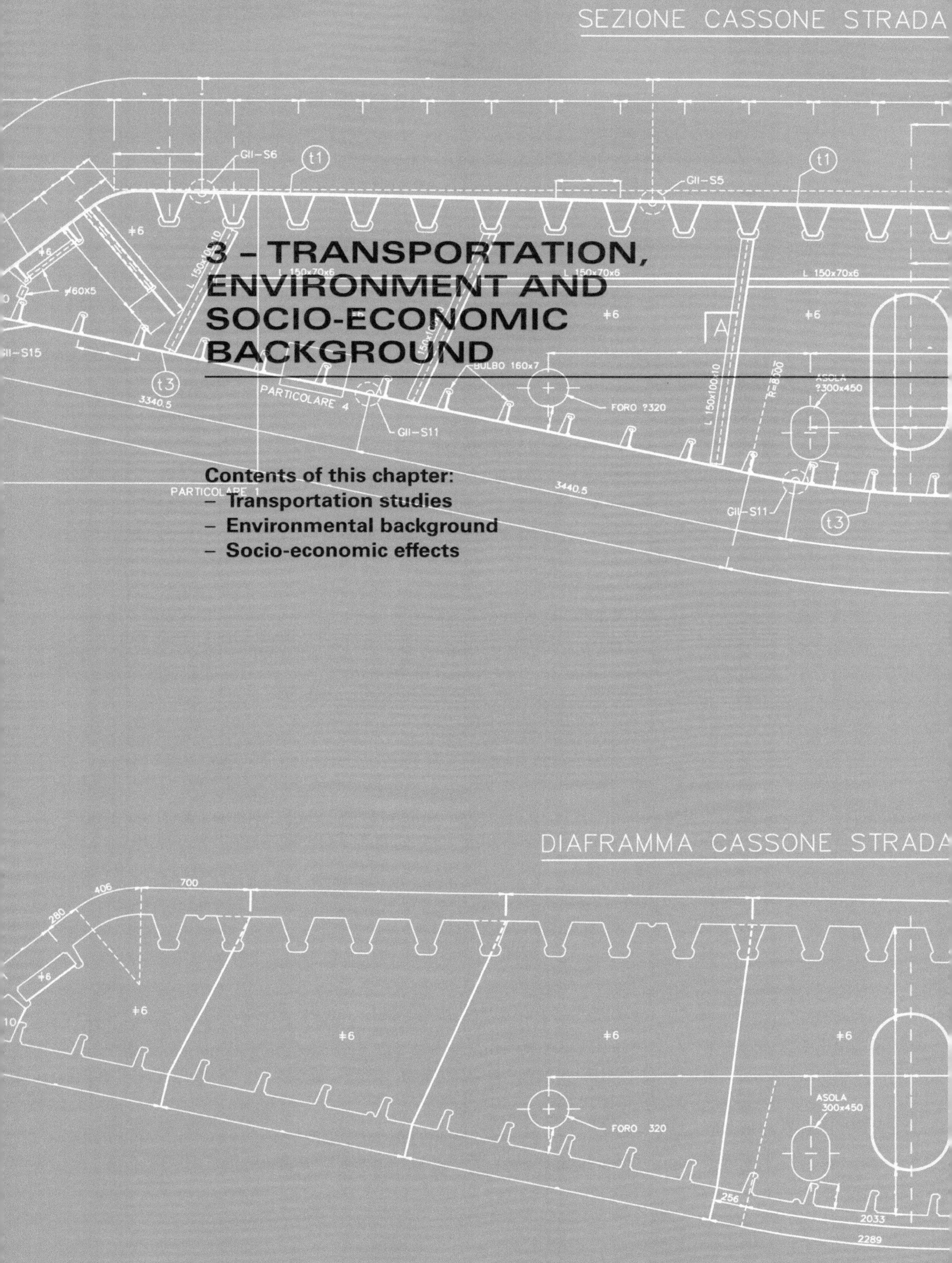

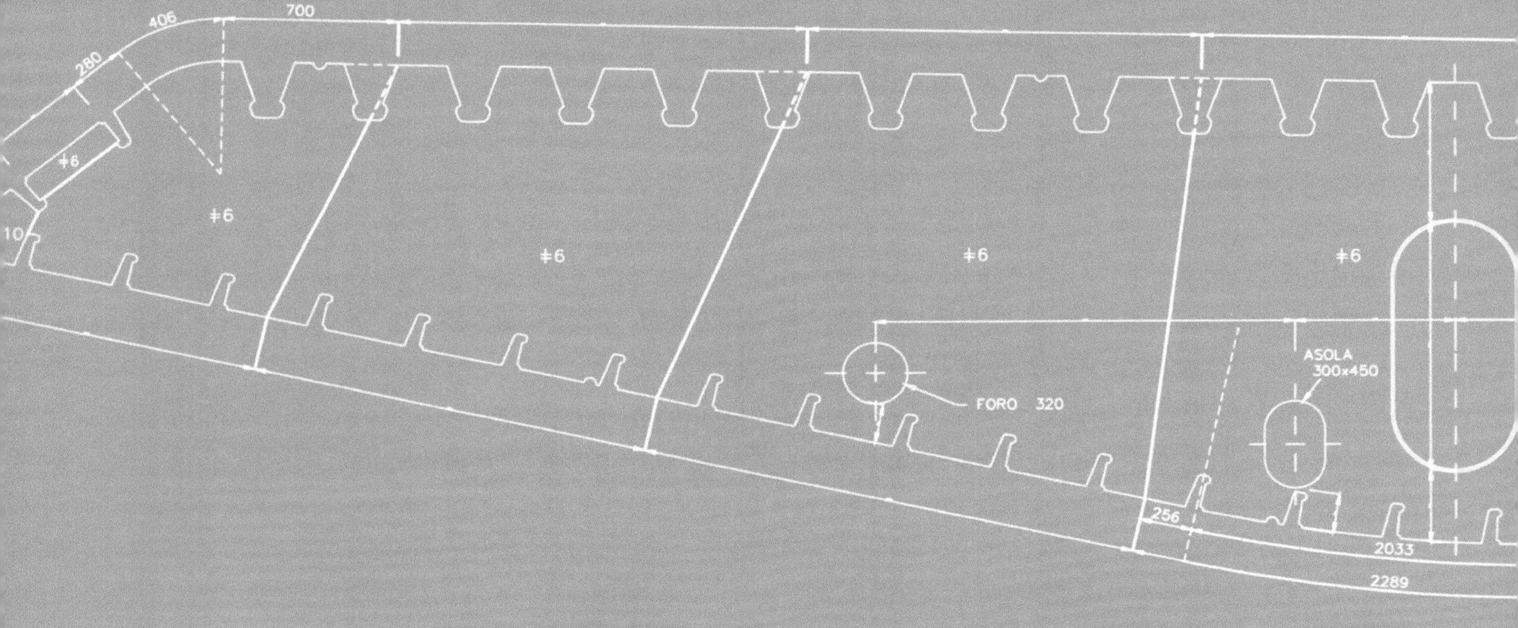

This chapter elaborates on some of the key issues which have already been briefly discussed relating to the evolution of the Project to date.

Evolution of transport demand since 1995 and an analysis of current Sicily to mainland passenger and freight mobility are reported in order to give an idea of the scale and complexity of these studies and related surveys which have no precedent in Italy for any single project.

The environmental and socio-economic background is also described, including results of Environmental Impact Assessment procedures completed in 2003.

Furthermore this chapter summarizes the socio-economic effects of the crossing together with findings of the cost-benefit analysis and strategic environmental assessment which have been carried out in order to have a scientific comparison between the two alternative methods of crossing the Strait (bridge or ferries) and a realistic picture of the future modernization of the area which will be a consequence of the construction of the bridge.

3.1 Transportation studies

3.1.1 General

In the different studies and design activities, several surveys and analyses have been carried out on the passenger and freight traffic between Sicily and the Italian mainland focussed on the crossing of Messina Strait. The most recent studies, started in 2005, took into consideration all transport modes and all routes, with the following main objectives:

- Building up a consistent data-set of past traffic trends and analysis of the evolution of transport demand in the previous 10 years (1995–2005).
- Analysing the current traffic demand between Sicily and the mainland.
- Building up a comprehensive database on traffic supply and demand to and from Sicily including all relevant characteristics.
- Forecasting future traffic demand on the bridge in different transportation and macro-economy scenarios by multi-mode computer models.

The scale and complexity of these studies and related surveys have no precedent in Italy for a single project and a single region. The results thus form a significant and unique geo-referenced database for any future transportation study in Southern Italy.

3.1.2 Past trends

The evolution of the transport demand in the 10 years between 1995 and 2005 was reconstructed through data obtained from all available competent public and private sources. For this purpose, the national passenger and freight demand to and from Sicily was considered for the following different modes and routes:

- Ferries across the Strait (for tyred vehicles, trains and pedestrians).
- Ro-Ro ships (for tyred vehicles and pedestrians) to and from the north and centre of Italy from and to the Sicilian ports of Messina, Palermo, Trapani, Termini Imerese, Pozzallo and Catania.

- Aircraft (passengers) from and to the Sicilian airports of Palermo, Catania, and also to and from Reggio Calabria.

The results obtained give a detailed picture of the evolution of the freight and passenger demand to and from Sicily during the ten years to 2005.

The **passenger demand** shows a positive trend from 18 million passengers in 1995 to 21.3 million in 2006 with a 18% overall increase. The positive trend temporarily declines in years 2000–2002. Years 2004–2006 show an increase in the share pertaining to air travel and a decrease of the share pertaining to Strait ferries, particularly to ferries operated by the railway company. This mainly suggests that a modal transfer of the demand is in progress especially from train to aircraft (Figures 3.1 and 3.2).

The **freight demand** shows a positive trend from 15.2 million tons in 1998 to 19.2 million tons in 2006, counter-tendencies in 2002 and 2005 can be related to the weakness in the economy in those years. The period 2002–2004 shows a significant increase in the share pertaining to long route Ro-Ro ships and a decrease of the share pertaining to Strait ferries, particularly in years 2001–2003. This modal transfer appears mainly due to the (temporary) decay in service level of the Salerno-Reggio Calabria highway whose substantial upgrading, presently in progress, involves important works and related troubles for users. (Figure 3.3)

The **overall flow of vehicles** crossing the Strait by ferries is almost stable, as shown in the following table. The trend shows a slight decrease

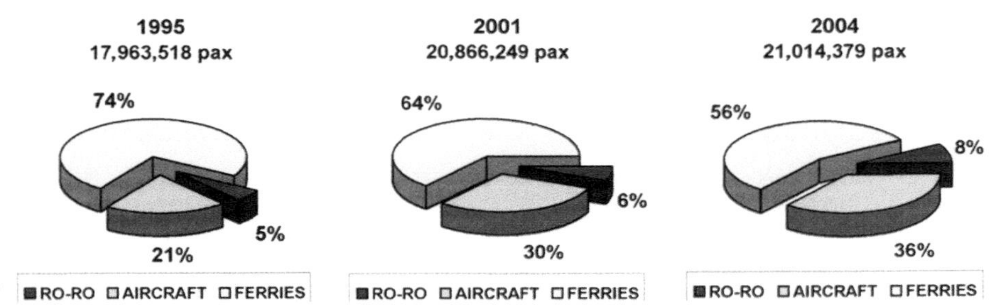

◄ Figure 3.1 10-year evolution of modal split for passenger traffic.

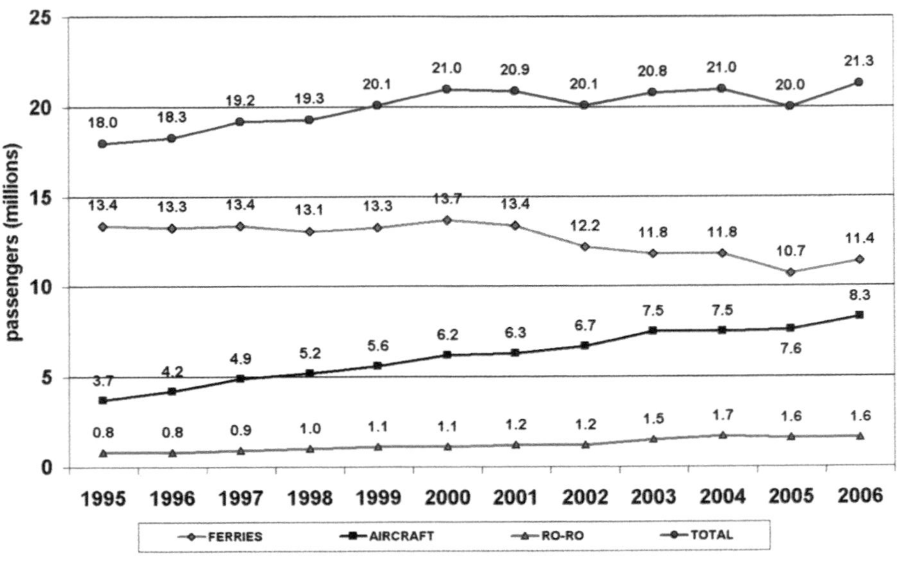

◄ Figure 3.2 Past trends of passenger traffic to/from Sicily.

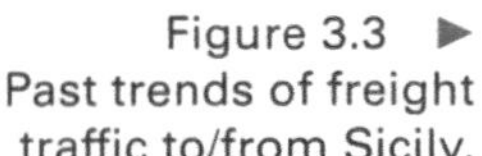

Figure 3.3 ▶ Past trends of freight traffic to/from Sicily.

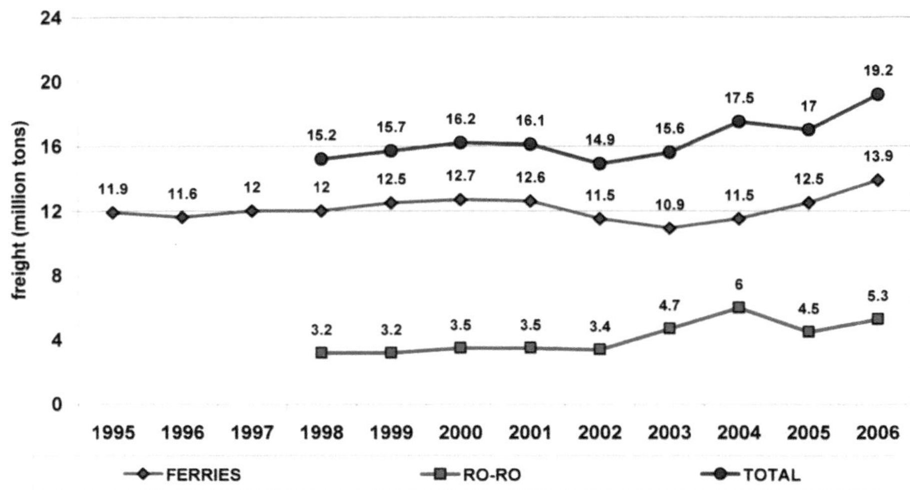

Table 3.1 ▶ Road vehicles crossing the Strait by ferry.

Million Vehicles Per Year	2000	2001	2002	2003	2004	2005	2006
Passengers	2.76	2.80	2.61	2.37	2.47	2.40	2.53
Freight	0.93	0.96	0.86	0.81	0.85	0.88	0.95
Total	3.69	3.75	3.47	3.34	3.32	3.28	3.48

in 2002–2003 and invariance in the following period, with 2006 showing a significant increase on the previous year (more than 5%). A yearly flow of about 2.5 million cars and 1 million trucks appears to be established.

3.1.3 Recent traffic (2005–2006)

The analysis of the recent Sicily to mainland passenger and freight mobility has been made through the following surveys:

- one-year automated traffic counts all vehicles crossing the Strait;
- about 23,000 Origin-Destination interviews and related counts at 2 airports (Catania and Palermo), 7 harbours (Messina, Palermo, Trapani, Termini Imerese, Pozzallo and Catania) and 5 railway stations (Palermo, Agrigento, Siracusa, Catania, Messina);
- about 14,000 Origin-Destination interviews and related counts at the ferry terminals of the Strait (for vehicles and trains);
- about 10,000 household telephone interviews in Reggio Calabria and Messina districts.

The interviews collected, in addition to O-D data, of all the socio-economic information needed for the multi-mode forecasting model, such as trip purpose and frequency, travel time and cost, access/egress modes to air and train terminals, vehicle type, number of truck drivers, etc.

The total volume of transport demand to and from Sicily in 2005, on all routes and by all modes (excluding maritime cargo transports) proved to be about 21.7 million passengers per year and 16.9 million tons of freight per year.

As a result of the modal evolution in the 1999–2005 period, the following current modal split was identified:

2005 modal split	Road	Railway	Aircraft	Ship
Long-distance passengers	25%	11.5%	54%	9.5%
Long-distance freight	55%	15%	–	30%

◀ Table 3.2
2005 modal split.

Average filling coefficients (i.e. the number of people or tons of freight per vehicle) for long-distance trips are higher than those used as the average standard relationship between overall transport demand and vehicle numbers according to published data: +6% for passenger vehicles (2.18 passengers per car) and +12% for freight vehicles (9 tons per equivalent truck).

All in all the transport across the Strait has been strongly affected by two main negative factors:

a. In 2003 average economic growth rate stood at 1.5% instead of 2.8% expected in the Government Financial Planning Document, and the infrastructure improvement program of southern Italian networks, including the extension of high speed railways south of Naples and the upgrading of the Salerno-Reggio Calabria highway, did not move forward as expected.
b. The service level of the Strait ferry system continues to reduce (i.e. longer crossing time, reduced frequency, higher fares) due to reasons including:
   - Compulsory diversion of truck ferries to the port of Tremestieri (to divert trucks from the centre of Messina) which implies a much longer sea route compared to Messina – Villa San Giovanni.
   - Poor economy and inefficiency of the system, particularly for rail ferries.
   - Increasing congestion of sea routes across the Strait.

The inefficiency of crossing the Strait by ferry significantly contributed towards the progressive loss of market share by rail transport.

3.1.4 Future traffic forecasting

As already mentioned, traffic volume forecasts have been carried out for the 2002 Preliminary Design and for the 2003 Concession Agreement. These forecasts have been based on four different scenarios coherent with the GDP growth foreseen by the national Economic and Financial Planning Document issued by the Government. Aggregate traffic data obtained by authorities or drawn from published literature have been used.

The predicted growth of overall transport demand in the 2005–2042 period was estimated to be about +95%, as the intermediate value between low and high growth hypotheses.

Considering the importance of this issue for the future of the project, new forecasting of traffic is in progress based on the detailed traffic data obtained through the above-mentioned surveys.

The multi-mode forecasting model takes into account overall Sicily to mainland mobility as shown in Figure 3.4.

Forecasting starts from detailed demand levels for the different routes, modes of transport and vehicle types obtained from the surveys. The aggregated transport demand is then increased in proportion to the economic

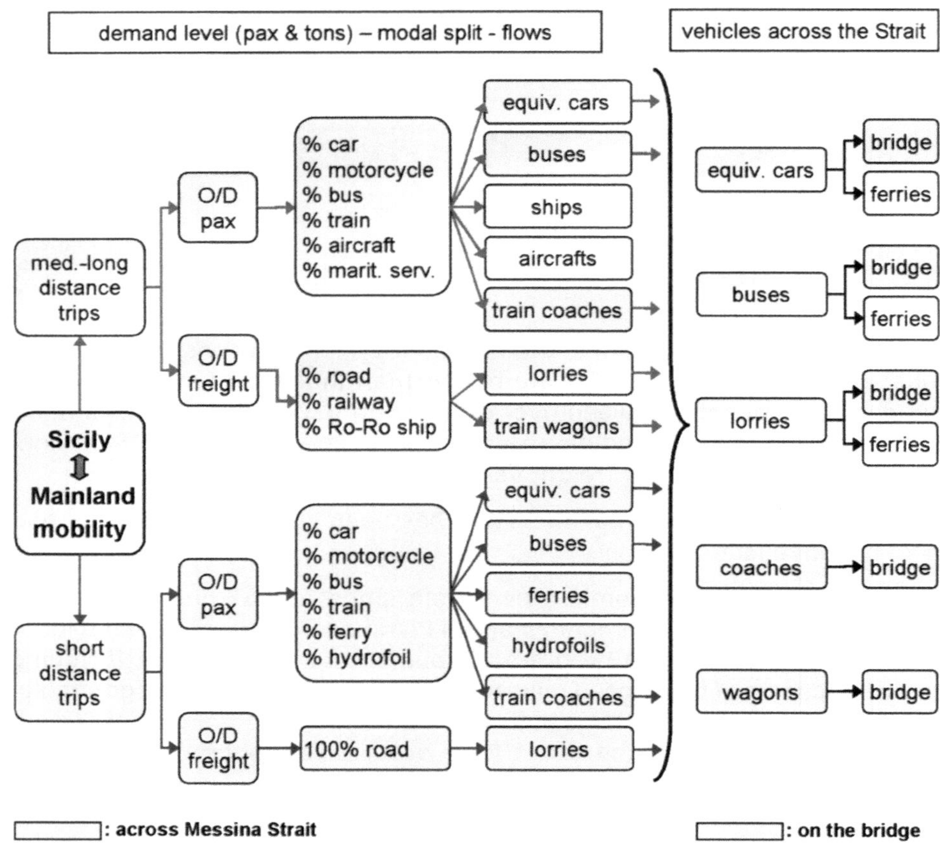

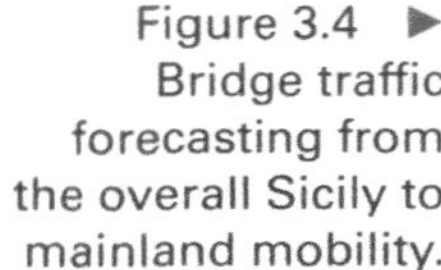
Figure 3.4 ▶ Bridge traffic forecasting from the overall Sicily to mainland mobility.

development represented by the anticipated GDP. The modal choice and the route choice are obtained by behavioural models based on the "random utility" theory.

In coherence with previous studies, the model has been set for two different macro-economy scenarios combined with two transportation scenarios:

- Two different trends of passenger and freight transport demand based on "high" and "low" growth rates of southern Italy GDP respectively. Anticipated growth rates are coherent with the most recent national Economic and Financial Planning Document.
- Two different settlements of transportation systems, favourable and unfavourable for the bridge respectively, as summarized in the box below. Both transportation scenarios consider the up-to-date programs for realization/upgrading of infrastructure and transport services in southern Italy and relevant evolution of service levels, up to the 2025 time horizon. In both scenarios a regional metro railway operates across the bridge.

The "unfavourable" transportation scenario

- ferries operated by RFI (railway company) are dismissed;
- private ferry companies compete with the bridge as much as they can, regardless of their cost/revenue balance;
- private ferries offer 6 hourly trips in each direction;
- ferry fares are 20% lower than bridge tolls;
- high speed boats for pedestrians offer 2 hourly trips in each direction.

The "favourable" transportation scenario

- ferries operated by RFI are dismissed;
- private ferries offer 2.5 hourly trips per each direction;
- ferry fares are equal to bridge tolls (unvaried in real value);
- high speed boats for pedestrians offer 0.5 hourly trip in each direction.

3.2 Environmental background

3.2.1 Environmental impact study

On 20 June 2003, the Ministry of the Environment's Special Committee for Environmental Impact Assessment announced their approval of the Environmental Impact Study carried out by a consortium formed by Systra SA, Bonifica Spa, Systra-Sotecni Spa and Ast Sistemi Srl. on behalf of Stretto di Messina S.p.A.(SdM).

The approval decree contains recommendations and guidelines that have to be implemented by SdM during the final design and construction phases.

The Environmental Impact Study (EIS) focused on the following main issues:

1. Assessment of the existing state of the environment in the Strait area and verification of the "minimum impact" condition;
2. Cost-benefit analysis of the two methods of crossing the Strait (bridge or ferry) in accordance with European Directive 42/2001/CE that provides principles and procedures for the Strategic Environmental Assessment (SEA);
3. Multi-criteria analysis in order to obtain a technical optimisation of the fixed link solution;
4. Individuation of possible mitigation measures.

According to Italian technical regulations (Decree DPCM 27/12/88) the EIS includes the analysis of the state of and interaction between the following factors: marine environment, human beings, fauna and flora, soil, water, air, climate, the landscape, noise & vibrations and ionizing radiations.

The *evaluation of the potential adverse effects* of the different alternatives and technical solutions has been carried out for each environmental factor, including:

- physical changes in the area (topography, land use, changes in water bodies);
- use of natural resources such as land, water, materials or energy;
- use, storage, transport and handling of hazardous substances;
- production of solid wastes during construction;
- risks of contamination of land or water;
- social changes (e.g. demography, traditional lifestyles, employment);
- vulnerability of important or sensitive areas (eg. for breeding, nesting, resting, overwintering or migration of birds);
- potential effects on marine or underground waters;
- congestion of transportation routes on or around the construction sites;
- visibility of the works to local people;
- vulnerability of any areas or features of historic or cultural importance;
- existing and future land uses;
- landslides, erosion, flooding, and extreme or adverse climatic conditions.

The best available techniques, such as environmental modelling, analysis of existing cartography and new thematic maps, advanced landscape analysis and photographic simulations have been applied, especially concerning the evaluation of changes to the landscape.

In fact, the outstanding scenic interest of the Messina Strait has demanded very special attention. While the protection of a designated scenic area has an intrinsic value, the bridge offers a unique opportunity to develop an area that has suffered for its role as a corridor experiencing a prolonged, aggressive and chaotic urbanization of its coasts in recent decades.

The protection and enhancement of landscape quality, character and local distinctiveness will be a priority during the final design phase in order to restore and strengthen the typical features of the Strait by designing appropriate optimization and mitigation measures.

The EIS included the comparison of the two methods of crossing the Strait (by bridge or ferry) in accordance with the European Directive 42/2001/CE that provides principles and procedures for the *Strategic Environmental Assessment* (SEA), even though in 2002 this Directive had not yet been put into effect by the Italian Government.

The purpose of the SEA Directive is to ensure that environmental consequences of certain plans and programmes are identified and assessed during their preparation and before their adoption.

SdM adopted a scientific approach including 64 indicators:

- 30 environmental indicators (effects and impacts);
- 10 on transport efficiency;
- 24 on economic and territorial sustainability.

The bridge solution was seen to be highly preferable to the upgraded ferry solution, as it achieves the following objectives:

Figure 3.5 ► Visual impact assessment by 3D interactive modelling: view of the bridge from Sicily.

- substantial reduction in exhaust gas emissions;
- important time savings in crossing the Strait (time savings for railway passengers are an average of 2 hours and for cars and trucks approximately 1 hour);
- large reductions in urban area congestion;
- higher degrees of socio-economic integration of urban areas along the Strait;
- positive effects on the economy and employment.

Technical optimisation of the Preliminary Design, together with the careful selection of construction methods and procedures for handling and disposal of materials, led to a significant 25% mitigation of the environmental impact, compared to the previous preliminary design. This improvement is a result of various actions including:

- improvement of the road connections in Sicily, by optimising the suspension bridge profile so as to allow the approach viaduct ("Pantano viaduct") to be reduced in length by 400 metres, and concealing in a tunnel a 350 metre stretch of road that was particularly impacting on the surface;
- overall optimisation of the road connections on both sides reducing the length of viaducts by 1,600 metres, minimising the Sicily toll plaza area and eliminating the Calabria one;
- direct connection of the bridge with the planned high-speed railway line, eliminating approximately 21 km of tunnels;
- requiring the handling of materials during construction to be by sea transport rather than by road, to mitigate interference with local traffic;
- precise location of material stockpiling in remote areas of low population density.

◄ Figure 3.6 Visual impact assessment by 3D interactive modelling: view of Pantano viaduct.

3.2.2 Environmental, territorial and social monitoring

As described in detail in Chapter 4, according to the recommendations and guidelines of the Special Committee for Environmental Impact Assessment, SdM will implement a specific environmental, territorial and social monitoring plan, not only during the construction works (in opera) but also during the preparation of the final design (ante operam) and during the operational phase of the fixed link after its completion (post operam).

Furthermore for the first time in Italy, a comprehensive monitoring plan of environmental, territorial and social impact will be implemented not only in the reduced area directly affected by construction activities (construction sites, pit areas, disposal areas, etc.) but also outside the construction sites in a wide area of the Strait.

During the executive design phase, the monitoring plan of the wide area will provide important information about territorial vulnerability and a comprehensive knowledge of the environment in order to provide the necessary mitigation measures and to properly design any possible environmental and social compensation measures.

3.3 Socio-economic effects

The application of adequate social and economic science and research is critically important to resolve the challenge of achieving a sustainable development of the region by constructing infrastructures that modernize the area while at the same time producing positive effects on population lifestyle and the environment. Furthermore, close coordination between social science, environmental science and engineering disciplines is necessary to optimize a project such as the link between Sicily and the mainland that will have an immediate and enduring impact in social, economic and environmental terms.

For these reasons independent advisors have analysed and reported on the potential effects to allow appropriate planning by SdM.

As noted above, the economic feasibility (cost/benefit analysis) has been evaluated on the basis of four development scenarios which take into consideration high and low scenarios for GDP growth for Southern Italy and favourable and unfavourable transport growth predictions. The conclusions demonstrate the economic feasibility of the project, even under the "low" GDP growth scenario and with unfavourable transport growth.

The results prove the considerable economic feasibility of the bridge project:

- the Economic Net Present Value is always positive;
- the benefits exceed the costs in all scenarios considered;
- and the Economic Internal Rate of Return has a range between 9% and 12%.

Both the construction and operation phases have been analysed in order to evaluate the economic and social impacts. The analysis of the construction phase has estimated the direct, indirect and induced impact of both the bridge and the connecting rail and road infrastructures. In the operational phase, the assessed positive impacts will generate structural changes in the economy of the surrounding areas.

These evaluations show that in both the construction and operational phase there will be a positive impact on the local economies of Messina and Reggio Calabria, as well as for the wider Sicily and Calabria regions.

The economic impact of the construction phase has been estimated using the so-called "sector interdependencies" methodology through a multi-regional model of the Italian economy. Overall, the direct, indirect and induced economic impact of the construction phase is estimated to be significant – approximately €6 billion – close to or exceeding (as often happens) the total investment cost. More than 50% of the economic impact will be felt in Calabria and Sicily and about 75% will be in southern Italy. The direct economic benefits in terms of added value in the various areas amount to 32.7% of the investment cost. The indirect and induced benefits are even greater at 48% and 42% of the investment respectively.

The effects on the labour market during the construction phase will also be positive. It is estimated that the direct and indirect occupational level will increase by approximately 40,000 units along the regions of the Strait.

The operational phase has been evaluated based on two main assumptions relating to the effective value of a permanent and continuous link (open 24 hours a day, 365 days a year), providing a facility to cross the Strait in only three minutes. The first one represents the great improvement in quality of service, resulting in positive effects on the development of the area and the quality of life. The second represents the contribution made by the bridge in reducing the gap in infrastructure levels as a result of the planned improvements to road and rail links in southern Italy. In other words, the bridge will act as a development multiplier, bringing extraordinarily positive results to the industrial and economic systems of southern Italy.

These evaluations have focused on the impacts on different sectors. First of all, trade and business activities will feel the most significant positive impact. By linking the two shores, the bridge will enlarge the market not only by facilitating greater trade activities between companies, but also by creating new business opportunities and promoting greater integration between the regional economies, leading to an increase in employment.

Secondly, the positive impact on tourism is at least twofold:

- the impact of better connections, facilitating both short and long distance accessibility, and
- the impact of the bridge itself as a tourist destination, attracting visitors from all over the world.

The urban impact lies in the reduction of traffic congestion, the result of which will be an improvement in quality of life, an improvement of the urban image and its resulting benefits in terms of added commerce and tourism benefits. Furthermore the construction of the bridge will allow and promote the renovation of existing harbour areas, currently used by local ferries, and increase the international cruise and leisure market.

It will also trigger the potential for very interesting projects of re-qualifying and developing Messina's waterfront, currently occupied by railways and other unsightly functions. Removal of facilities and activities currently devoted to train dis-assembly, loading, unloading and re-assembly, as well as eliminating the "barrier" between the town and the sea, currently created by the railway (that will become underground), will allow the re-development of a highly valuable urban area (more than 40 hectares adjacent to the town centre). (Figure 3.7)

Figure 3.7 ▶ Detail of Messina centre and harbour. The future underground railway connection to the bridge is added in black. The valuable waterfront area currently devoted to train shipping is clearly visible in the lower zone of the image, at southern margin of the town centre.

The new crossing will also facilitate the creation of an integrated urban area comprising Messina and Reggio Calabria, linked by the bridge, resulting in better cultural, health, commercial and leisure facilities, as already mentioned in Chapter 1.

When considering the socio-economic value of the project, we should not disregard the importance of scientific research related, for instance, to construction materials, techniques and technologies. The high-tech content of the bridge will provide an important added value for industry and techno-scientific research, as has been observed in other countries where major bridges have been constructed.

Furthermore, both the construction and operation phases of the crossing can induce the development of "advanced technology poles" for research and technical support in respect of safety and the environment.

A further socio-economic impact will concern the so-called "credibility effect" consequence: that is the positive influence on the business community, including a greater interest from external investors, connected to the fact that a large investment is being made in the south of Italy with such a high profile project.

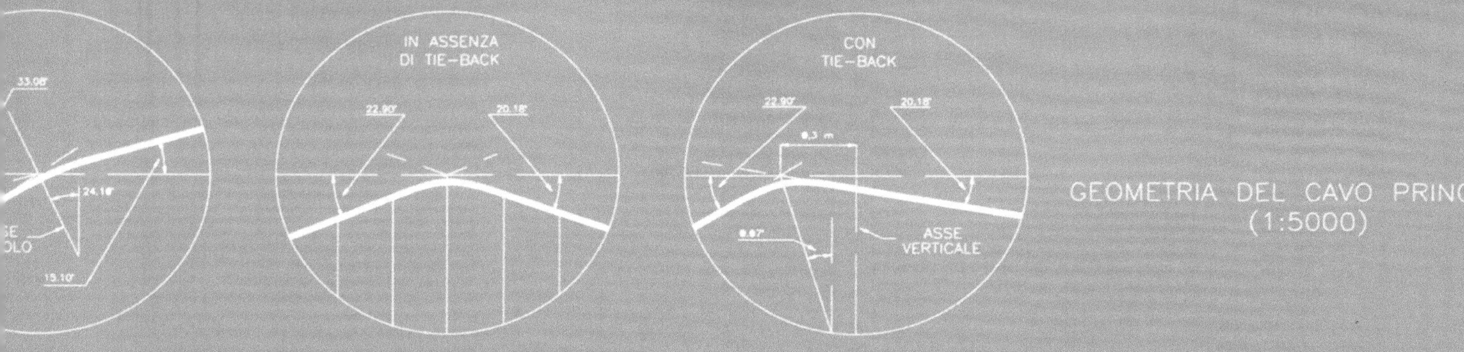

4 – TERRITORY AND ENVIRONMENT

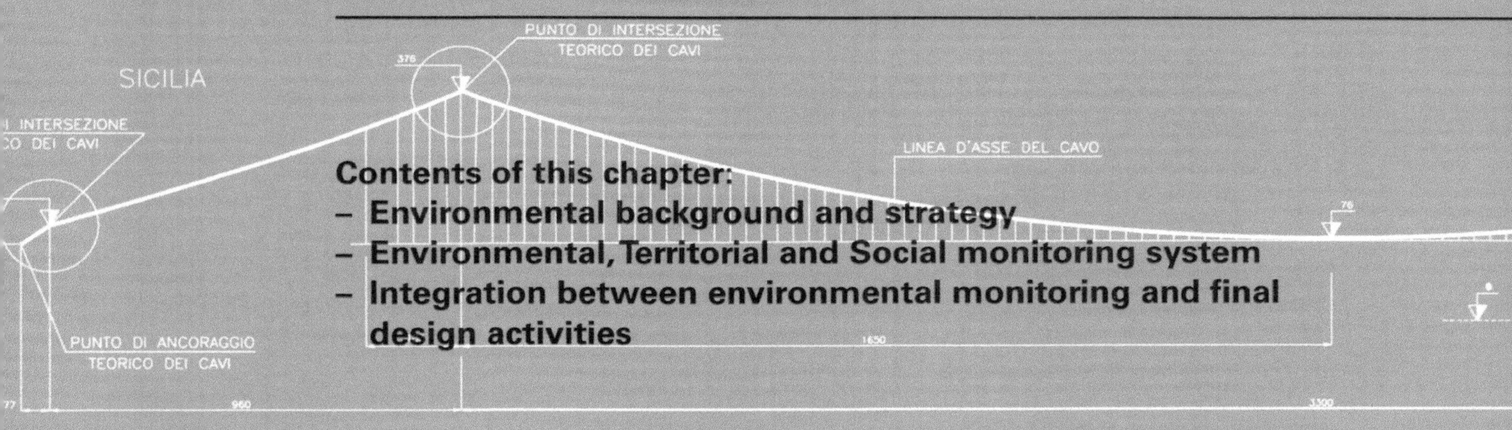

Contents of this chapter:
- **Environmental background and strategy**
- **Environmental, Territorial and Social monitoring system**
- **Integration between environmental monitoring and final design activities**

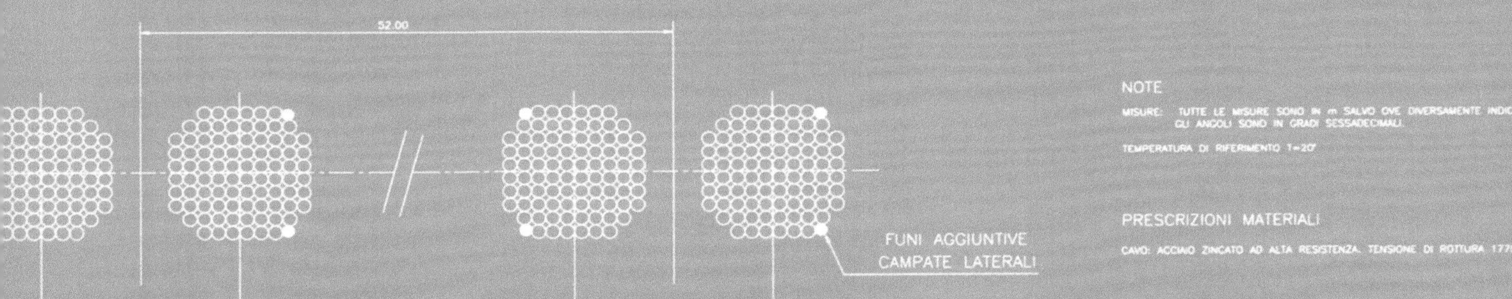

PRINCIPALE	CAMPATA CENTRALE	CAMPATA SICILIA	CAMPATA CALABRIA
(mm)	5.38	5.38	5.38
DEI 4 CAVI	44352	44352+928	44352+488
) DEI 4 CAVI	88	88+2	88+2
	504	504/464	504/244
QUATTRO CAVI (m²)	1.008	1.029	1.019
IE (m²)	0.01146	0.01146/0.01055	0.01146/0.0055
CAVO (m)	1.236	1.249	1.243
E (m)	0.135	0.135/0.131	0.135/0.093
5.38 mm PER 1 DEI 4 CAVI (Km)	148284.3	46426.1	38694.3
PER 1 DEI 4 CAVI (t)	26461.790	8284.880	6905.113

SI ASSUME COME METODO DI TESSITURA DEL CAVO L'AIR-SPINNING
DIAMETRI CORRISPONDONO AD UNA PERCENTUALE DEI VUOTI PARI AL 22% PRIMA DELLA COMPATTA-ZIONE E PARI AL 18% DOPO LA COMPATTAZIONE

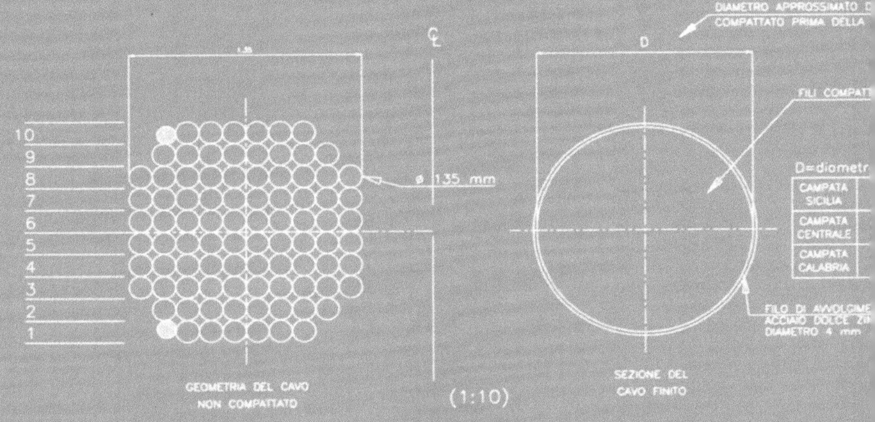

This chapter describes the environmental control and monitoring methods of the Project. These have been developed specifically, representing an innovative approach, especially for Italy, and will generate a unique and valuable knowledge bank for a wider understanding of state of the environment of the whole Strait area. This chapter describes the Environmental, Territorial and Social monitoring system designed by SdM in order to ensure the independence and objectiveness of the monitoring and to represent the evolution of the environment in a scientific, objective and reliable way.

The complexity of the project and the environmental context of the Messina Strait demanded the design of a monitoring system which guaranteed that the environmental data became not only a control element to be used by SdM, the Environmental Monitor and the PMC, but also a design element to be used by the General Contractor.

Furthermore this chapter summarizes the innovative methodologies applied to evaluate the interaction and possible interferences between the project and bird migrations and marine mammals.

The Social Monitoring Plan adopted to monitor socio-demographic, micro/macro-economic impacts, social perceptions, and the degree of consensus among local communities and national stakeholders (which are often not properly monitored) is also described.

4.1 Environmental Strategy: an innovative approach

4.1.1 How to build the consensus

The importance of environmental matters in the decision-making process of public investment on infrastructure projects has recently increased, significantly influencing the destiny of such projects. As a matter of fact, many projects of proven environmental value get stuck, even after several attempts, because of difficulties in determining the works location or disagreements with stakeholders on the environmental protection and mitigation measures needed to make the project environmentally sustainable and tolerable for the affected region.

The so-called NIMBY complex (= "not in my back yard"), which is often fuelled by environmental concerns, no longer occurs only during the selection of the location of a new infrastructure project or a new plant facility which is considered to be a benefit for the community. It has in fact recently spread to all construction phases, even when all the necessary authorizations have been granted. The glaring effects of this phenomenon are evidenced by the frequent delays incurred by projects under construction due to the need to respond to requests for variations in or even cancellation of construction works, and handle the legal implications.

On one hand, the model which is presently used in Italy to represent the community during planning, permitting and design of important infrastructures, often proves to be inadequate. On the other hand, environmental matters are often handled with too much emotiveness or superficiality. The opposition which emerges once the authorization and planning approval are obtained, is rarely based on accurate environmental studies capable of producing a reliable and objective evaluation of the potential negative effects.This can hinder a real sustainable development of the affected region.

This is the reason why it is necessary to consider environmental matters as a key element in the design and construction of important projects. As a matter of fact, it constitutes a fundamental technical and communication mechanism for those companies or authorities engaged in important projects which, despite being intended as community utilities, impose a strong environmental and social impact.

On this basis it was clear from very early on that the environmental issues connected with the Messina Strait Bridge had to be dealt at a very high level in order to gain the support of stakeholders and the wider population and to reach the final objective of a sustainable development of the area.

The requirements imposed by Italian and European environmental legislation constituted only the starting point for voluntarily developing the means adopted by Stretto di Messina S.p.A (SdM) for managing the environmental aspects of the project.

SdM decided to monitor and supervise the General Contractor's activities and to apply specific and independent procedures during both the final design and the construction phases.

The approvals process for the project was attacked on environmental grounds largely because it followed the special administrative procedure introduced in Italy by the "Legge Obiettivo", which established a preferential procedure for infrastructures of prevailing national interest. (See box)

The debate on what, for many, was the assumed devastating effect of the project on the environment did not always focus sufficiently on the substantial environmental control methods which are an inherent part of the project and which will constitute an important justification of the environmental sustainability of the bridge.

For this reason, this chapter will not deal with the procedures for Environmental Impact Assessment which have already been widely described in several articles and publications supporting or attacking the bridge, but will, on the contrary, describe the environmental control and monitoring methods which represent an innovative approach, especially for Italy, and which will generate a unique and valuable knowledge bank for the project.

Environmental Impact Assessment under "Legge Obiettivo"

As described in Chapter 2, a new legislation was approved by the Italian Parliament in 2001 to facilitate the implementation of large-scale public infrastructures and strategic industrial investments (Law no. 443 of December 27, 2001, known as the Legge Obiettivo). This law gave the Government powers to identify and influence those infrastructure projects of significant strategic importance in the national interest, and also allowed for the financing of public infrastructure projects through the contribution of private capital. It also defined specific procedures for streamlining the Environmental Impact Assessment and approval processes. The Legge Obiettivo was implemented by Legislative Decree no. 190 in August, 2002.

Procedures for the approval by CIPE (the Inter-ministerial Economic Programming Committee) of an EIA relating to the Preliminary Design are set down. Consent is granted by a majority vote for approval by the heads of the relevant autonomous regional or provincial authorities who give their opinion after consulting the municipal authorities on whose territory the works will be carried out. The approval by CIPE of the EIA procedure determines the environmental acceptability of the works, and effectively grants planning and building permission on behalf of those municipal authorities affected by the project.

Subsequently, the Final Design is subject to a further environmental approval procedure, to check that the project still satisfies the necessary requirements and that any changes or conditions imposed upon it by the approval of the Preliminary Design have been made.

4.1.2 Strategy and objectives

From the environmental point of view the project strategy and objectives addressed the following considerations:

- The independence of the environmental monitoring activity is fundamental for any major project such as the Messina Strait Bridge. Thus this activity was granted to a third party, independent of the General Contractor.
- Although the Messina Strait is an exceptionally environmentally sensitive area, the available data was not considered adequate for a thorough understanding of the issues. To redress this lack of knowledge it was necessary to implement environmental monitoring which focused not only on the sites immediately affected by the construction works, but also on developing a wider understanding of the environmental, territorial and social impacts of the project on the whole Strait area.
- The construction of an important structure such as the suspension bridge constitutes an extraordinary opportunity to invest significant resources in the study of the state of the environmental of the area. Thus a monitoring network and database was conceived which would be connected to the existing networks and interfaced with local and national environmental agencies to set up a widespread permanent system for monitoring the environmental and territorial risks.
- The proposals went beyond the traditional approach of evaluating the environmental impacts at the end of the process – the so called "end-of-pipe" approach – which tends to be limited to beautification and re-greening measures after the event. Instead, the process establishes the need to force the work production to routinely face the environmental aspects in order to integrate them within all decision-making processes during the final design and construction optimization phases.
- The environmental sustainability of a major structure such as the bridge needs tangible actions to be carried out by the selection of appropriate design and engineering solutions, the optimization of construction methods, the selection of compatible materials, and the design of adequate mitigation and compensation measures where necessary.

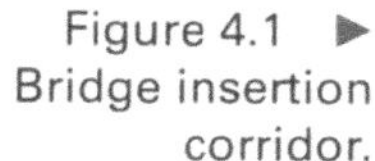

Figure 4.1 ▶ Bridge insertion corridor.

These subjects will be briefly dealt with in the following, describing the measures taken by SdM to ensure the above-mentioned goals are achieved during the final design and construction stages.

4.2 Wide area monitoring

4.2.1 A double environmental monitoring system

The first challenge to be faced involved designing an Environmental Monitoring System (EMS) able to obtain a dynamic assessment of environmental vulnerability of the whole area, during the design, construction and operation phases.

To achieve this it has been necessary to use monitoring techniques that not only allow the final design to take into account the real "ante operam" (i.e. before the works commence) undisturbed conditions, but also to plan appropriate mitigation or corrective actions during and after construction.

These intentions led to the implementation of a monitoring plan including a wide area of the Strait, the so-called "Wide Area Monitoring". These activities, which do not include the monitoring activities that will be carried out by the General Contractor on the construction sites, amount to about €30 million, a record sum in Italy in terms of environment-dedicated resources within an infrastructure design budget.

The main targets pursued by SdM during the planning of the EMS were:

- To ensure the independence and objectiveness of the monitoring;
- To represent and describe the evolution of the environment in a scientific, objective and reliable way;
- To allow a real understanding of the region's response to the introduction of the infrastructure over a wide area and not just the insertion corridor;
- To determine the undisturbed condition of the environment by improving and extending the available existing information on the region, particularly in view of the sensitivity of the Messina Strait area;
- To make the "ante operam" monitoring data available already during the final design phase, thus allowing the proposals to be optimised, fully respecting the environmental and regional constraints;
- To plan and implement adequate construction controls to prevent the occurrence of possible environmental damage during construction;
- To verify the efficacy of the adopted environmental mitigation and compensation measures.

The complexity of the project and the environmental context of the Messina Strait demanded the design of a monitoring system which guaranteed that the environmental data became not only a control element, but also an input for the final design. The environmental data thus become a design element to be used not only by the environment specialists but also by the General Contractor and by the PMC.

The preparation of the technical documents for the tender was developed right from the start to achieve two goals:

- To ensure that the environmental monitoring and control activities were thorough and carried out with an extremely high level of accuracy.
- To develop a monitoring system that could be well integrated into the design and construction procedures and involve both the General Contractor and the PMC in the monitoring activities.

The first innovative step was the creation of the role of the "Environmental Monitor". SdM voluntarily took this decision since no law or environmental regulation imposes a requirement to carry out such a double environmental monitoring on construction sites and on a much wider area. (Figure 4.2)

The preparation of the tender documents for the selection of the Environmental Monitor was in many ways far more difficult than for the General Contractor or the PMC since there were no previous examples or models of such wide area monitoring in Italy.

Important guidelines on the monitoring activities to be carried out were given by the Environmental Impact Assessment Commission (part of the Ministry of Environment) who issued "Guidelines for environmental monitoring". These guidelines refer only to construction sites and to the corridors which are directly affected by the works.

Until now in Italy, the assessment of the interaction between infrastructure and environment has been limited almost entirely to the direct effects arising from the choice of alignment and by the construction works. Once the Environmental Impact Assessment stages are completed, the solutions that could lead to more sustainable or optimum results or that can verify the environmental effects during construction are seldom explored.

On this project, however, the Environmental Monitor will act over a wider area and influence a larger number of elements than the General Contractor. Therefore the knowledge and experience of the Environmental Monitor must include not only environmental monitoring and surveys and the assessment of landscape aspects, but also the specific activity of social impact monitoring.

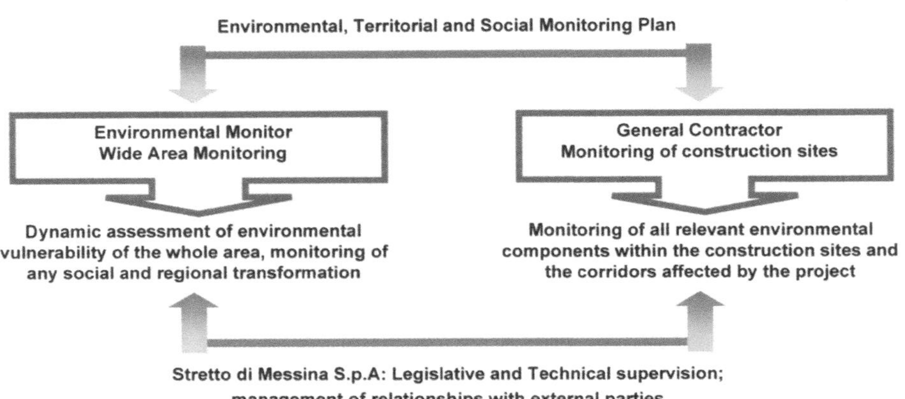

Figure 4.2 ▶ Double monitoring system.

4.2.2 Areas of monitoring and range of environmental components

The Preliminary Design and the Environmental Impact Study together define the areas directly affected by the works: alignment corridors and construction sites. However, the definition of the so-called "wide area" and the environmental components to be monitored required an in-depth analysis of the various previous studies and a careful assessment of the technical, environmental and territorial literature concerning the Strait area.

The Messina Strait is, without any doubt, one of the most important areas of Southern Italy from the environmental, tourist and cultural point of view. It is also characterized by several important industrial and protected areas, such as the Special Protection Areas recently identified by European and Italian legislation. The environment of the Strait has also already suffered for its role as a corridor experiencing a prolonged, aggressive and chaotic urbanization of its coasts in recent decades.

The whole area is potentially subject to natural events which can be of extreme intensity. These are not limited to strong earthquakes, but also include strong meteorological phenomena made worse by often very steep slopes and landslides, and by the influence of nearby volcanic activity.

The need to define the wide area during the tender demanded a thorough study of all these critical aspects. The analysis of the available information and the study of existing maps of environmental and territorial vulnerability led to the definition of the wide area. (Figure 4.3)

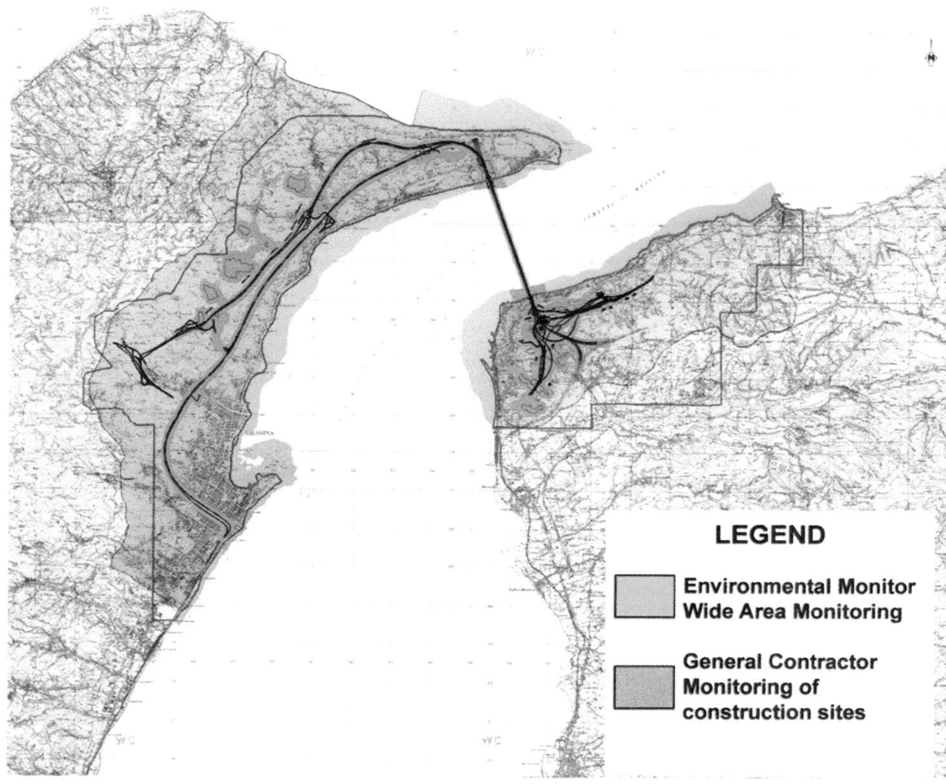

◄ Figure 4.3
Definition of the "Wide Area" (in green) and construction sites and corridors (in blue and black).

The environmental, land and social monitoring plan which emerged from this study is, without doubt, one of the most extensive ever developed for Italian infrastructure projects, owing to the number of assessed components and the nature of the applied investigation methods and techniques. (Figure 4.4)

The fact that the surveys carried out independently by the General Contractor and the Environmental Monitor will be comparable and overlapping will help to ensure the transparency and reliability of the collected data. In order to ensure proper comparability and compatibility of results, the same sampling protocols and analysis techniques will be adopted by both parties, and the same quality indicators and methods for data handling and processing will be used. All these aspects, which were defined in the tender documents, will permit a more integrated interpretation of the data and easier and more productive processes for supervision and control by SdM.

Data collection carried out both on the work sites and over the wide area will be both discreet and continuous. Discreet measures will be carried out

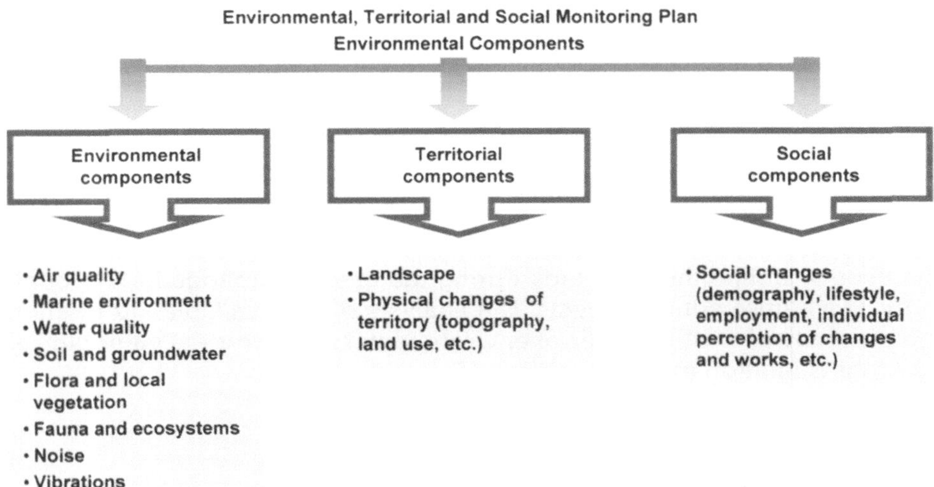

Figure 4.4 ► Monitoring components.

periodically, for a pre-determined duration, in connection with particular phenomena or activities of specific interest, using removable units or equipment or by portable data logger. Continuous measures will be carried out over long periods using fixed monitoring stations equipped with automatic data loggers and sensors.

It is important to emphasise that the completeness of the adopted monitoring plan is not characterised merely by the number of analysed environmental components but also by the number of monitored parameters for each component. The innovation in the plan arises partly through the inclusion of not only those parameters which are normally measured during the construction of road and rail infrastructures but also of several extra parameters which allow a fuller understanding of environmental developments and changes.

Monitoring of the **marine environment** will deal with all the normal environmental components (waters, benthos, deposits, the biota) and foresees the assessment of the following analyses and parameters:

- Seawater chemical, biological and physical analyses: Turbidity, transparency, temperature, dissolved oxygen, saturated dissolved oxygen level, pH, salinity, microrganisms (BOD5, total dissolved solids) nutrients (total phosphorus, Orthophosphate phosphorus, total nitrogen, ammonia nitrogen, nitric nitrogen, nitrous nitrogen), chlorophyll (Chl-a, Chl.b), ecotoxicological tests, coliform bacteria, faecal streptococcus, Escherichia coli;
- Sediment chemical, biological and physical analyses: Organic matter, granulometry, pH; redox potential; main heavy metals (As, Cd, total Cr, CrVI, Hg, Ni, Pb, Se, Va, Zn);
- Ecotoxicological and biodiversity tests – Benthos on soft and hard substratum;
- Plancton;
- Bioaccumulation analysis;
- Diving for "visual census";
- Ecosonar recording – diving;
- Measurements of sea currents;
- Measurements of solid transport;
- Continuous monitoring of physical and chemical characteristics of seawater by multi-parameter sensors.

A similar approach was adopted for the other environmental components by selecting those parameters that not only allow compliance with environmental regulations to be checked, but also permit the evaluation of the "quality status" of each component. The monitoring plan applies the definition of "environmental quality status" provided by the most advanced European legislation and technical references.

4.2.3 A dynamic monitoring system

The Ministry of Environment guidelines for environmental monitoring define the need to carry out monitoring during three different project stages: ante operam (i.e. before the works commence), in opera (i.e. during the works), and post operam (i.e. during the operational phase). (Figure 4.5) Thus the target of the three stages applied in the Messina Strait Bridge environmental, territorial and social monitoring plan is as required by law.

In addition, the technical specifications developed by SdM to control the activities of the Environmental Monitor and the General Contractor specified the scope of each stage in relation to the main target of integrating the environmental issues within the entire "production" cycle of the project, including final design, construction and ongoing management of the crossing.

The importance attributed by SdM to the real and comprehensive knowledge of the environmental and territorial conditions to ensure the environmental sustainability of the project is demonstrated by the fact that the financial resources required for the ante operam monitoring, excluding environmental activities of the General Contractor, amount to almost one third of the whole monitoring activity anticipated over the whole construction period.

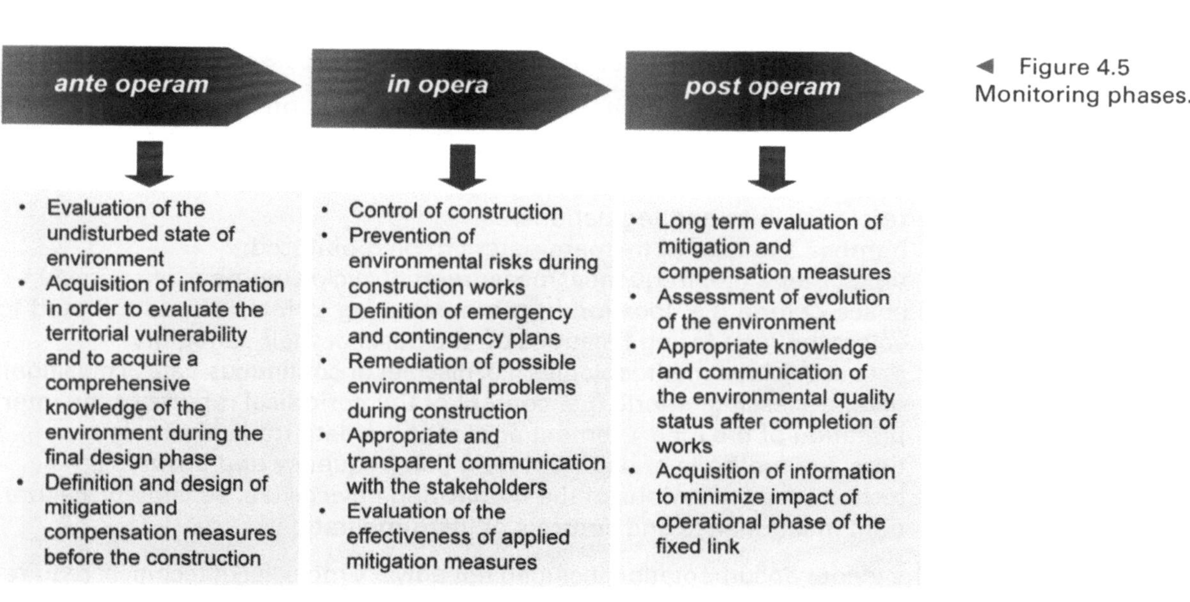

◄ Figure 4.5 Monitoring phases.

4.2.4 Defining the environmental specifications

A fundamental aspect of the tenders has been the inclusion of the technical specifications which define detailed and unambiguous monitoring procedures and methods both for the environmental, territorial and social monitoring of the wide area and for the environmental activities to be carried out by the General Contractor. This was particularly true in view of the lack of consolidated experience and knowledge of such wide area environmental, territorial and social monitoring which is not limited to particular sites or construction activities.

On one hand, Italian experiences of environmental monitoring for infrastructure construction projects, such as those of TAV[1] during the construction of the recently-completed sections of high speed railway, represent an important body of knowledge and demonstrate the need to combine traditional quality assurance techniques with accurate environmental monitoring. On the other hand, it also highlights the importance of defining more and more advanced monitoring methodologies in order to minimise environmental risks and ensure the suitability of any mitigation measures to be adopted.

The environmental specifications have been defined in detail depending on sensitivity and vulnerability to the envisaged impacts of the project, and on the changes likely to be induced by the construction. Those specifications are aimed at ensuring:

- that the different environmental activities respect the environmental laws and regulations applicable to such operations (verification of environmental conformity);
- the control of suitability of the measures applied by the General Contractor;
- the evaluation of environmental impacts on the whole Strait environment during the construction works;
- proper understanding of the relationship between the disturbances and changes induced by the project and the different environmental factors so as to enable the optimisation of the final design and a clear definition of the mitigation measures to be taken;
- the application of the best available techniques in terms of environmental survey, measurement and analysis;
- the application of international standards in terms of control and internal validation of the monitoring and evaluation processes.

The minimum requirements of the tender specifications cannot be described in detail here. However, it is possible to group these requirements, which are specific for each environmental component, as follows:

- reference laws and regulations;
- number and type of the parameters to be monitored;
- time criteria (i.e. frequency, measurement cycles, number of surveys);
- space criteria (i.e. location of the monitoring points, criteria adopted to select the monitoring locations on the basis of their suitability);
- data acquisition methodologies (i.e. discreet or continuous data acquisition);
- data processing criteria (i.e. content of the periodical reports of the interpretation of the measurement and/or of the data transmission);
- time and methods of data reporting (i.e. frequency and formats);
- technical specifications of the monitoring devices (i.e. sensitivity, calibration, maintenance and controls of instrumentation).

The tender documentation included not only a simple list of technical requirements, but also contained a complete Monitoring Plan. Apart from containing the requirements and technical specifications, the tender documentation also provided clear guidelines to the approach to be followed in carrying out the activities and in data interpretation, so as to obtain a full understanding of the environmental, territorial and social condition of the area.

[1] TAV – "Treno Alta Velocità SpA" is special purpose entity owned by RFI (the owner of Italy's railway network) for the planning and construction of a high-speed network in Italy.

To attain this goal, the monitoring carried out by the Environmental Monitor and the General Contractor will make use of mathematical models, properly calibrated according to site conditions, which will allow a careful assessment of the effects that the project could induce on each environmental component.

Table 4.1 illustrates, as an example, the activities foreseen in the technical specifications for air pollution monitoring during construction.

◀ Table 4.1 Air pollution monitoring requirements.

Air Pollution Monitoring			
Parameter/Activity	**Monitoring Period**	**Frequency**	**Monitoring points**
Sampling and chemical analysis of the following pollutants: (NO_x, SO_2, CO, BTEX)	15 days	2 times per year	84 (43 in Sicily and 41 in Calabria)
Mobile monitoring station (All parameters foreseen by Italian legislation on air quality)	30 days	Continuous	4
Installation of permanent monitoring stations All parameters foreseen by Italian legislation on air quality	Construction period	Continuous	3 (2 in Sicily and 1 in Calabria) plus 2 still to be defined
Monitoring of weather conditions by permanent stations	Construction period	Continuous	3 (2 in Sicily and 1 in Calabria)
Traffic intensity by automatic sensors	Construction period	Continuous	5 (3 in Sicily and 2 in Calabria)
Emissions inventory	Construction period	At the beginning of construction	–
Simulation of air quality conditions by mathematical models	Construction period	4 times	–

4.3 An innovative approach to evaluate the interaction between infrastructures and ecosystems: monitoring birds and marine mammals

SdM carried out two ante operam studies to analyse possible interferences between the project and bird migrations and to evaluate any potential effect of the bridge on marine mammals.

These activities represent a very particular and perhaps controversial aspect among the complex environmental issues connected with the project, and for this reason they have attracted the attention of certain opponents to the bridge. Nonetheless, they illustrate and develop the trend towards an important cultural change, much needed in Italy, to scientifically and objectively investigate the risks associated with planned new infrastructure projects and implement proper mitigation measures where necessary.

Many projects carried out today in Italy, have a severe potential impact on birdlife and marine mammals living in the Mediterranean region. The problem of the bird mortality due to accidental impacts with man-made structures or high tension power lines, telecommunication towers, aerial structures in general, transport and airport infrastructures, is rarely taken into account properly during the planning and design of such projects.

Similarly potential risks for marine mammals, and marine fauna in general, are rarely considered adequately during marine works such as dredging and land reclamation, port construction, laying of submarine lines and other coastal works, in spite of the possibility of damage due to sound waves or to sudden changes of seawater quality.

For the Messina Bridge these issues were carefully investigated with the assistance of recognised international experts by the application of advanced instruments rarely used before in Italy or in Europe, and other in-depth studies will be carried out during the final design.

The conclusion of these studies will be verified by the Ministry of the Environment during the approval process for the final design, so the results reported here cannot be considered as final results. Nonetheless they constitute an important knowledge base for the future evolution of the project, and a starting point for all those activities having a potential impact on bird and marine mammal life in the region.

4.3.1 Monitoring of marine mammals

The study on marine mammals (cetaceans) included the following activities:

- Collection of original and comprehensive data on the presence, distribution, number and use of the habitat of the marine mammals living in the Strait.
- Intense research programme (visual and acoustic) carried out on a sea area of about 2,300 km^2. (A distance of 8,795 km was covered in 125 days.) (Figure 4.6)
- Space analysis and modelling of the presence and distribution of the different marine mammal families.

Different important species of marine mammals protected under the EU Habitats Directive were regularly recorded in the Strait waters, confirming the importance of the area for the conservation of marine fauna, notwithstanding the heavy disturbance already existing due the intense shipping between Sicily and mainland.

During the survey 80 marine mammals were observed, belonging to six species (striped dolphin, bottlenose dolphin, sperm whale, grampus, ziphius, and the common dolphin). These species are easily disturbed by human activity and it was very interesting that 70% of the sightings occurred in the Ionian Sea. (Figures 4.7 and 4.8)

Important considerations, which are not usually taken into account during marine works, emerged from this monitoring activity.

The study revealed that, during the construction phase, the potential impacts on cetaceans must be carefully limited. It will be necessary to make particular efforts to minimize the disturbance from pressure and sound waves generated by submarine works, but at the same time there are many possibilities to implement mitigation measures by the application of the more advanced techniques.

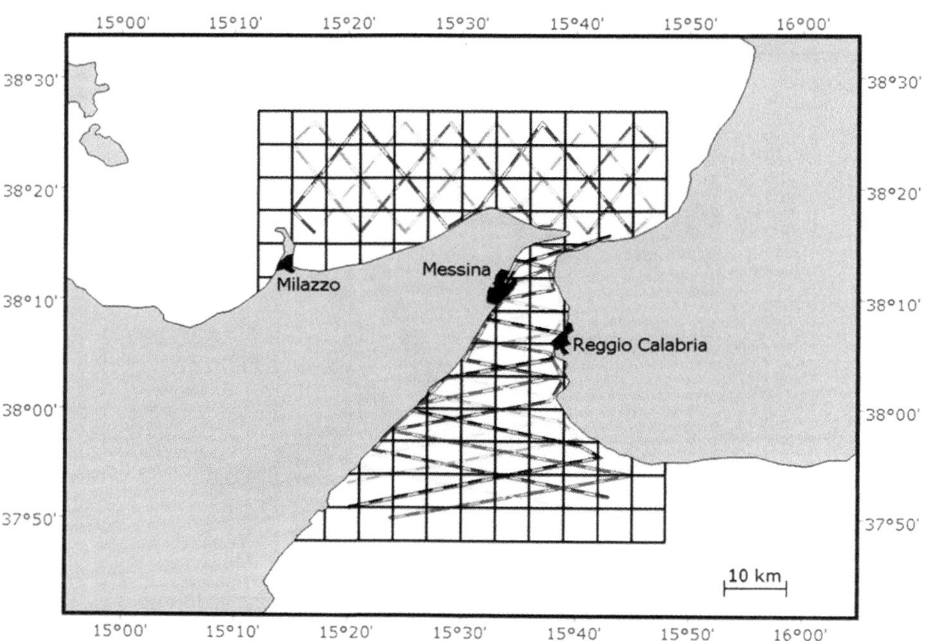

◄ Figure 4.6 Marine mammals monitoring: sea routes.

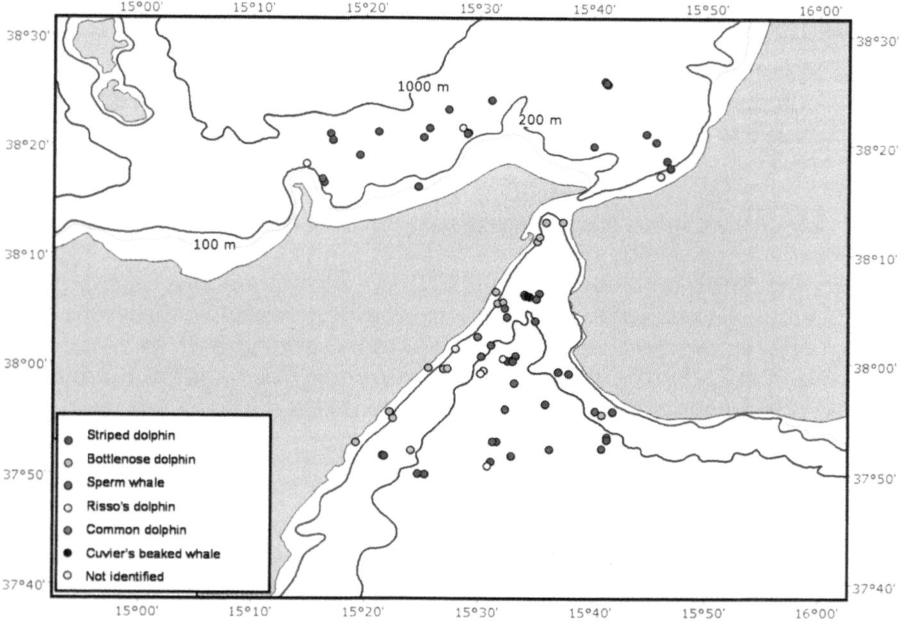

◄ Figure 4.7 Marine mammal monitoring: map of observations.

For the Messina Bridge this risk is limited to the early stages when the temporary service jetties necessary to facilitate the main construction works will be constructed. No other submarine work is foreseen since the main towers of the bridge are located on the shore and it is forbidden for the General Contractor to deposit waste materials at sea. Dredging operations and marine disposal of soil and dredged materials may affect seawater quality and could cause negative effects on the marine flora and fauna, so these are explicitly forbidden by the technical specifications. Furthermore special monitoring protocols will be implemented during the works to avoid interference between cetaceans and shipping due to the construction.

The study also considered the potential risks of seawater pollution during the works and the possibility of problems produced by the shadow of the

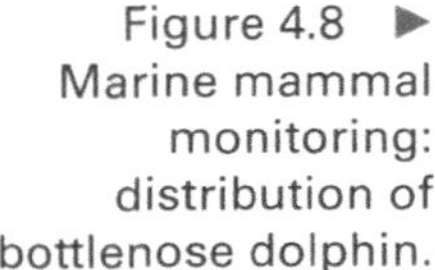

Figure 4.8 ▶ Marine mammal monitoring: distribution of bottlenose dolphin.

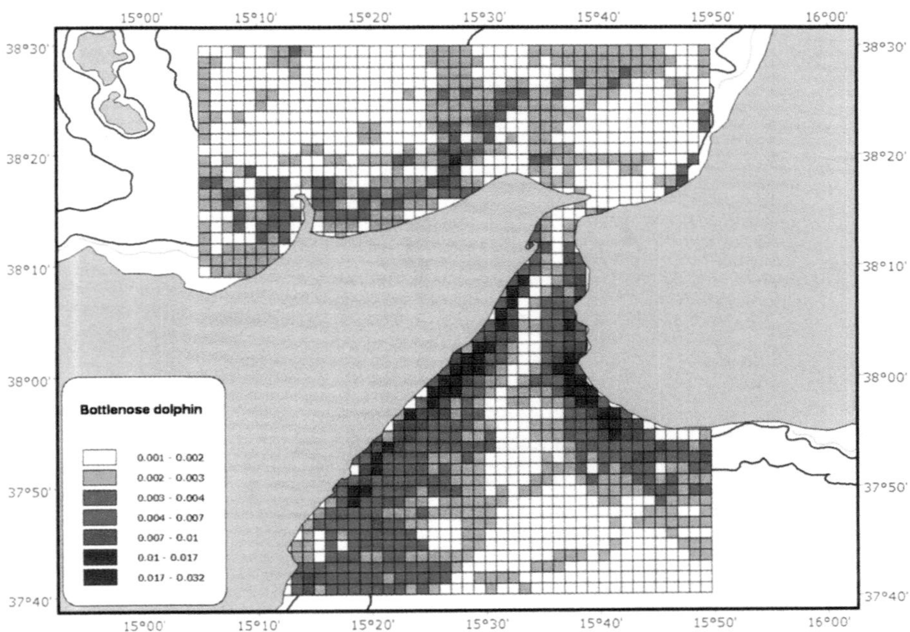

bridge once completed. Some people expressed concern that the shadow of the deck on the water surface could disturb sea life.

On the basis of previous international experience with long span bridges and the outcome of the present studies, no proven negative effect for cetaceans due to the bridge shadow has yet been detected.

4.3.2 Monitoring of bird migration across the Messina Strait

The Messina Strait is an important European and Italian site from a birdlife point of view. As a matter of fact the monitoring revealed the presence of 321 bird species in the area of the Strait, representing 64% of all bird species living in Italy. Furthermore the area supports rich national biodiversity, which well represents the whole of Italian bird life.

The collision of migrating birds against man-made structures is a familiar and well studied concern, but has rarely been considered seriously in Italy within the context of the design and construction of rail or road infrastructure or of buildings in general.

The published history of human-caused bird fatalities is long, and there is a large volume of literature. Whilst many people have seen or are aware of birds colliding with buildings (in particular windows) or vehicles, most people are not familiar with other anthropogenic agents of bird mortality. There are known and documented design features of structures that can increase bird mortality, and in some cases combinations of those features work together to attract and sometimes kill birds.

Banks in 1979[2] estimated that approximately 10 billion birds die annually in the USA from all causes. Anthropogenic causes (that is those caused by manmade

[2] Banks, R. C. 1979. Human related mortality of birds in the United States. Special Scientific Report, Wildlife No. 215. Washington, DC: Fish and Wildlife Service, U.S. Department of the Interior.; 16p.

influences) accounted for 196 million bird deaths, less than 2% of the total mortality rate. Of these, only 1% of human-influenced bird fatalities were attributable to collisions with tall structures, whereas 61% was due to hunting, 29% to vehicles, 2% to impact with window glass and 7% to other causes.

Structures reported to cause bird kills are mainly telecommunication or radio transmission towers, chimneys or stacks, large glazed windows, power transmission and distribution lines and airports. The causes of bird collisions with man-made structures are normally considered to be invisibility (eg. power lines or guy-wires) particularly at night, deception (caused by glazing in buildings) or confusion (caused by light refracted or reflected by mist).

Most of these structures, which are distributed all over Italy, are rarely subject to special requirements for bird life protection, not least because they are subject to a faster authorization procedure than larger projects such as the Messina Bridge.

A comprehensive literature review clearly demonstrates that bridges do not usually cause significant bird mortality. Nevertheless, considering the ecological significance of the Strait area, SdM decided to study this phenomenon by applying the most advanced survey methods.

The activities implemented by SdM for this project consisted of:

- A compilation of surveys and literature covering the most significant incidences of breeding, moulting, staging, wintering and migrating birds so as to assign species into categories of national and regional importance which could potentially be affected by the construction and operation of the fixed link.
- Biological and conservation analysis carried out according to traditional methods during the migration season, including a preliminary assessment of potential impacts posed by the bridge on relevant species.
- Instrumental analysis carried out using a radar technique, applied for the first time in Italy, to estimate the species and number of birds migrating across the bridge area, their seasonal distribution and their flying height. (Figure 4.9)
- Estimate of the influence of weather conditions on the concentration and flying height of migrating birds.
- Risk assessment to estimate the possibility of bird collisions with the bridge so as to make a preliminary assessment of potential mitigating actions.

The application of a risk assessment approach to this problem is particularly widespread in Northern Europe, particularly in connection with offshore structures and wind turbines. It enables a scientifically determined estimate of the environmental effects of a project and the planning of necessary mitigation measures.

In the Messina Strait area, diurnal migration mainly involves small passerines (short-distance migrants and swallows in particular) and raptors. In addition, some resting waders, water birds, and particularly gulls may pass the area of the bridge on diurnal commuting flights.

In fine weather with good visibility, these birds will see the bridge from far away. It is reasonable to assume that most diurnal migrants will climb to cross over the bridge where it is relatively low, or will pass either side of it at its landward edges. Collisions are expected to be rare under such conditions.

Figure 4.9 ► Radar techniques used to estimate species and number of birds migrating across the bridge area.

Figure 4.10 ► Position of radar and monitored area. The monitored area extended over a 4–5 km radius for small birds and up to 8 km radius for big birds and flocks of birds.

The situation changes in poor weather conditions with reduced visibility (eg. drizzle, fog, mist) when the potential hazard increases. While agile birds may be able to react quickly at short distance, large heavy birds with low manoeuvrability are exposed to a higher risk of collision. The problem is most severe in cases of low hanging clouds, when the migrating birds might be concentrated at the level of the bridge and the water vapour or drizzle reduces visibility. On the other hand under such conditions, which are very rare around the Messina Strait, the intensity of bird migration is reduced, particularly in raptors.

Nocturnal migrating birds comprise 91% of passerines, and 9% of continuously flapping birds like waders and water birds. Visibility is generally reduced at night, so the risk of collision is increased. Nonetheless, radar and infrared surveys carried out on windmills located far from the coast (Deholm & Kahler, Desholm et al. 2006) and on the Öresund Bridge connecting Sweden and Denmark (Nilsson & Green 2002) showed that with good visibility birds avoid large man-made structures even at night.

The study allowed the definition of appropriate mitigation measures to be taken in order to minimize collisions in the worst conditions. The colour of the structure should be bright in order to assure its visibility, even without additional lighting. Bright metal-surfaces or a pale colour (eg. white or light grey) is suggested to increase the reflection of natural and artificial night-light from the surroundings, and thus improve visibility. Adequate lighting of the bridge will help to increase detection by birds at night and avoid attraction phenomena.

Finally the use of a simplified avian risk collision model, which constitutes an important innovation for the evaluation of potential risks for birdlife, allowed the number of potential collisions to be estimated. The preliminary results demonstrate a very low predicted collision rate in relation to the total number of transits. An even more detailed risk assessment will be carried out during the final design.

The selection of mixed techniques here described, associated with a complex interdisciplinary analysis to obtain the input data for the risk model, represents a first and unique experience in Italy in this field and provides an important knowledge base to facilitate decision making in relation to the final design of the bridge.

4.4 From environmental and territorial monitoring to social monitoring

The construction of a large project always takes place in the context of possible conflicts of interests in relation to environmental issues, territorial changes and to the perception of the project and of its effects on the life of the local community.

Traditional monitoring methods do not always permit a complete analysis of the social perception of a project. The environmental concerns often conceal hardships and rejections connected to social aspects which are often not properly monitored. This is why many people's perception of environmental risk is frequently at odds with the real assessed environmental risk.

The Messina Bridge project falls into this type of category, as evidenced by its past and present history. In fact the bridge synthesises several different values:

- an instrumental value connected to its function as a major piece of transport infrastructure,
- a symbolic value due to its importance and cultural significance, and
- a social value due to the changes it will introduce into the social structure, spatial relationships and the distribution of influence in the region.

On this basis it was decided to include the study and analysis of this important social component among the activities to be carried out by the Environmental Monitor, thus introducing an innovative approach for major infrastructure projects.

The integration of environmental and social issues already constitutes a sound management principle for many organisations, companies and multinational corporations that adopt structures and management methods based on Corporate Social Responsibility schemes. However, to apply this kind of integrated approach within the context of an infrastructure project in Italy is a new initiative, and SdM faced many challenges in introducing such an innovation through the definition of a specific Social Monitoring plan for the Messina Strait area.

The model adopted by SdM comprises three distinct activities intended to allow a comparative analysis of instrumental, symbolic and social impacts:

- Socio-economic monitoring, aiming at defining socio-demographic, micro-economic and macro-economic impacts that focus on the instrumental and social levels.
- Monitoring of social perceptions, aiming at defining impact perceptions and the degree of consensus about the project among local communities and national stakeholders.
- Monitoring of both local and national media to assess the temperature of public opinion and provide a continuous barometer of the degree of consensus so as to be able to detect in advance any potential shift in project perceptions.

Furthermore the Environmental Monitor has to carry out a comparative analysis of socio-economic indicators and social perceptions together with the results derived from measuring the environmental components.

The main challenge is to develop an adequate technical and scientific rationale to justify the selection of suitable compensation measures which, although fulfilling the real needs of the community, may not strictly constitute an environmental and social compensation for any perceived losses among the community as a result of the project.

The bridge project produces different social impacts:

1. From the instrumental point of view:
   - It constitutes an element of an integrated transportation system which contributes to a general improvement of the Italian and Sicilian network, by improving and updating the transport infrastructure in one of the most deprived areas of Italy;
   - It is an integral and fundamental part of the infrastructure system necessary for the socio-economic development of the area;
   - It can contribute to the improvement of the environment of areas subject to degradation, and to the re-development of Messina and Villa San Giovanni and the reduction in traffic congestion.

Figure 4.11 ▶ Landscaping and simulation of new urban development.

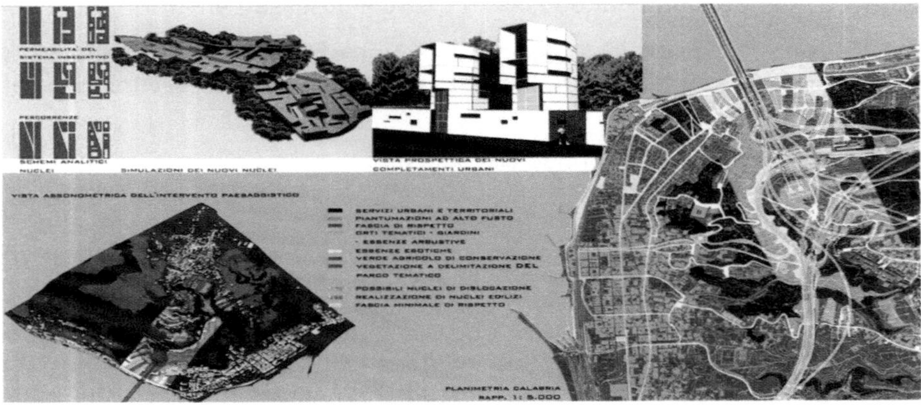

2. From the symbolic point of view:
   - Its location holds a special place in the collective national consciousness, not restricted to Sicily and Calabria. The Messina Strait has a special identity and is a place of memories embodying myths and legends as well as significant historical events and community values;
   - It influences the perception of the space and traditions of the island of Sicily in particular and even the concept of an island in general;
   - It influences the representation of the visibility and role of the State in southern Italy;
   - It influences the perception of environmental risk;
   - It influences the representation of the landscape;
   - It increases positive and negative expectations on the potential development of the territory, on the quality of life and on opportunities for groups and individuals;
   - It embodies ideas, desires and fears which are not consciously expressed;
   - It influences the identity model, conceived as the combination of shared values, emotions and meanings that define the community.

3. From the point of view of the social structure:
   - It creates new social subjects, with new specific interests that can influence decisions, targets and strategies;
   - During the construction phase it will generate problems of integration between the local population and visiting seasonal workers;
   - It influences the availability of and access to educational and recreational services and other facilities;
   - It influences the distribution of welfare and the map of social hierarchy;
   - It influences social mobility.

4.5 Integration between environmental monitoring and final design activities

A distinctive element of the procedures developed by SdM within the project is the nature of the integration between the environmental monitoring and the final design activities.

The General Contractor must deepen its environmental knowledge, initially by a critical examination of the Preliminary Design and the studies already carried out, and then through development of the design technologies and anticipated environmental mitigation measures, in order to verify the suitability of the adopted solutions or identify other design improvements. Furthermore, the General Contractor must optimise all construction phase activities on the basis of the results of environmental monitoring.

In addition to environmental monitoring of the construction sites, the General Contractor has to carry out a series of in-depth environmental studies to verify the design assumptions embodied in the Preliminary Design, such as:

- to determine the location of the preliminary and final sites for the disposal of soil and other materials arising from excavations, including the necessary associated landscaping and environmental restoration.
- Hydro-geological and geochemical studies of the Pantano area in Ganzirri, Sicily, comprising modelling of surface and groundwater circulation

in order to prevent potentially damaging impacts due to the construction of the Sicily tower foundation.

- Noise studies including phonometric measurements and the development of an ambient acoustic model of the area.
- Hydro-geological study of the areas where tunnel boring is planned.
- Study of relevant ecosystems according to the Habitat Directive (92/43/EEC).
- The study of the environmental insertion of the works.
- The study of the environmental restoration of the all sites affected by the works, including those selected for the final disposal of waste materials.
- Environmental investigations of the areas involved in the works, checking compliance with the relevant regulations on soil and groundwater contamination.

On the other hand the activities carried out by the Environmental Monitor in the wide area constitute at the same time both an input and a "calibration element" for the above mentioned activities of General Contractor.

For large projects and works of national interest regulated by the Legge Obiettivo, which include the Messina Bridge project, the Environmental Impact Assessment and the environmental approval procedure are carried out on the Preliminary Design. This does not mean, as some have claimed, that the Final Design loses its environmental value, but on the contrary that it plays a fundamental role in confidently achieving project sustainability through proper engineering solutions, the selection of construction materials and the optimization of construction stages.

During the approval of the final design the Ministry of Environment will assess whether the proposals comply with all the requirements and environmental stipulations identified in the Environmental Impact Assessment.

With this in mind, SdM have defined clear technical and environmental specifications for the General Contractor who is required to:

- Carry out suitable design optimization concerning engineering solutions, construction methods and materials selection;
- Design adequate mitigation measures where necessary;
- Select and design adequate and appropriate compensation measures after an accurate and scientific environmental evaluation of the positive and negative effects induced in the area.

In particular, the design of mitigation and compensation works will be carried out within an overall balanced framework that takes account of all the inputs deriving from the various different environmental activities involved in the procedure.

As far as design optimization is concerned, priority will be given to construction techniques characterized by:

- Minimum environmental and landscape impact;
- Minimum consumption of materials and energy;
- Maximum use of renewable resources;
- Maximum use of low toxicity materials;
- Minimum requirement of resources for maintenance and management after construction.

All possible opportunities for using local and regional resources and materials should be taken in order to reduce transport requirements during construction and thus minimise disturbance and disruption to the local community as well as the emission of pollutants into the atmosphere.

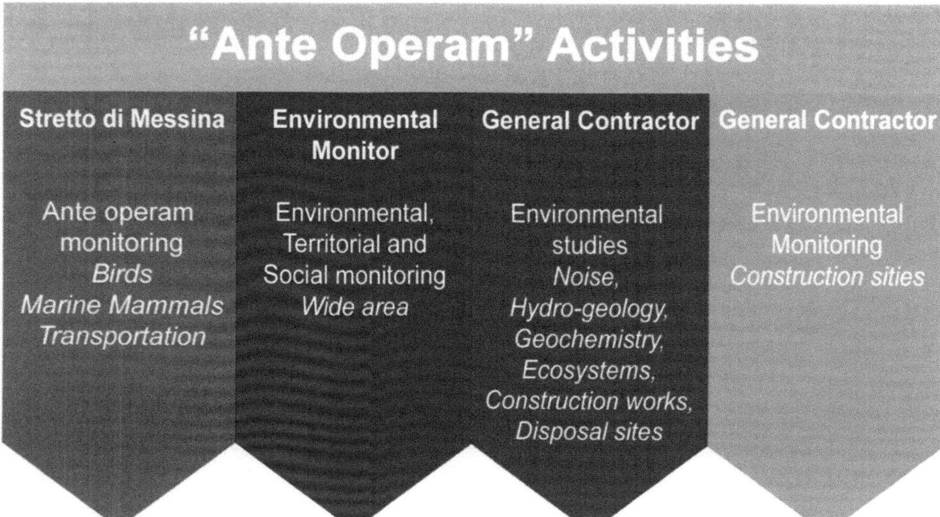

◄ Figure 4.12 Environmental inputs to the Final Design.

◄ Figure 4.13 Landscaping and mitigation measures.

Similarly, particular attention will be paid to the possibilities of recycling and recovering materials and embodied energy, and to the opportunities for deriving benefit from the careful disposal of waste materials deriving from the demolition of existing structures and from construction activities.

4.6 The role of PMC within the environmental management of the project

SdM gave the PMC a key role in achieving the integration of environmental matters within the "production" phase of the project, from final design to completion of construction. The PMC must supplement its engineering and technical staff with professionals specialized in environmental, territorial and social issues in order to carry out all the required control functions and specific activities of it. To achieve this target, SdM described in the tender specifications for the PMC all the required activities to ensure the proper interaction between the activities of the General Contractor and the Environmental Monitor.

In this respect, the most important tasks of the PMC include:

- ensuring proper integration and co-ordination between the General Contractor's and the Environmental Monitor's monitoring plans;
- providing an integrated Environmental, Territorial and Social Monitoring Plan and ensuring that it is incorporated into the final design for approval by the authorities;
- providing specialized support to SdM to check that the design activities of the General Contractor meet the necessary environmental prescriptions in respect of the Environmental Impact Assessment and the requirements of CIPE, and also comply with the tender specifications regarding environmental optimization and mitigation, including the selection of adequate compensation measures.

References and Further Readings

[1] Nimby Forum ARIS (2006). Infrastrutture, energia rifiuti: L'Italia dei si e l'Italia dei no.

[2] Banks, R. 1979. *Human related mortality of birds in the United States*. United States Fish and Wildlife Service, Special Scientific Report-Wildlife No. 215: 1-16.

[3] Shenzhen Western Corridor – Investigation and Planning, Environmental Impact Assessment Report (2001) – *Bird Collision with Manmade Structures with Reference to the Proposed Shenzhen Western Corridor.*

[4] National Environment Research Institute, Ministry of Environment Denmark (2005) – *Construction of a fixed link across Fehmarnbelt – Preliminary risk assessment on birds.*

[5] Federal Ministry of Transport, Building and Urban Affairs, Germany, Ministry of Transport and Energy, Denmark (2006) – *A Fixed Link across the Fehmarnbelt and the Environment, Environmental Consultation Report.*

[6] Wallace P. Erickson et al. (2005) – A summary and Comparison of Bird Mortality form Anthropogenic Causes with an Emphasis on Collisions. USDA Forest Service Gen. Tech. Report.

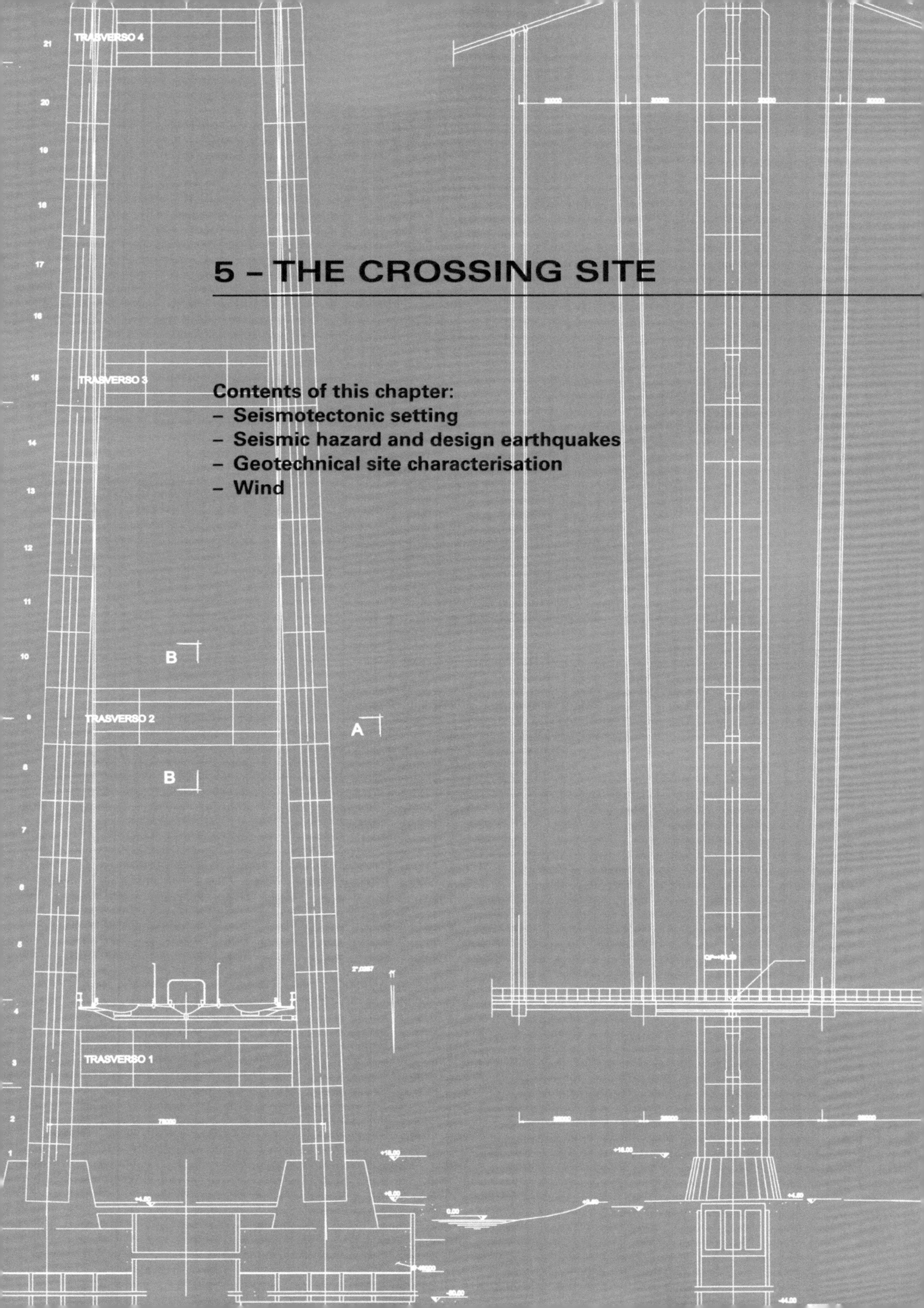

5 – THE CROSSING SITE

Contents of this chapter:

- **Seismotectonic setting**
- **Seismic hazard and design earthquakes**
- **Geotechnical site characterisation**
- **Wind**

This chapter addresses the main natural phenomena which critically influence the design of the bridge, namely the incidence of very strong earthquakes, the geotechnical ground conditions and the nature of the wind.

The Messina Strait is known to be a zone with strong tectonic activity responsible for significant earthquakes, including the devastating Messina earthquake of December 1908. This chapter describes the vast amount of research into that earthquake and other seismic events in the region, providing an insight into the complexities involved in analysing and assessing such phenomena, and explains the background to the evolution of the design spectrum and other criteria to be used in the bridge design.

The ground conditions affecting the nature of the two tower foundations and the cable anchorages are described, together with the extensive ground investigations and in-situ and laboratory tests which have been carried out to determine the foundation design parameters.

Finally, the chapter describes the local investigations and measurements of speed, direction, turbulence intensity and other characteristics of the wind, leading to the specification of the extreme wind conditions to be used in the design.

5.1 Seismotectonic setting

5.1.1 Introduction

The fundamental goal of this chapter is to provide a summary of the seismotectonics of the Messina Strait with special reference to the 28th December 1908 earthquake, the bridge Design Earthquake. We summarize what is known about the earthquake, describe the source model that has been used for assessing the bridge design seismic action (see Section 5.2. *Seismic hazard and design earthquakes*) and discuss some short- and long-term implications of tectonic activity in the area.

5.1.2 The Messina Strait, a very special object of research

For geologists and tectonicists the Messina Strait is certainly one of the most interesting sites of the entire Mediterranean. The spectacular evidence for fast uplift and the sedimentary suites associated with rapid erosion and deposition processes have attracted geographers and geologists for well over a century. The occurrence of the catastrophic 1908 earthquake and of the ensuing tsunami further spurred the interest of Italian and foreign scientists from various disciplines, also because of the availability of a number of seismographic recordings and observations of elevation changes induced by the earthquake. It took many years, however, for scientists to understand the source of the 1908 earthquake in relation to the main geological and tectonic features of the Strait. The earthquake was generated by a blind fault, thus preventing direct investigation of its causative source, and the resolving power of available instrumental data turned out to be rather limited.

Due to these circumstances, the seismological community had to wait until the beginning of the 1980s for modern analyses of the earthquake source to be published. In the ensuing 20 years, more detailed analyses of the earthquake, the tsunami and the permanent modifications of the topographic surface led to a unified and largely accepted model for the recent evolution of the Strait. The model is dominated by the interaction between the uplift of southern Calabria, a large-scale process related to the complex geodynamics of the Central Mediterranean, and the 40 km-long normal fault responsible for the 1908 earthquake. The match between the recent evolution of the area and surface deformation induced by the earthquake suggests that its causative fault dominates the seismotectonics of the Strait. This conclusion is corroborated by the limited historical seismicity of the area, where the seismic cycle appears to be controlled by the repetition of the maximum-size event.

We provide a summary of research on the seismotectonics of the Strait from its very beginning, pointing out the main controversies and the established facts. Our analysis moves from early studies of active tectonics in the area, proceeds through investigations of the 1908 source using instrumental data, and ends with a discussion of the evolution of the Strait in relation to the 1908 source and of the available evidence to estimate future earthquake recurrence.

5.1.3 The 1908 earthquake: early source models

The 1908 earthquake did not come unexpected, as it took place in the middle of perhaps the most seismically active area of the entire Italian peninsula (Figure 5.1). It was indeed a large event, at least by European standards (Figure 5.2), assessed as magnitude M = 7½, the largest ever recorded for a European earthquake. According to Baratta [1] the severity of the shock appalled Giuseppe Mercalli to the point that he decided to extend his own Mercalli intensity scale to the XI degree. As pointed out by contemporary workers such as Omori [2] and later confirmed by modern analyses, however, much of this damage was to be explained by poor building styles and strong site effects rather than by the actual severity of the earthquake.

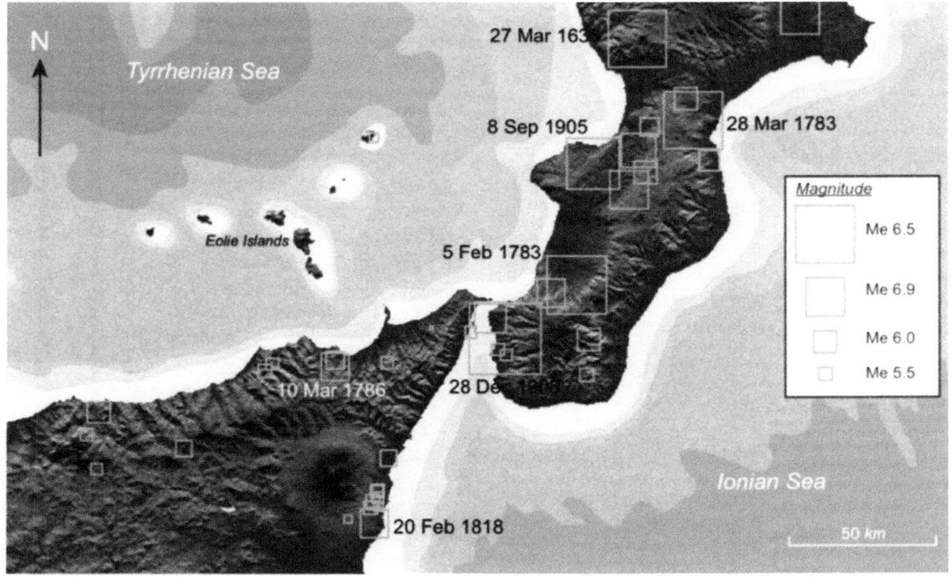

◄ Figure 5.1 Historical seismicity of the Messina Strait and surroundings.

From Catalogue of Strong Earthquakes in Italy [5].

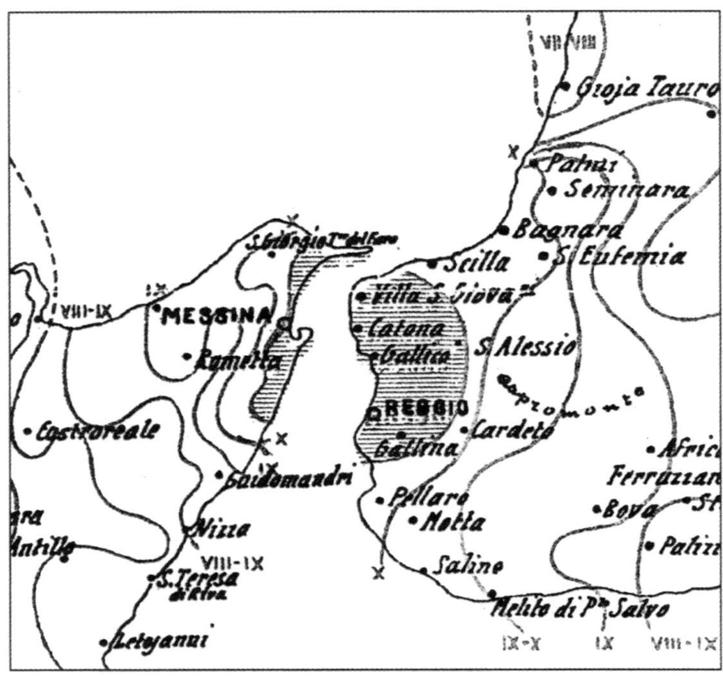

Figure 5.2 ▶ Intensity pattern of the 1908 earthquake. The hatched area suffered the strongest ground shaking. Notice the different decay of intensity on either side of the Strait and the strong asymmetry of damage towards Calabria.

From Baratta [1].

The international scientific community was immediately involved in the investigation of the effects of the earthquake and of the ensuing tsunami. Instrumental information available to the early investigators included more than 110 regional and teleseismic seismograms of the main shock [3] and about 100 elevation changes from re-measuring two levelling lines fortuitously surveyed shortly before the earthquake [4] (Figure 5.3).

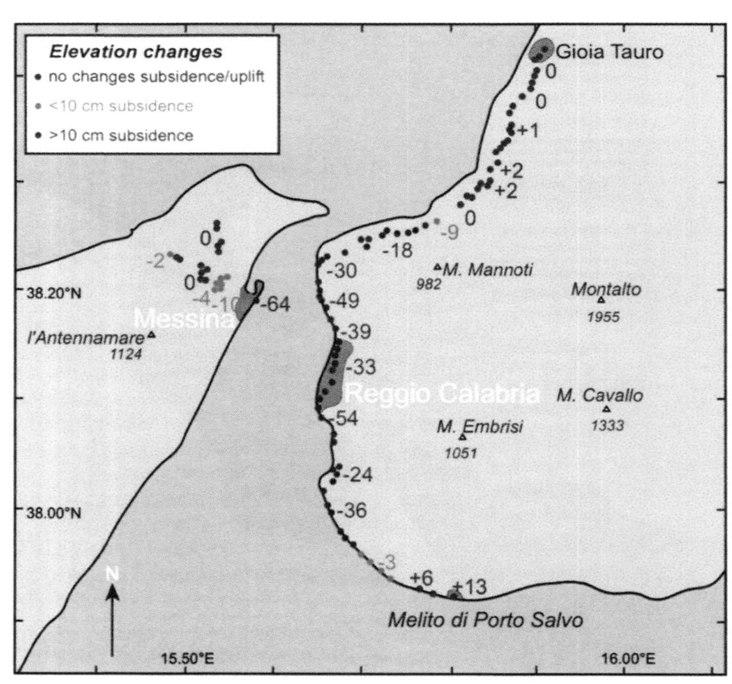

Figure 5.3 ▶ Elevation changes (in cm) measured by Loperfido [4] following the 1908 earthquake. The largest genuinely tectonic subsidence was recorded in Reggio Calabria. The Messina harbour also experienced subsidence up to 70 cm, but the observed values are suspected to reflect also non-tectonic deformation due to settling of coastal deposits.

Field observations were best represented by the macroseismic survey performed by Baratta [1] (Figure 5.2). Even though the magnitude of the earthquake should have ensured their existence, modern readers of this classical report found no reference to ground ruptures that could positively be ascribed to surface faulting. Because of the comparable level of damage in Messina and Reggio Calabria, various investigators suggested that the source was located in the middle of the Strait, a hypothesis that would also account for the lack of surface faulting [6, 7]. This was seen to be consistent with results obtained early on by Omori [2], who used Robert Mallet's method for locating the earthquake based on observations of collapsed monuments to suggest that the shock nucleated halfway between Messina and Reggio Calabria. With a similar approach, Omori located the source of the tsunami in the Strait off Reggio Calabria, not far from the earthquake epicentre.

In search of a reliable source mechanism

Based on the re-examination of seventeen first motion polarities, Schick [6] proposed that the 1908 earthquake occurred by pure normal faulting on a N15E°-trending, 70° west-dipping plane beneath the axis of the Strait (fault "A" in Figure 5.4). This solution, however, was inconsistent with Loperfido's elevation changes, that show significant subsidence of the coastal plains on both sides of the Strait and mild uplift of the adjacent ranges. The pattern of vertical motions caused by the earthquake was indeed unusual; even though the levelling lines extended well beyond the area of largest damage (compare Figures 5.2, 5.3), more than 95% of the signal indicated subsidence. To reconcile his prediction of coseismic uplift with the observed subsidence of the Calabrian shore, Schick assumed that the earthquake was generated by a unilateral sinking force superimposed on double-couple dislocation, dropping the crust even in the footwall of a steep normal fault.

Bottari et al. [7] re-examined the macroseismic field of the 1908 earthquake by applying the Medvedev-Sponheuer-Karnik (MSK) intensity scale. Based on the pattern of the higher intensity isoseismals they hypothesised that the earthquake had been generated by a NE-trending steep normal fault dipping to the NW and located under the Calabrian side of the Strait ("B" in Figure 5.4).

Several investigators attempted traditional elastic dislocation modelling of the dataset to infer the earthquake fault parameters. Mulargia and Boschi [8] proposed a model with two N15°E-trending normal faults dipping in opposite directions, the principal of which dips at about 35° toward the east. Capuano et al. [9] proposed a rupture mechanism based on a single 356°-trending, 39° east-dipping, oblique-slip fault ("C" in Figure 5.4). Using a slightly more sophisticated variable-slip approach, Boschi et al. [10] proposed motion on a single, N15°E-striking, 30° east-dipping pure normal fault, with dynamic slip only below 4 km and concentrated into two main slip patches ("D" in Figure 5.4): the main patch was located near the assumed epicentre, while the other was found near the northern end of the fault. Based on a similar approach, De Natale and Pingue ("E" in Figure 5.4) [11] investigated the expected slip distribution for the fault geometry proposed by Capuano et al. and

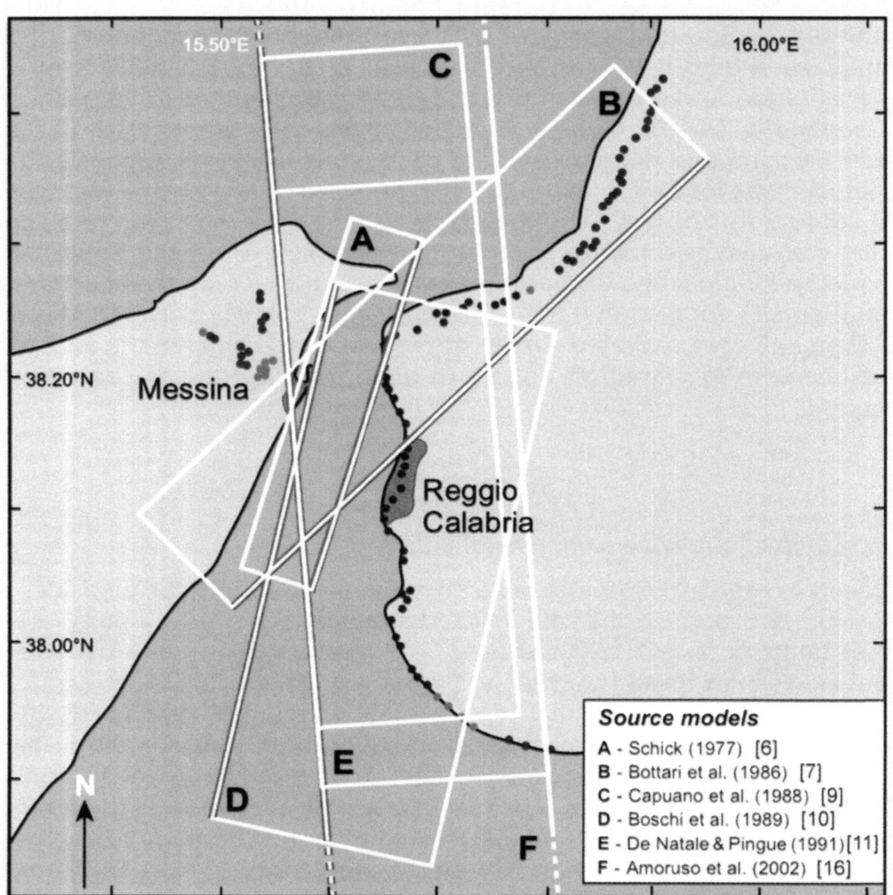

Figure 5.4 ► Models of the 1908 source. The rectangles represent the surface projection of the earthquake causative fault according to various workers. A thicker line marks the upper edge of the fault plane, thus indicating its dip direction. Models C, E and F share the same strike but differ in length. In particular, model F extends beyond the box edges.

located the region of largest moment release nearly coincident with that recognised by Boschi et al. [9, 10, 11]. Most of these models proposed an earthquake magnitude in the range M = 6.8 to 7.1. This range represents a lower bound, since geodetic data can seldom resolve all crustal deformation induced by an earthquake, especially in coastal areas.

In contrast with most coseismic models of the 1908 earthquake, tectonic schemes of the Strait invariably called upon Grabe-style faulting to explain its geological setting [12, 13, 14]. Most of these schemes envisioned the formation of the marine terraces frequently observed in the region (see discussion below) by the action of flights of steep normal faults partitioning a large amount of extension across the Strait. For some structural geologists, the 1908 earthquake was not necessarily a primary manifestation of the stress field that had resulted in the long-term evolution of the Strait. Numerous seismic sources having different geometry would participate in determining the observed geological structure. This would make the recurrence of significant earthquakes virtually random, and the deterministic prediction of the associated ground shaking impossible.

Despite a full decade of intense study, the end of the 1980s did not bring a significant growth of consensus about the 1908 earthquake source. The models based on instrumental data, and particularly on elevation changes, envisioned a 40–80 km long, NNW to NE-striking, dominantly dip-slip fault gently dipping to the east (Figure 5.4). This fault would extend between 3–5 and 12–20 km depth and would therefore be positively blind. However, most structural geologists (e.g. [14, 15]) maintained that the earthquake source had to be found among the many faults exposed on either side of the Strait, and particularly on the Calabrian side. (Notice that the east-dipping fault would project onto the Sicilian side). Due to the scatter in the investigators' opinions, further goals such as evaluating the full extent of the 1908 source region and the earthquake repeat time just seemed impossible to attain.

5.1.4 The 1908 earthquake: reference source model

The 1990's and the early 2000's brought about several new observations and analyses that helped reconciling the discrepancies delineated above.

Valensise and Pantosti [17] started off by proposing that the pattern of elevation changes following the 1908 earthquake resembles the landscape and recent evolution of the Strait, and hence that the area's long-term evolution could simply reflect the repetition of several 1908-type earthquakes. This hypothesis and its implications for earthquake recurrence will be discussed in the next section.

Following a thorough reassessment of the seismograms relating to the 1908 earthquake, Pino et al. [3] obtained a reliable source-time function and hence an estimate of its magnitude and rupture length. Their slip model is very similar to that obtained by Boschi et al. [10] and rather similar to that proposed by De Natale and Pingue [11] (Figure 5.5). Although the quality of the data did not allow Pino et al. to determine the focal mechanism of the earthquake by waveform modeling, the match with two separate models obtained from leveling observations confirms, on a totally independent basis, that most if not all seismic moment was released within the Strait. The seismological model sets an upper bound for the magnitude at M = 7.1,

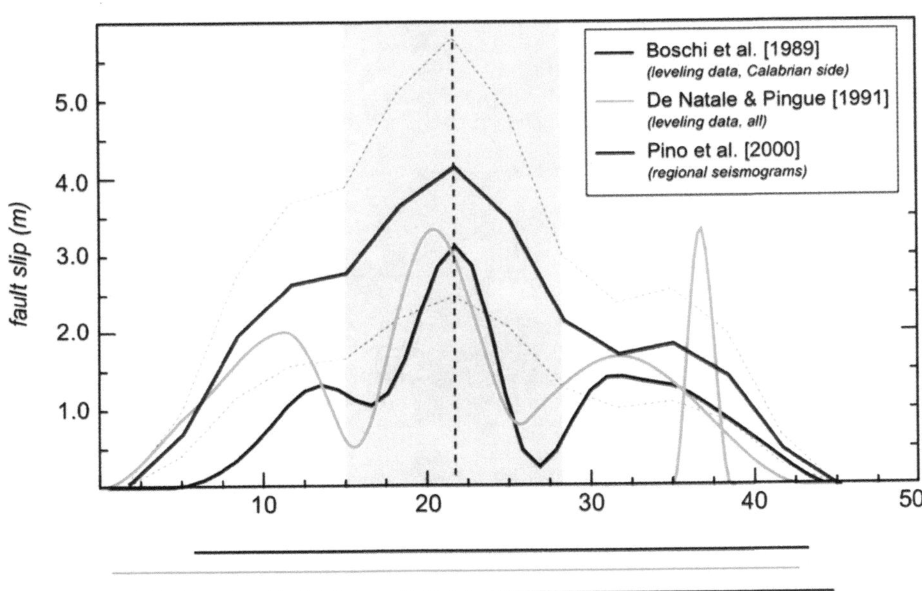

◄ Figure 5.5 Comparison of slip models obtained by different investigators (see colour-coding). The lines below the diagram indicate the inferred rupture length. All models suggest a fault length of ~40 km and an average slip of 1.5–2.5 m, consistent with an M = 7.0 to 7.1. North is to the right.

the same obtained by Schick [6]. This figure compares well with the 6.8 to 7.1 estimate obtained using leveling data. The seismological analysis also showed that the rupture propagated from south to north, and hence that it must have nucleated in the southern portion of the Strait, as already suggested by Omori [2]. Predominantly unilateral rupture propagation is seen in approximately 80% of large shallow earthquakes ruptures [18] and has significant implications for the estimation of ground motion at the fault ends, as discussed in Section 5.2.

Amoruso et al. [16] performed a non-linear joint inversion of seismological (polarity of first arrivals) and geodetic (elevation changes) data to improve the source model for the 1908 earthquake. Their preferred model fault ("F" in Figure 5.4) strikes nearly N-S, dips to the east at an angle ~40°, and exhibits transtensional slip with a significant right-lateral component. Slip is concentrated in the southern half of the fault.

Michelini et al. [19] used first arrivals from the same historical seismograms to derive the hypocentre, i.e. the nucleation point of the earthquake. Their best location falls near the southern end of the fault models proposed by Capuano et al., Boschi et al. and De Natale and Pingue [9, 10, 11], albeit with significant uncertainty, and at a depth that is comparable with the depth of the base of the fault (12–15 km). Remember that in most normal faulting earthquakes the rupture nucleates near the base of the fault and then propagates upwards.

Better seismological and crustal deformation data (essentially using GPS) and analyses have allowed more reliable investigations of the stress and strain fields. The Messina Strait experiences frequent low-energy earthquakes in the upper crust and occasional events in the upper mantle. Upper crustal earthquakes were recently analysed by Neri et al. [20], who found that they are best explained by an extensional stress field oriented WNW-ESE (Figure 5.6). This regime is compatible with most models of

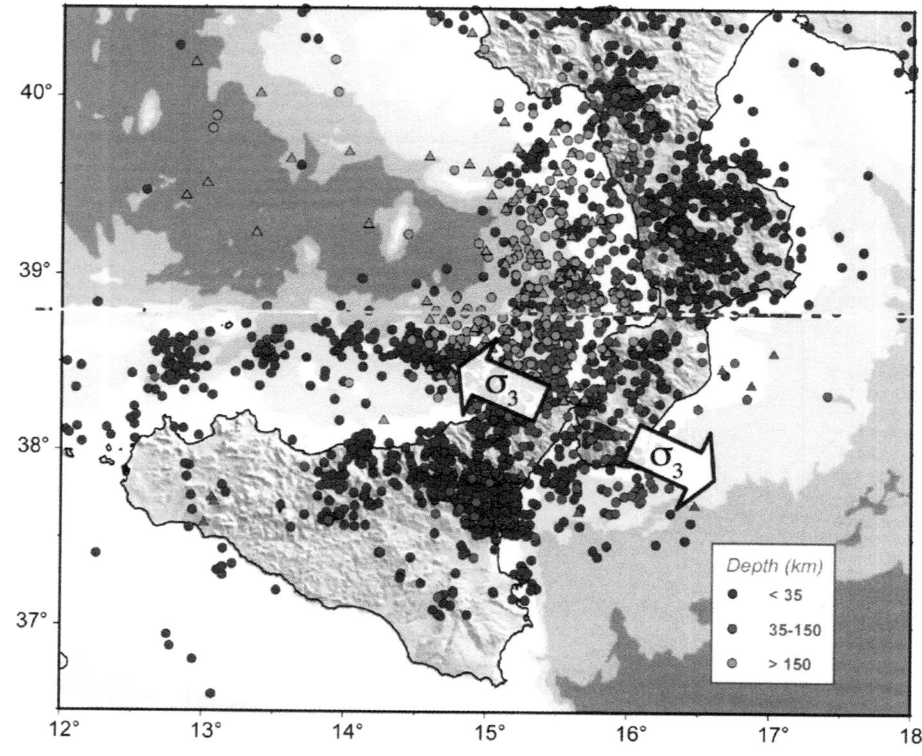

Figure 5.6 ► Seismicity of Sicily and Calabria from the catalogue of the Italian National Seismic Network for the period 1984–2001 (circles) and deeper earthquakes (h > 50 km) for the period 1964–1984 (triangles). Notice that background seismicity concentrates around rather than within the Messina Strait, where much elastic strain was relieved by the 1908 earthquake.

the 1908 source and with tectonic evidence [14]. However, it implies pure extension on faults oriented NNE-SSW (models A, B and D in Figure 5.4), and oblique faulting (extensional with a component of right lateral slip) on faults oriented N-S or NNW-SSE (models C, E and F).

The Messina Strait was one of the first sites in Italy to be closely monitored for horizontal strains by a relatively dense geodetic network. The oldest surveys [21], however, did not return significant or reliable estimates of ongoing deformation, probably due to a combination of feeble strain rates and limited resolving power of the land-based geodetic techniques available in the 1970s and '80s.

Things were to change following the introduction of modern GPS technology, but earlier attempts to assess the relative motion of the two sides of the Strait did not supply evidence for ongoing deformation beyond the data noise level [22]. It took a few more years for the GPS network to be able to detect sizable strains (Figure 5.7). D'Agostino and Selvaggi [23] documented WNW-ESE extension across the Strait at a rate of 2–3 mm/yr, consistent with most models of the 1908 earthquake and with the tectonic and seismological evidence discussed above.

The rather diverse evidence described in this chapter was used to identify a reference source model for the 1908 event suitable for deriving earthquake scenarios and ground shaking predictions (Table 5.1). The model is based on that originally proposed by Boschi et al. [10], satisfies most of the instrumental observations and agrees with large scale recent geological and landscape features (see following section).

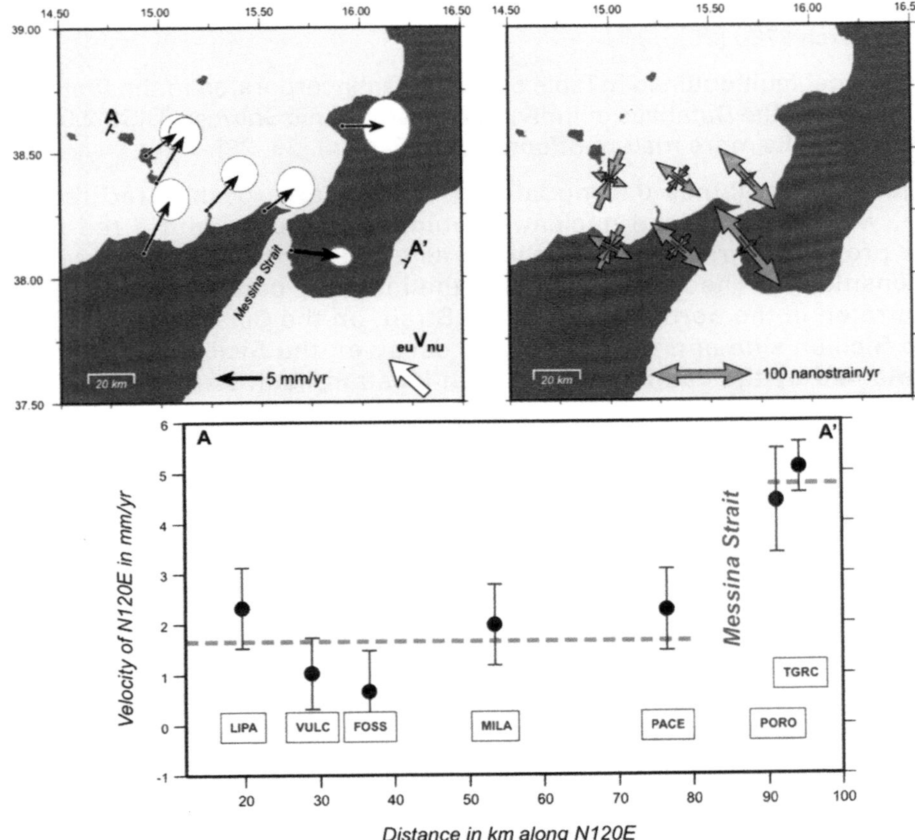

◄ Figure 5.7 (*top left*) GPS velocities in a Nubia reference frame. (*top right*) Principal axes of the horizontal strain rate tensor (in yellow) and associated 1sigma errors (red bars). (*below*) Velocity in the N120°E direction as a function of distance along the same direction (see trace of profile in figure above left). The error bars represent one standard deviation.

All figures are from D'Agostino and Selvaggi [23], redrawn.

Table 5.1 ▶ Summary of fault parameters for the reference model of the 1908 earthquake.

Strike (degrees)	20°	*Orientation from north*
Dip (degrees)	29°	*Angle between fault plane and the horizontal*
Rake (degrees)	270°	*Corresponds to pure normal faulting*
Length (km)	40.0	*Along strike of the fault*
Width (km)	20.0	*Along dip of the fault*
Min depth (km)	3.0	*Depth of upper edge of fault plane*
Max depth (km)	12.7	*Depth of lower edge of fault plane*
Slip (m)	1.42	*Average over the entire fault plane*
M	7.0	*Moment magnitude*
M_0 (Nm)	3.5×10^{19}	*Seismic moment*
Longitude Latitude	15.378 N 37.939 E	*Coordinates of southernmost corner of fault projection onto the surface*

The location of the proposed source model is consistent with the inferred rupture nucleation and with the damage pattern. The overall extent of the fault is constrained by geodetic and seismological data; it does not extend beyond the northern end of the Messina Strait, as suggested also by the lack of sizable tsunami effects north of the Ganzirri peninsula [24], and is therefore entirely confined within the Strait proper. Its strike is constrained by geological and geomorphological observations discussed by Valensise and Pantosti [17]. Its southern end coincides also with a known transverse structural feature that may have generated the magnitude M = 5.6 earthquake on 28th March 1780 [25].

The model fault outlined in Table 5.1 has been incorporated in the first public release of the Database of Individual Seismogenic Sources (DISS 2.0) [27] and later in its more mature offspring (DISS 3.0.4) [26, 28].

Figure 5.8 combines the model fault (shown as a dashed red rectangle) with the proposed nucleation point [19] (shown with a red star), the proposed directivity [3], (shown as a red arrow) and the observed intensities [1]. The main features of the intensity pattern (damage concentrated in the northern part of the Strait, on the Calabrian more than the Sicilian side and faster intensity decay on the Sicilian side) is well explained by the combination of fault location, fault strike and rupture directivity.

Figure 5.9 shows the elevation changes and the horizontal strains that are expected to be caused by motion on the fault outlined in Table 5.1. According to the reference model, the 1908 earthquake somewhat surprisingly caused minimal changes in the relative position of the two sites for the bridge foundations; less than 10 cm in elevation and a few cm in distance.

5.1.5 The recent evolution of the Messina Strait: clues to the 1908 earthquake source

Several elements of the recent geology and landscape suggest a close relationship between the processes of the formation of the Messina Strait

◄ Figure 5.8 Synoptic view of the 1908 earthquake rupture history and of the associated damage. The region of largest intensity is outlined in green/blue, the surface projection of the reference fault in yellow.

From Baratta, [1], modified; see text for discussion.

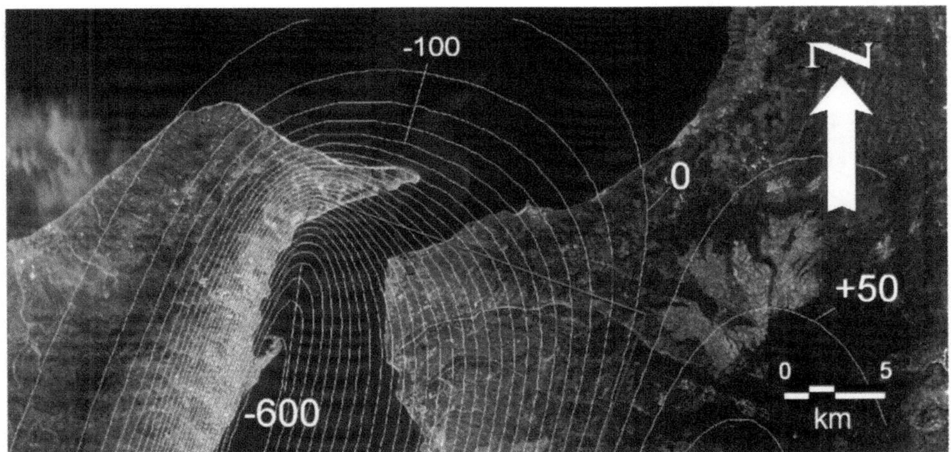

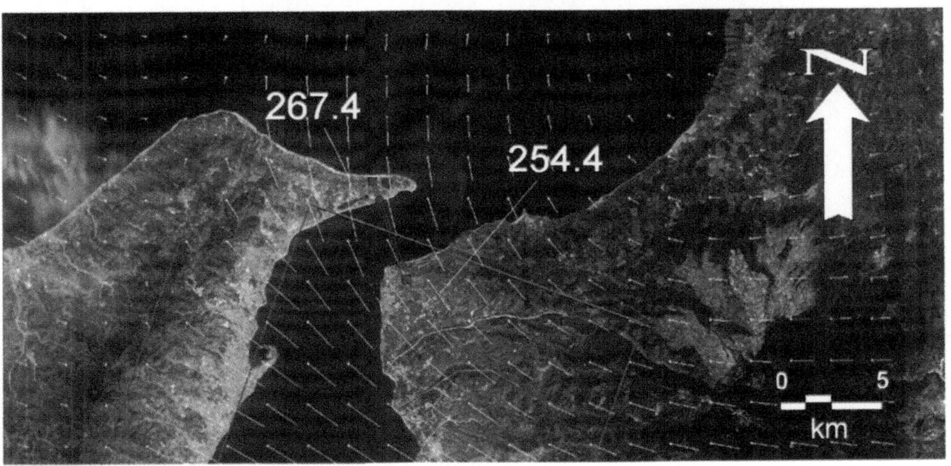

◄ Figure 5.9 Simulation of elevation changes (top, shown with contours) and horizontal motions (bottom, shown with vectors pointing opposite to the yellow dot) associated with the 1908 earthquake. The calculation assumes elastic dislocation following uniform slip on the model fault outlined in Table 5.1. The model predicts significant subsidence along the axis of the Strait, the Sicilian side being steeper than the Calabrian side, and mild uplift of both its shoulders. Contour interval in top map is 25 mm. Horizontal motions are shown in mm. The bridge site is shown in blue.

and the source of the 1908 earthquake as described by the reference model discussed above. Proving this relationship would also prove that sustained slip on the 1908 fault is a powerful landscape generator and would ultimately support the appropriateness of the reference fault model.

Consider first the similarity between the topography of the Strait and coseismic elevation changes predicted by the model. These describe a "coseismic trough" that is parallel to the Peloritani range in Sicily and to the Aspromonte range in Calabria (Figure 5.9). At both terminations of the fault the ends of the trough coincide with major structural discontinuities: the drop in elevation of the Peloritani range, the Ganzirri Peninsula and a sharp change in the direction of the Calabria coastline to the north, and a distinct change in the elevation and trend of the crest of the Peloritani range to the south. A section across the central part of the Strait shows that the footwall flank of the "coseismic trough" is roughly twice as steep as the hangingwall flank. Similarly, the average topographic gradient of the Sicilian side of the Strait (0.15) is roughly twice as large as that of the Calabrian side (0.08). Notice that the model faults striking N-S or NNW-SSE (e.g. C, E and F in Figure 5.4) violate these basic constraints as they cut through the Ganzirri peninsula and would hence destroy it after a few tens of seismic cycles. Similarly, model faults dipping westward (A, B in Figure 5.4) would produce uplift or subsidence with a pattern that is totally inconsistent with that of current topography.

Proving that repeated slip on the reference fault may generate the most youthful geological features of the Strait, however, requires a long-term evolution model that accounts for all major landscape features of the Strait (Figure 5.10). For instance, 1908-type faulting may account for subsidence

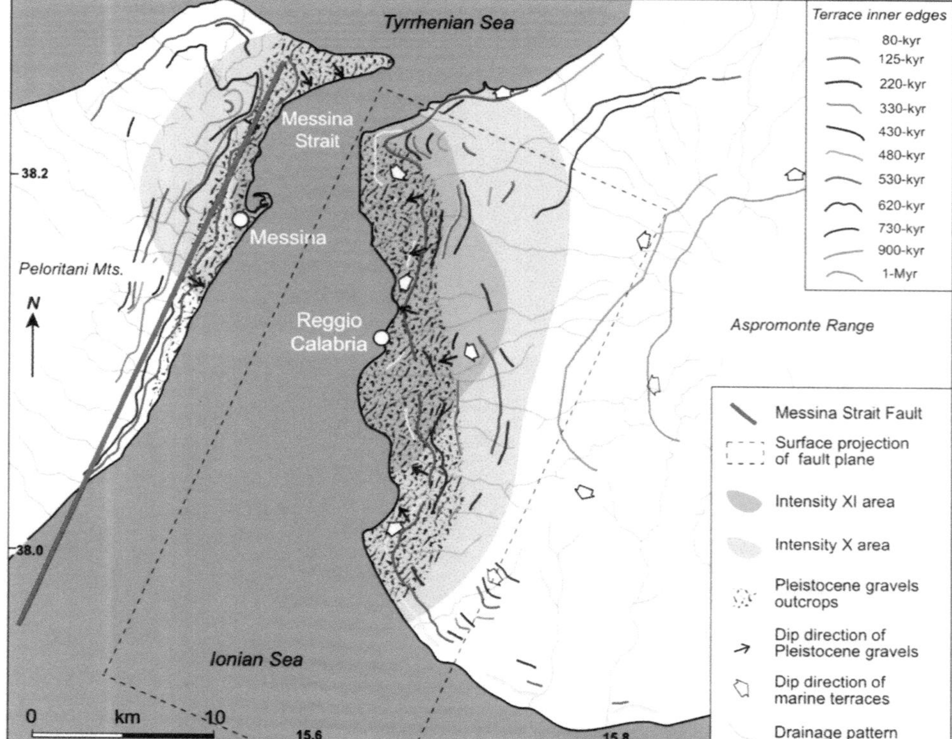

Figure 5.10 ▶ Summary of geological and geomorphological features of the Messina Strait in relation with coseismic elevation changes generated by the 1908 earthquake.

From Valensise e Pantosti [17], redrawn and modified.

of the Strait, but not for the creation of high topography in the Aspromonte and Peloritani ranges, which requires an independent and large-scale engine.

The restless uplift of the Calabrian Arc

All Calabria and northeastern Sicily have been known for the wealth of elevated coastal landforms and characteristic coastal deposits since the end of the 19th century. One of the most outstanding of these marine features is found in Aspromonte, facing and with the same orientation as the Messina Strait; a 20 km-long, up to 5 km-broad, flat, gently seaward(westward)-dipping wave-cut platform carved into crystalline, Late Tertiary and Early Pleistocene rocks forming the Calabro-Peloritano massif (Figure 5.10). Its inner edge (fossil shoreline angle) is found at an elevation of 1,150–1,200 m. Because the platform is assumed to be younger than 1.0–0.8 Myr based on the age of its youngest bedrock [13], this "upper terrace" demonstrates uplift at an average rate up to 1.4 mm/yr.

Fast uplift is also directly evidenced by several observations of littoral deposits containing Strombus Bubonius, a fossil that is a faithful marker of a sea level highstand dated 125 kyr (corresponding with the Marine Isotope Stage 5e), scattered around Reggio Calabria at elevations in the range 100 to 160 m (see Bordoni and Valensise, [29], for a summary). Several other wave-cut platforms can be recognised and dated with respect to the age of the 125 kyr terrace and of the Upper terrace and have been the object of intense investigations in the past two decades (Figure 5.10).

Hence, simple but well constrained observations from marine terraces suggest that uplift has been continuous at least during the past 0.7 Myr. Similar evidence, although with substantially slower rates, can be found virtually anywhere in Calabria and northeastern Sicily, indicating that this quite fast uplift affects the whole region.

Valensise and Pantosti [17] reconstructed in detail the inner edge of the 125 kyr marine terrace between Scilla and Lazzàro (Figure 5.11), and found that it ranges in elevation between 100 and 170 m, corresponding to uplift rates in the range 0.9 to 1.4 mm/yr. They also argued that the present elevation of this important geological marker matches the pattern of 1908 subsidence along the Calabrian shore of the Strait. In other words, the elevation fluctuations of the 125 kyr marine terraces are the result of repeated slip on the 1908 fault: the marine terrace is lowest where it approaches the region of largest 1908 subsidence, and highest outside the region affected by 1908 faulting, i.e. outside the Strait proper. This circumstance and the smoothness of the elevation trend are important indications that none of several mapped faults crossing it (e.g. the Reggio Calabria fault of Monaco and Tortorici [14]) have been active in the past 125 kyr, as they would have sharply displaced the marine terraces leaving a distinct signature. Similarly, the lack of displacement in the trend of 1908 elevation changes across these faults and the absence of reports of coseismic surface breaks comprise a further indication that they are no longer active, not even as secondary features capable of sympathetic slip.

The superposition of 1908-type faulting onto regional-scale uplift has important outcomes for the evolution of a number of geological occurrences, all of which are closely interconnected as they belong to the same geomorphological and sedimentological domain (refer to Figure 5.10):

- relatively recent sedimentary rocks are preserved only in those portions of the Strait that experienced the least uplift in the past 125 kyr (long-term) and the largest subsidence in 1908 (modern), whereas only older crystalline rocks crop out elsewhere;
- more specifically, the Mid-Upper Pleistocene Messina Gravels, a young clastic deposit that is characteristic of the Messina Strait, occur only in areas of least long-term uplift, and in fact outline them rather accurately;
- drainage divides tend to mimic the shape of the region of least long-term uplift on both sides of the Strait;
- drainage on the Calabrian side has rotated from an outward radial pattern near the top of Aspromonte Range to an inward flowing pattern near the modern shoreline, following the trend of both long-term and modern subsidence;
- similarly to drainage, the shape of the shorelines has evolved over time from outward convex on top, to outward concave near the modern shoreline, thus reflecting the inception of depocentres (low-lying sites of largest deposition) in the Strait.

Therefore, repeating the strain pattern induced by the causative fault of the 1908 earthquake does produce a good fit of the youngest geological and landscape features of the region. This mechanism, however, does not explain the overall configuration of the Strait. Good quality seismic reflection data [30] indicate that the Strait is a wide north-northeast-trending late Miocene trough subdivided into a series of small perched basins that developed on top of major thrust sheets of crystalline basement. Individual thrust-top basins are filled by late Miocene to early Pliocene sediments, and all are unconformably overlain by a complete late Pliocene-Holocene basin-filling sequence. According to Monaco et al. [31], these basins developed within the accretionary wedge associated with Ionian subduction and are not related with the rifting of the Tyrrhenian sea; their setting suggests a complex interplay between compressional and extensional tectonics at least until the Early Pleistocene.

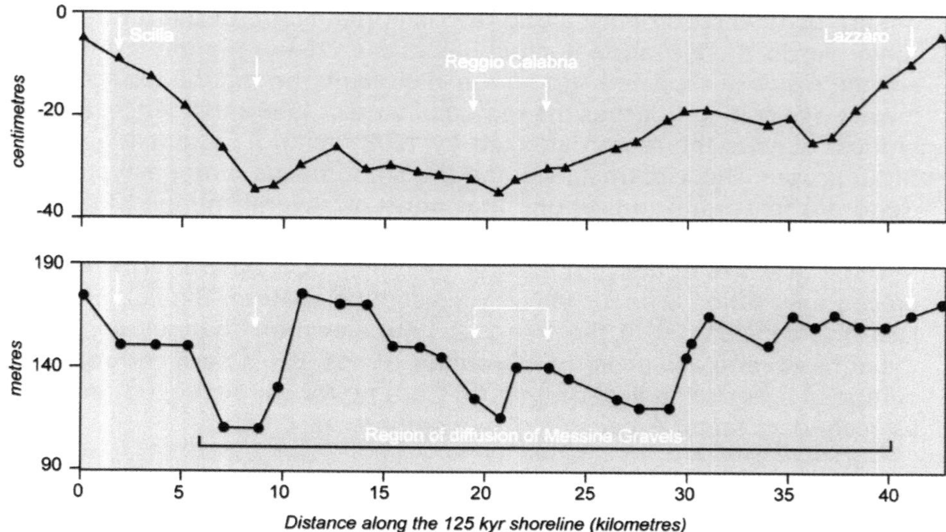

Figure 5.11 ▶ Comparison between 1908 coseismic subsidence (top) and the elevation of the 125 kyr terrace inner edge along the Messina Strait from Scilla (to the north; far left in figure) and Lazzàro.

From Valensise e Pantosti [17], redrawn and modified.

If the 1908 fault slips at about 1.4 mm/yr (that for the given fault location and geometry corresponds to a maximum fault-related subsidence rate of 0.7 mm/yr along the axis of the Strait) and started operating at about 0.7 Myr, as suggested by Valensise and Pantosti [17] and indirectly confirmed by Monaco et al. [31], one would expect to see fault-related landforms with a maximum topographic relief of about 500 m. Instead, both the topographic relief between the onland sides of the Strait and the sea floor and the structural relief of the crystalline basement are in excess of 2,500 m. Only a small portion of this relief can be accounted for by the broad uplift of the entire region which, according to Bordoni and Valensise [29], has a wavelength in the order of 100 km and exhibits a culmination right at the Messina Strait. This implies that the agreement between the presence of a large normal fault beneath the Messina Strait and its graben-like configuration is at least partially fortuitous, and that a structural trough separating Sicily from mainland Italy existed prior to the inception of the 1908 fault.

5.1.6 Earthquake recurrence in the Messina Strait

In the previous section we discussed the similarity between long-term and modern strains in the Messina Strait. Valensise and Pantosti [17] argued that this similarity in itself proves that the 1908 is a "characteristic earthquake" for the Messina Strait; ie. a large earthquake that has a tendency to repeat with about the same magnitude and rupture length without leaving much room for different-size events. They used the same similarity to propose a recurrence interval of about 1,000 (+500 –300) years for 1908-type earthquakes. Millenary recurrence intervals are found for most large Italian faults and are expected given the relatively low rates of tectonic deformation in most of Italy [27]. In fact, the causative source of the 1908 earthquake turns out to be one of the most frequently re-activating of the entire Italian peninsula.

Guidoboni et al. [32] investigated earthquake recurrence in the Messina Strait using both historical and archeological evidence. Historical data show no evidence for earthquakes comparable in size with 1908 during the most recent part of the first millennium A.D. and the entire second millennium; that is to say, over a period in excess of 1,000 years. Historical sources for this period are reasonably complete, as testified also by several accounts concerning smaller earthquakes that occurred in the Strait.

The archeological approach was based on the awareness that the 1908 earthquake caused enormous territorial upheaval and left signs in the settlements that are still largely recognisable today. Guidoboni et al. hence hypothesised that the Messina Strait area, which has been densely populated even in ancient times, may similarly retain evidence of one or more much older "upheavals" of the settlement network, and that this evidence may be recognised through a careful analysis of archaeological observations. They found that the settled area around the Messina Strait contracted substantially around the middle of the 4th century A.D., when many sites were abandoned or relocated, a contraction that can hardly be justified by the contemporary economic and military setting. Specific archaeological findings within the cities of Messina and Reggio Calabria also suggest a serious decline of the region during the same period.

The archaeological hypothesis is hence in good agreement with available historical, palaeoseismological and geological evidence. It suggests that a large earthquake, perhaps similar to the 1908 event, took place in the

Messina Strait around the middle of the 4th century A.D., possibly in the interval 350–363.

5.1.7 Conclusions

The 28th December 1908 earthquake in the Messina Strait is the oldest Italian earthquake for which a significant number of instrumental records is available. Similarly to several other Italian cases, it was generated by a blind fault that would have been very hard to detect and characterise based on geological and tectonic observations only. Thanks to the wealth of instrumental data this fault has been investigated in great detail. The residual scatter among the solutions proposed by different investigators is rather small, and in any case small enough for deriving a reliable design earthquake and for the ground motion predictions discussed in Section 5.2. In fact, instrumental observations of the 1908 earthquake gave geologists and tectonicists a unique opportunity to understand the recent history and current dynamics of the Messina Strait. Overall, investigations of the 1908 earthquake in the context of the region's active faulting comprise one of the world's best examples of Seismotectonics, a young discipline at the boundary between Seismology and Tectonics.

What is in the future of the Messina Strait, aside from the unavoidable repetition of large earthquakes similar to 1908? Valensise and Pantosti [17] used the comparison between long-term and 1908-related strains to estimate an average slip rate of 1.4 mm/yr for the fault responsible of the 1908 earthquake, and hence a maximum subsidence rate of 0.7 mm/yr along the axis of the Strait. While this estimate agrees well with the most recent results from GPS observations, it also implies that 1908-type faulting cannot fully explain the regional-scale uplift of the Strait, that according to the same investigators proceeds at 0.9–1.4 mm/yr depending on the specific location. In other words, the 1908 fault can make the Strait progressively broader and deeper, but in the long run it will not prevent it from evolving into a narrow sound and ultimately into emerged land. A simple extrapolation of the trend shown in Figure 5.10 suggests that the Calabrian shoreline is slowly migrating towards Sicily at a rate that is 10–15 times larger than the estimated extension rate (the Sicilian shoreline is also progressing towards Calabria, but in a more subtle fashion due to the steeper topography).

Geology is governed by basic laws, such as Lyell's motto "*The Present is the key to the Past*" from his Principles of Geology (1830). But it is legitimate also to reverse this principle and project it into the future ("*The Past is the key to the Future*"), following what David Hume had already written in 1777: "*...all inferences from experience suppose... that the future will resemble the past...*". Geology thus tells us that not later than 200 kyr from now Sicily's insularity will come to an end. With the bridge, man can achieve the same result much faster.

5.2 Seismic hazard and design earthquakes

5.2.1 Introduction: historical and recent seismicity of the Messina Strait zone

After briefly recalling some salient aspects of the historical earthquakes and the current seismicity of the Strait region (more extensively discussed in 5.1), this section describes first how the main design earthquake for the bridge was determined, in the form of an elastic response spectrum for horizontal and vertical motion. Subsequently, an independent check of the validity of the design spectra is illustrated, based on advanced modelling of the source and of the ground motion generated in numerical simulations of the 1908 earthquake.

While the reader is referred to Sections 5.1.3 and 5.1.4 for some salient features of the latter event and to Baratta [1] for a detailed picture of its impact on the built and geological environment, it is of interest to note that relatively large earthquakes in the Strait region have been few and far between, as shown by the epicentre map of historical earthquakes in Figure 5.1. Over possibly more than 1500 years, only one other strong earthquake occurred in the Strait region proper, i.e. the 6th February 1783 shock, with a magnitude substantially smaller than 1908. The 1783 earthquake was apparently not caused by rupture of the main seismogenic fault of the Strait. If we enlarge the zone of interest to Southern Calabria, the only other relatively large event (M = 6.5) took place just the day before on 5th February 1783, with probable origin on a fault bordering the Gioia Tauro Plain, to the NE of the Straits. As to the past activity of the Strait fault itself prior to 1908, some authors have conjectured that it may have ruptured in a possibly major earthquake around the middle of the 4th century A. D. [32], as mentioned in Section 5.1.6.

The distribution of the recent, instrumentally recorded seismicity of the Strait area, is displayed for a 20 year period in Figure 5.12, where magnitudes have been added with respect to Figure 5.6. Such distribution does not exhibit specific spatial patterns; if anything, the low level seismic activity (proportional to the size of the symbols) present in the Strait is suggestive of quiescence. Thus, while the historical and current seismicity picture does not contradict the indication given in Section 5.1 of a characteristic earthquake with a return period of 700–1500 years that ruptured most recently in 1908, the quantitative simulation of the source process of such an event must rely heavily on the corroboration of tectonic and geologic elements.

5.2.2 Bases for the determination of the main design earthquake for the bridge

The Design Specifications for the bridge (See Table 7.2) prescribe that the design seismic action for Ultimate Limit State (ULS) verifications be corresponding to a return period of 2000 years. This may be compared with

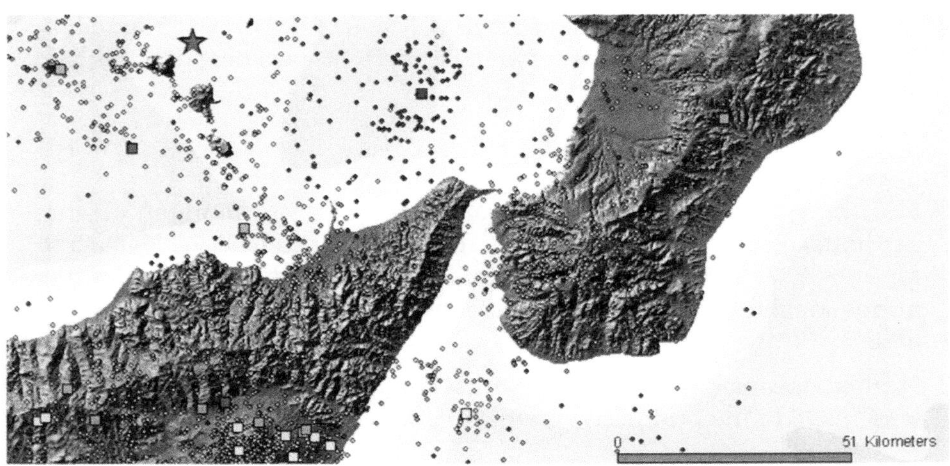

◄ Figure 5.12 Distribution of the instrumentally monitored seismic activity in the Strait and surrounding areas for the period 1981–2002 (the size of the symbols is proportional to earthquake magnitude).

From http://legacy.ingv.it/CSI/ (INGV – Instituto Nazionale di Geofisica e Vulcanologia – Italy).

the current Italian Structural Design Code [33] indication of 2475 years for Collapse Prevention Limit State verifications of structures with large dimensions and a nominal design life not less than 100 years. For the Bridge, a useful economic life of 200 years has been assumed in design. The 2000 year requirement poses a challenge, since the catalogue of Italian earthquakes is considered complete, at the larger magnitudes, for not more than 1000 years.

The design earthquake had to be quantified primarily in the form of a 0.05 damped elastic response spectrum of pseudo-velocity (PSV), separately for the horizontal and vertical components of ground motion. In addition, a set of appropriate spectrum-compatible accelerograms, obtained through modifications of the frequency content of real strong motion records, had to be provided, but these will not be discussed here. The design spectrum was obtained through the following main steps:

1. Determination of the relationship between macroseismic intensity (MCS scale) at the project site and the return period,
2. Conversion of MCS intensity into Peak Ground Acceleration (*PGA*) through a suitable empirical correlation, and
3. Definition of the shape of the spectrum by consolidated empirical rules, and calibration with spectral envelopes from a set of recorded strong motion accelerograms.

Approach for deriving the design spectrum

Step 1 was carried out using a hybrid statistical-probabilistic approach, which allowed the introduction into the formulation of the earthquake source zones shown in Figure 5.13. These were regarded as seismotectonically homogeneous by the project "Geosismotettonica" Advisory Group of the time, and differ in a few respects from the Seismic Source Zones nowadays used for the seismic hazard mapping of Italy [34]. The seismicity in each zone was described by the complementary distribution function of epicentral intensity I_0, that is

$$\tilde{F}_{I_0} = 1 - F_{I_0}(i).$$

Various tests carried out for determining such a distribution function, showed that the zones could be subdivided into two homogeneous groups or "areas", conventionally called A (including zones 5, 6, 7, 8, 12 in Figure 5.13) and B (including zones 2, 3, 4, 9, 10, 11, 13). The function $\tilde{F}_{I_0}$ for areas A and B, represented in Figure 5.14, was described by the doubly exponential function

$$\tilde{F}_{I_0} = \exp\left[\exp(\alpha \cdot 6 + \beta) - \exp(\alpha i + \beta)\right] \tag{1}$$

valid only for $I \geq \text{VI}$, and with parameters α and β calibrated on the earthquake catalogue data for the two areas in question. Note that for area A, which includes zones 7 and 8 controlling the hazard at the bridge location, the function $\tilde{F}_{I_0}$ is considerably more severe than for area B.

Further, the epicentral intensity I_0 of an event in any of the source zones was linked to the bridge site intensity I through an attenuation relation calibrated on the intensity maps of a vast number of regional earthquakes. The form of the attenuation relation is

$$I = I_0 e^{-a(\theta)[d - r(I_0,\theta)]} \quad (2)$$

where d = epicentral distance (in km), and a and r are parameters depending on the direction θ according to an elliptical law, separately calibrated for each source zone in Figure 5.13. The radius r of the epicentral intensity area takes values as high as 20 km for $I_0 \geq$ X, with a strong influence on the final results. From (1) and (2) the cumulative probability distribution of site intensity was obtained in the form

$$F_I(i) = P\ [I \geq i \mid I \geq \text{VI}]. \quad (3)$$

To establish a relationship between site intensity and return period $T_I(i)$, only earthquakes generating $I \geq$ VI at the project site were considered, and the well known relationship

$$T_I(i) = \frac{T_I(\text{VI})}{1 - F_I(i)} \quad (4)$$

was used, with T_I (VI) = return period of site intensity VI = 16 years, obtained after testing different estimation procedures. The resulting seismic hazard curve predicts return periods of 70, 200, 600, and 2200 years for intensities VIII, IX, X, and XI, respectively.

For Step 2 a new intensity-*PGA* correlation was adopted, based on available Italian records and macroseismic observations, and on other proposed correlations, as shown in Figure 5.15. The new correlation was determined by constraining it: (*a*) to be linear in the log-linear plot, (*b*) to take, for I = XI, the highest of the *PGA* values considered in an independent study by Professor G. Veneziano and, (*c*) to envelope all available significant data from Italian earthquakes.

The final curve of *PGA* vs. return period is shown in Figure 5.16 and provides the basis of the design values:

- *PGA* = 0.58 g for the ULS verification return period, horizontal motion;
- *PGA* = 0.32 g for the operational earthquake (elastic limit) verifications, at 300 year return period, both horizontal and vertical motion.

The reliability and degree of conservatism of the previous *PGA* values can be checked *a posteriori* with the values of horizontal peak ground acceleration prescribed by the current Italian Structural Design Code [33], expressed as *PGA* = Sa_g, where S is a factor accounting for local site conditions. As shown in Figure 5.16, if the subsoil category C of the norms is assumed, median values of Sa_g equal to 0.29 g, 0.34 g, and 0.47 g are obtained for return periods of 300, 500, and 2000 years respectively. These are significantly smaller than the design *PGA*s for the bridge, notably for the 2000 year return period. The norms in question provide slightly different values of the design parameters for the Sicily and Calabria foundation sites, but only the latter are used in the comparison in Figure 5.16 since they are more conservative.

The format of the design seismic action for the bridge is that of a velocity (S_V) response spectrum, defined by the piecewise linear curve (polygon) and the parameters shown in Figure 5.17, on a two-way logarithmic diagram, for

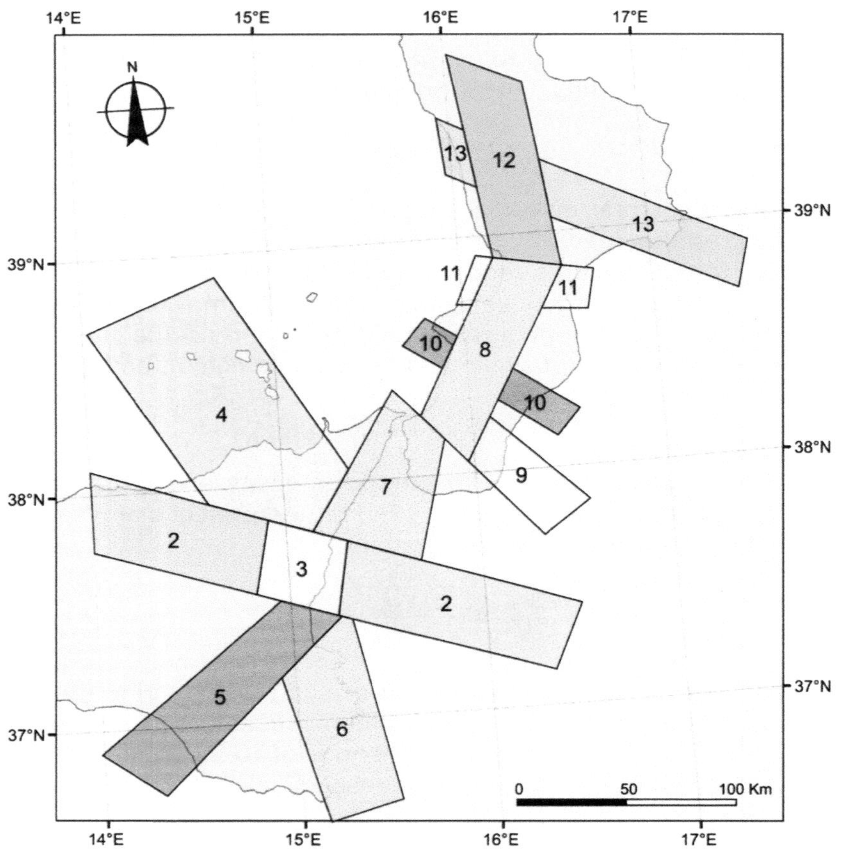

Figure 5.13 ► Seismogenic zones used for the seismic hazard study at the project site.

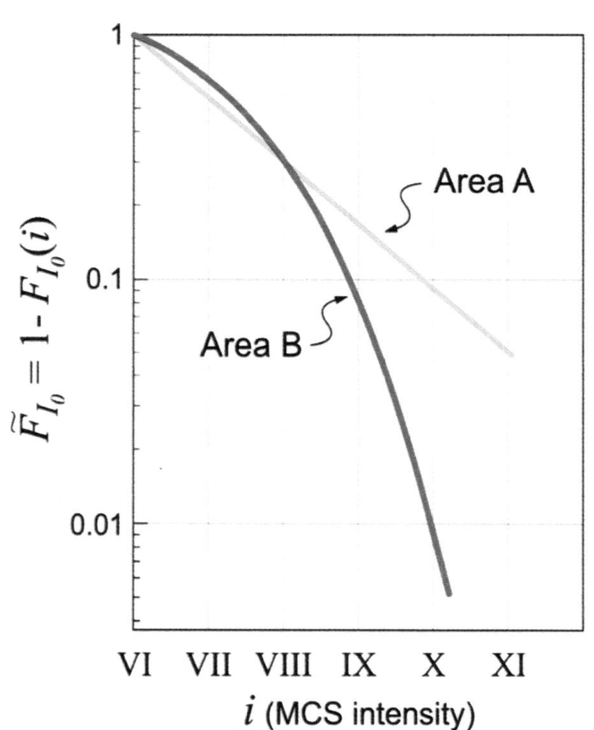

Figure 5.14 ► Complementary cumulative probability distributions of MCS intensity (i) for the two groups ("areas") of seismogenic zones, A and B, see Figure 5.13.

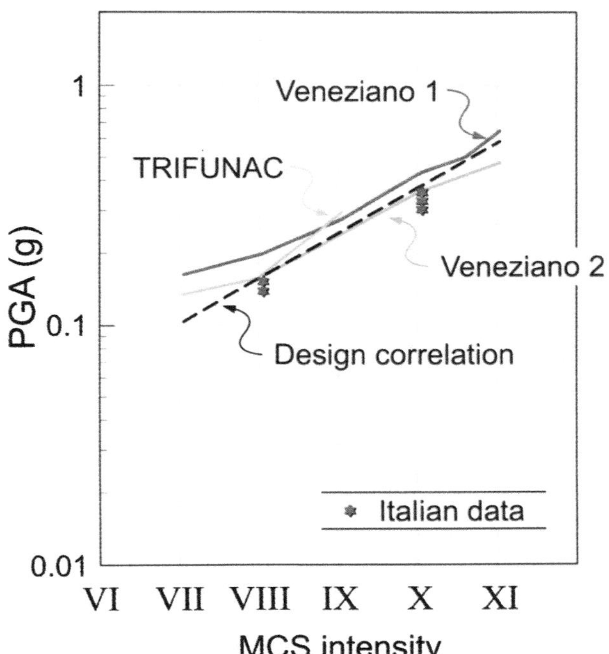

◄ Figure 5.15 Correlation between *PGA* and macroseismic intensity I.

horizontal motion and 0.05 damping factor. This spectrum, like the previous *PGA* values, did not take into account the specific ground conditions at the project site since the MCS intensity values used as primary descriptor of the ground shaking hazard were assumed as representative of "average" soil conditions, not necessarily rock.

Outline procedure for building the design spectrum

In Figure 5.17 the starting point of the velocity spectrum corresponds to the value $PGA = 0.58$ g in the (oblique) acceleration scale, associated with a vibration period $T = 0.03$ s. The first side ends at $T_1 = 0.1$ s, consistent with current seismic norms, with $S_v(T_1)$ such that the ordinate of the pseudo-acceleration spectrum $PSA = (2\pi/T_1)\ S_V(T_1) = 2.5 \times PGA$, or 1.45 g. The second side of the polygon, which corresponds to the constant acceleration plateau in the acceleration response spectrum, is thus extended up to $T_2 = 0.65$ s, i.e. a conservative bound with respect to current norms (typically giving 0.4 to 0.5 s for ground types A, B, and C of Eurocode 8). This determines the ordinate of the constant spectral velocity region, i.e. the third side of the polygon, at 147 cm/s. The fourth side was constrained by taking, rather conservatively, a ratio $S_d / PGA = 160$ (S_d = displacement (cm/s) spectral response) based on significant strong motion records of the 1980 Irpinia earthquake. This yields a spectral displacement = 92.8 cm, which determines the control period T_3 at 3.5 s. Based on extensive simulations of the source and of the near-field ground accelerations (synthetic seismograms), the constant S_d region was extended up to $T_4 = 10$ s. From then on the spectrum is decreased so as to reach the peak ground displacement value of 60 cm at $T_5 = 30$ s. The 60 cm displacement is consistent with the empirical rule, established decades ago for the design of nuclear power plants, by which slightly less than 1 metre of peak ground displacement (actually, 91 cm) entails 1 g of *PGA* [35].

The design velocity spectrum for vertical motion, not shown here, was obtained by applying reduction factors to the horizontal spectrum, with suitable changes in some of the control periods. Shown in colour in Figure 5.17 are also the 16-percentile, 50-percentile (median), and the mean statistical envelopes of the *PSV* spectra calculated from a set of 16 strong motion accelerograms recorded in extensional tectonic environments, broadly comparable to that of the Strait. These accelerograms were selected, after amplitude scaling to compensate for the differences in magnitude, as a basis for independent calibration of the design spectrum. A number of them were also separately subjected to modifications in their frequency content,

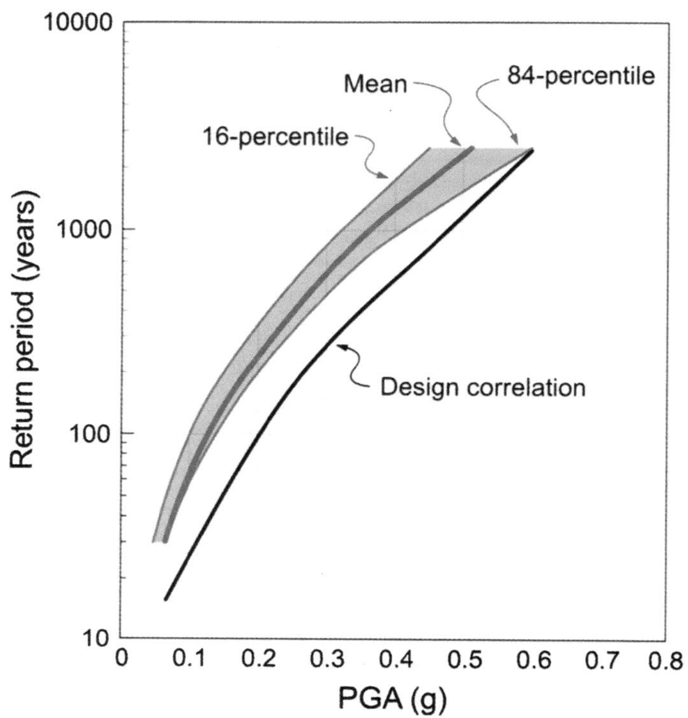

Figure 5.16 ▶ Design peak ground acceleration (*PGA*) as a function of return period at the project site (solid black curve) compared with the curve applicable at the same site according to current seismic norms of Italy on subsoil category C (median, with limits of shaded band corresponding to 16- and 84-percentile levels).

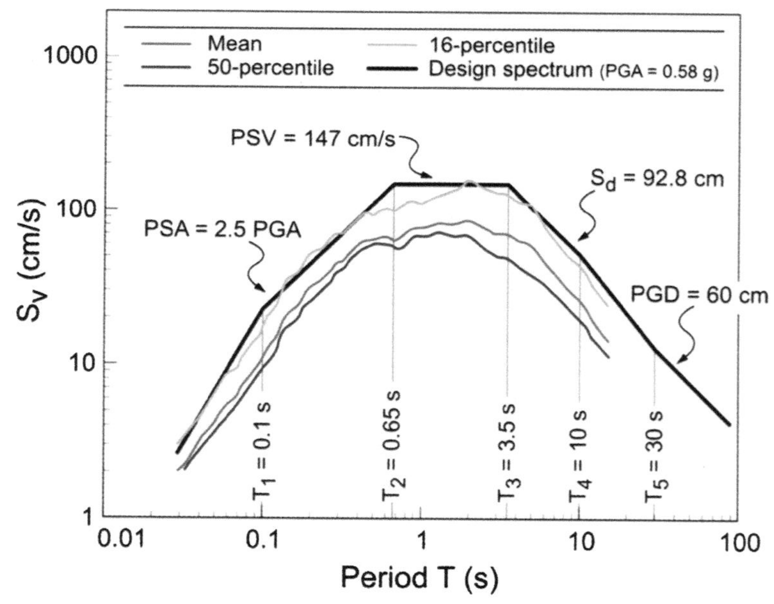

Figure 5.17 ▶ Design horizontal (pseudo-) velocity response spectrum, shown by black piecewise linear curve (with control periods T1…T5), and envelopes of set of scaled real accelerograms at different percentile non-exceedence levels.

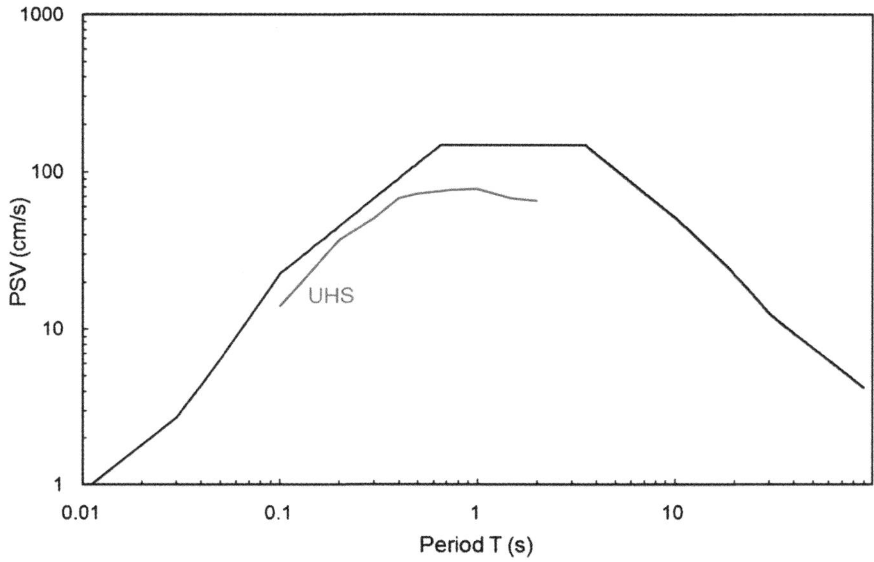

◄ Figure 5.18
The ULS elastic spectrum for the bridge (black curve) compared with the 2000 year Uniform Hazard Spectrum (UHS red curve) applicable on hard ground at the same site, according to the current seismic norms of Italy.

as already mentioned, in order to provide suitable time-history representations of the design earthquake for purposes of dynamic structural analyses. Figure 5.17 shows that the design spectrum closely matches the statistical envelope of the (scaled) records at the 16-percentile level of exceedence, which is quite useful to assess the amount of conservatism inherent in the design earthquake.

An independent check on the validity of the 2000 year design earthquake is displayed in Figure 5.18 which compares the spectrum just described (solid black curve) with the Uniform Hazard Spectrum (UHS) on hard ground for both the Messina and Reggio Calabria sites, as given by the current hazard zoning of Italy for the 2.5% in 50 year exceedence probability, i.e. 2000 year return period, in the interval of periods between 0.1 and 2 s.The comfortable margin of conservatism of the design spectrum, associated to 2000 year return period, should be noted (even though the ground conditions are not exactly the same).

5.2.3 Simulation of the 1908 earthquake ground motion

The re-assessment of the seismo-tectonic setting carried out for the project in the early 2000's (see Section 5.1 and the technical specifications prepared by SdM [36]) led to an improved understanding of the seismogenic fault of the Strait as the likely seat of characteristic earthquakes having the general features of the 1908 event. On this basis, an evaluation test of the design earthquake for the bridge was considered worth carrying out, by numerically calculating the strong ground motion that would be generated by a 1908-like earthquake, with particular emphasis on the range of intermediate and long vibration periods of most interest for the seismic response of the bridge. Such evaluation, or validation, was to rely on state-of-art methods for the simulation of the 1908 seismic source, and to simultaneously take into account the current knowledge on the regional Earth crust properties, as well as the specific ground conditions at the Sicily and Calabria foundation sites.

In the following sections, a description of the method used for the simulation of the ground motion at the two foundation sites is first given. Then the results obtained are shown in terms of acceleration and displacement time histories, and of response spectra, the latter being compared to the design spectrum for both horizontal and vertical motion.

5.2.3.1 Method

The surface seismic ground motion at the two foundation sites has been calculated by combining two different approaches for generating the motions in the low and high frequency ranges, respectively.

The main steps of the overall approach adopted, shown in the following flow diagram (Figure 5.19), are:

- definition of a site and source model for simulating the low frequency ($f < 1$ Hz) seismic motion;
- generation and processing of a time sequence of white noise, with zero mean, for simulating the high frequency motion ($f > 1$ Hz);
- combination of results from previous two steps.

Detailed input data and specific aspects of the implementation of the methods just outlined are provided in the detailed discussion at the end of Section 5.2. A critical influence in such implementation is exerted by the assumptions regarding the profile of wave propagation velocities in the uppermost few km of the Earth's crust (shown in Figure 5.20), the geometry of the active fault of the Strait with respect to the crossing site (Figure 5.21), and the distribution of the inferred relative displacement, or *seismic slip*, caused by the 1908 earthquake rupture on the same fault (Figure 5.22).

5.2.3.2 Results

The numerical simulations carried out with the hybrid deterministic-stochastic approach (described in more detail at the end of Section 5.2) yielded two sets of 3-component acceleration (and displacement) time histories, one for each foundation site, with 20 realisations each. Figure 5.23 illustrates the

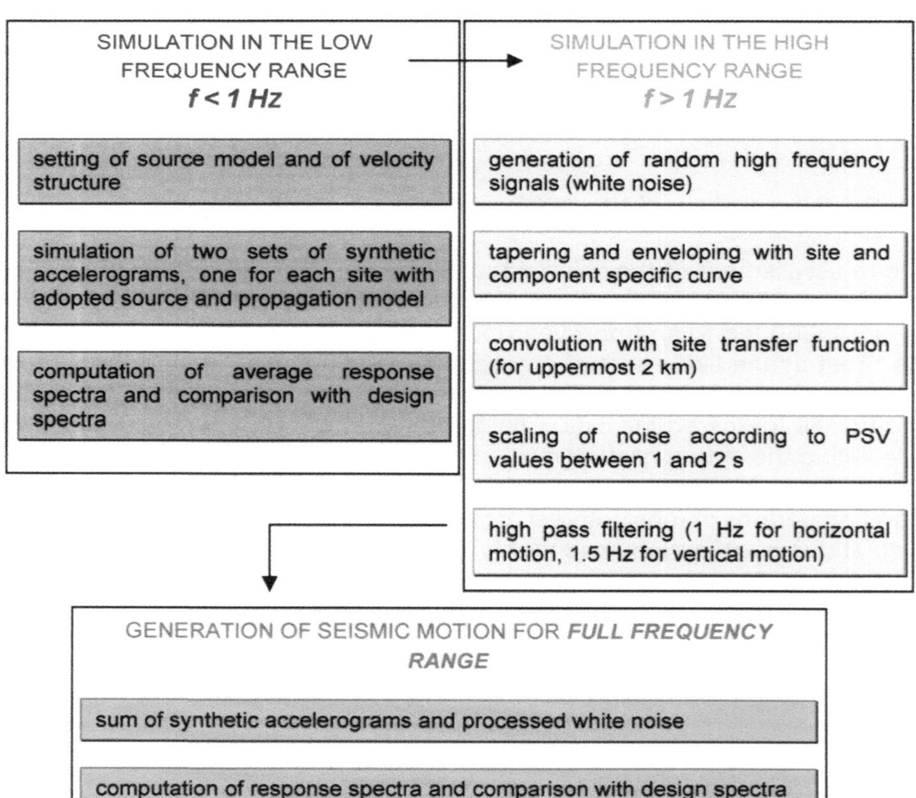

Figure 5.19 ▶ Outline of method of analysis.

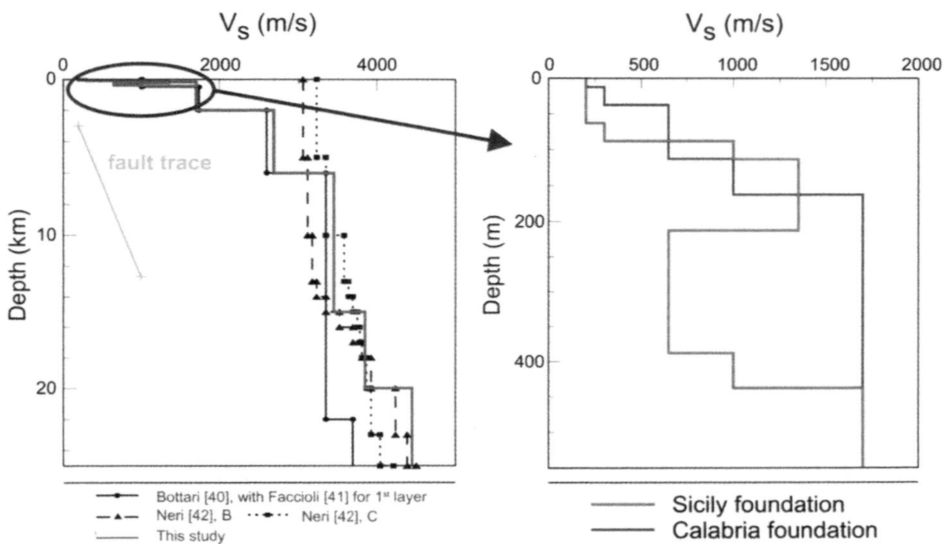

Figure 5.20 Crustal S-wave velocity model together with detailed profiles for the upper 500 m below the two foundation sites.

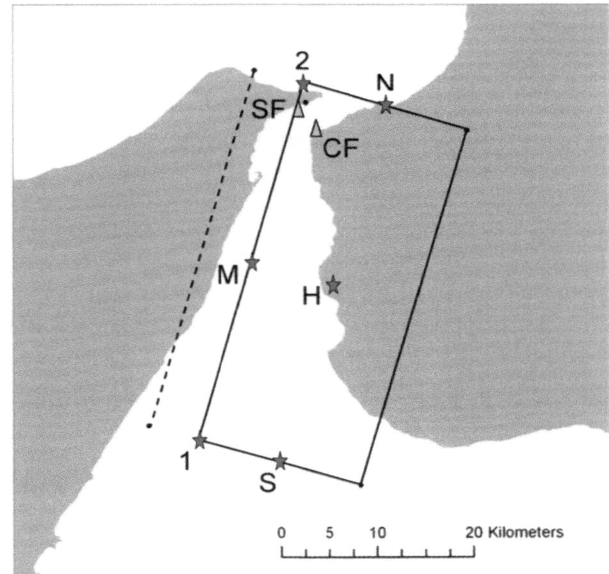

Figure 5.21 Plan view of modelled fault with position of the two foundation sites SF, CF (green triangles) and hypocentres (red stars) tested in the analysis.

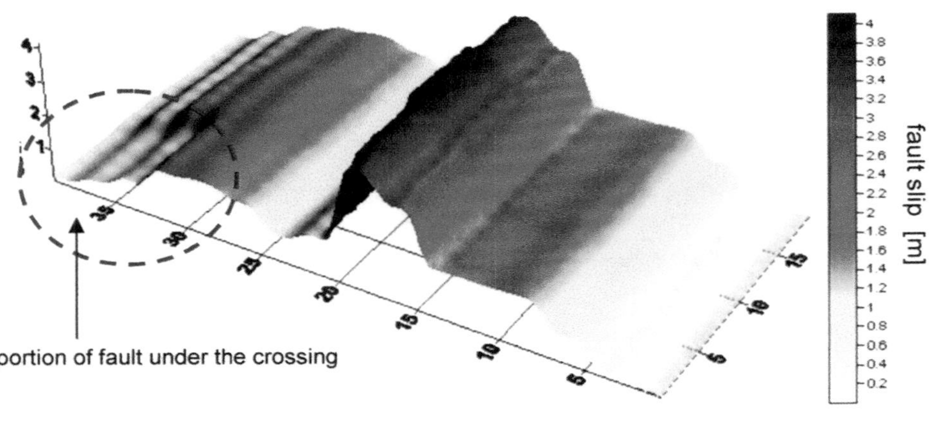

Figure 5.22 Adopted final distribution of relative seismic displacement (*slip*) on fault plane.

examples of two realisations (n. 8 and 17) in terms of acceleration, while Figure 5.24 shows, for the entire set of simulations, the mean ±1 standard error bands for the pseudo-velocity response spectra, compared with the design spectra, for horizontal and vertical motion.

All simulations exhibit physical features, especially in the low frequency range, dominated by the source model. The mean value for the horizontal peak acceleration is 0.24 g (with a maximum value of 0.578 g and a standard deviation of 0.09 g), while the vertical mean peak acceleration is 0.19 g (with a maximum value of 0.46 g and a standard deviation of 0.1 g). The synthetic accelerograms for the Calabria foundation site are richer in high frequencies and display higher peak values than the Sicily ones, for all components. The stochastic variability significantly differentiates each realisation; this can be seen, for instance, from the amplitude of the dispersion band of *PSV* curves, which is indeed higher at low periods, as physically requested.

The vertical components of *PSV* spectra are strongly influenced, at shorter periods, by the site specific amplification, albeit with low amplitudes.

Source rupture directivity has a significant influence on the amplitude of the two horizontal components, NS and EW, respectively almost parallel and perpendicular to the fault azimuth. They differ significantly in terms of mean values and dispersion bands of spectral ordinates, as expected. In particular, NS spectra dominate over EW ones in the long period range ($T > 3$ to 4 s) on both shores.

The simulated spectra are significantly below the design spectrum at low periods ($T < 1$ s) especially at the Sicily foundation, and this may not be completely realistic. However, at longer periods ($T > 1$ s for Sicily and $T > 5$ s for Calabria) the simulated horizontal spectra approach the design level in the NS components, while the vertical spectra are consistently low over the whole frequency range. It hardly needs to be stressed that the bridge dynamic response is dominated by the fundamental vibration period of the towers at around 3 s, a range where the design spectrum is well matched by the simulations, while the spectral ordinates at shorter periods have little influence.

In order to check the previous results, the number of stochastic simulations was subsequently extended from 20 to 60. The mean peak values and the acceleration time histories for the larger sample confirm the validity of the previous work. Horizontal components, in the time domain, display roughly the same features (with a slight increase, 11.3%, in mean peak values), while

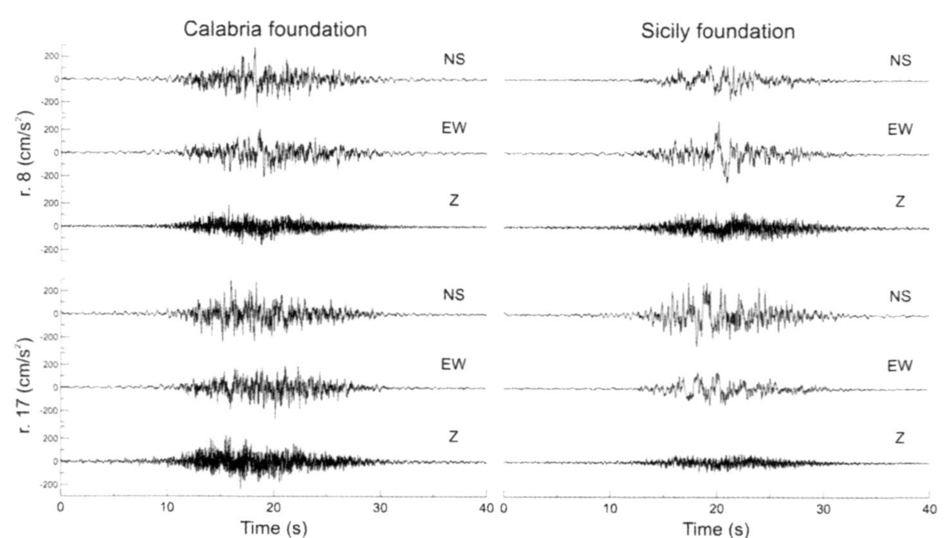

Figure 5.23 ▶ Final simulations for Calabria and Sicily foundations, two examples.

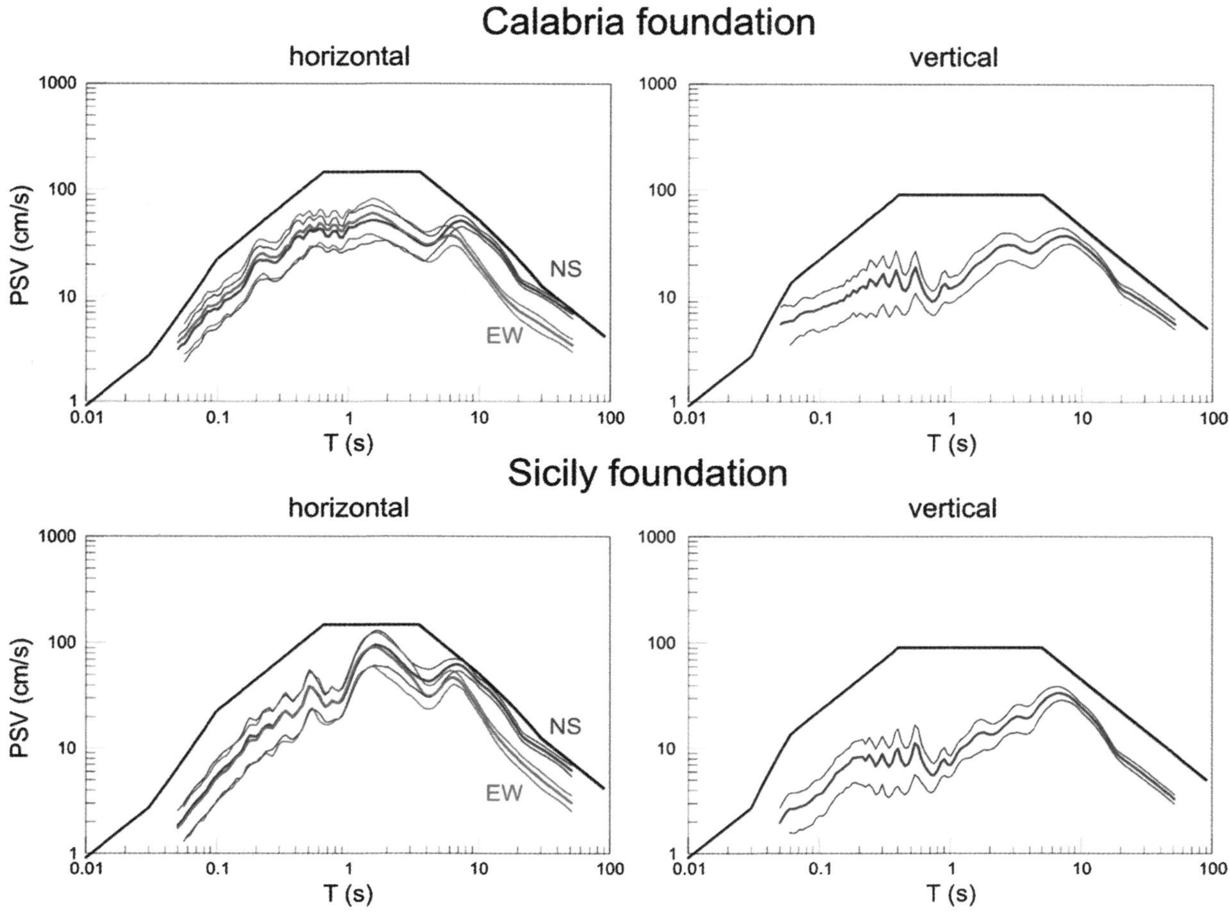

▲ Figure 5.24
Mean *PSV* response spectra at 5% damping from wide band accelerograms, for Calabria and Sicily foundations (mean ± s.e.). Design spectra are shown for comparison.

vertical components are affected by a greater variability in peak values (the mean increases by about 30%). Standard deviations of peak values increase more significantly for vertical components, as expected. The mean *PSV* spectra are almost identical to those in Figure 5.24, except for a small increase (less than 4%) in horizontal amplitudes between 1 and 2 s.

5.2.4 Conclusions

This section has shown how the main design seismic action for the bridge had been determined, around two decades ago, in the form of an elastic design spectrum, and the severity of the spectral levels proper. It has also demonstrated how the spectrum in question has resisted, so to speak, the challenge of time. In other words, it has been shown to be suitable even after checking against the results of later studies (including very recent code seismic hazard provisions) performed with updated inputs and more sophisticated tools of analysis. Of special significance among these later studies have been, on one hand, the successful simulation of strong ground motions likely to be generated by a 1908-like earthquake, and on the other hand the availability of probabilistic, uniform hazard spectra applicable near the crossing site.

These checks have been successful, proving that the design spectrum was determined with a remarkably well calibrated amount of conservatism, and that there is therefore no apparent need to introduce significant changes in seismic design criteria.

Detailed Discussion

Generation of low frequency ground motion: f < 1 Hz

Deep and shallow crustal profiles

A deterministic, analytical approach was adopted to compute the ground motion in the low frequency range, i.e. the method of Hisada [37, 38, 39] using an extended kinematic source. This method computes displacement and stress Green's functions for viscoelastic layered half-spaces, with explicit consideration of static (permanent dislocations due to fault rupture) and dynamic transient terms.

The application of the method requires knowledge of the propagation velocity structure in the Earth's crust (all the way to the surface) and of the source. The proper definition of these models and of all the computational parameters is crucial for the generation of realistic, physically credible time series at the ground surface. For this, after careful examination of the available information, several tests were performed to achieve a definitive choice.

Crustal velocity model

The selected crustal profiles rely both on published data, for the deeper structure, and on project documents for the near-surface structure.

The crustal model by Bottari et al. [40] is based on data of dipole electrical soundings and, for the deeper structure, on seismological observations and on their comparison with data from active seismics (explosion profiles). On the other hand, Faccioli [41], in his study for the feasibility evaluation of the crossing, used a simplified 2D profile based on the available in situ tests of the previous years (cross-hole, down-hole, deep seismic reflection profiles), with a bedrock layer consistent in its properties with Bottari's (cit.) first layer. The crustal *Vp* and *Vs* velocity structures by Neri et al. [42], derived from tomographic inversions of earthquake data, significantly improved the knowledge of the crustal structure of the Southern Tyrrhenian and Sicily and were used for the deeper layers (See Figure 5.20).

The velocity profiles in Figure 5.20 also show details of the uppermost 500 m below the Calabria and Siciliy tower foundation sites. The data are drawn from the SdM report [43], which is a 1 D and 2 D local amplification study for the computation of site-specific spectra. *Vp* and *Vs* values for in the uppermost 100 m or so were provided by cross-hole measurements at the foundation sites [44, 45].

The choice of including the uppermost layers of the local soil profile (0–2 km) in the source and crustal models allowed them to generate ground motions that already included site specific effects on wave propagation.

Source model

The source model adopted was that described in Section 5.1.4, slightly modified in its slip values according to the full-scale work of Pino et al. [3] on the 1908 earthquake and the SdM (2004) report. [36]

The modelled seismogenic fault of the Strait (summarised in Table 5.1), 40 × 20 km in size, lies between 3 and 12.7 km under the surface and dips towards ESE. The strike, dip and rake angles are 20°, 29° and 270°, respectively. The overall fault was subdivided into 32 × 15 subfaults of equal size, in order to suitably carry out the integration of Green's functions up to 2 Hz (maximum frequency simulated). Figure 5.25 shows a plan view of the fault, with the Sicily tower and Calabrian tower foundation sites (SF and CF) and all the hypocentres tested in the analysis.

The adopted slip distribution on the fault plane was slightly modified with respect to the reference source model, though achieving a final seismic moment ($M_o = \Sigma_{subfault\ j}[G_j \times slip_j \times Area_j]$) consistent with Pino et al. [3], i.e. 5.38 (±2.16) × 10[19] Nm. Figure 5.22 illustrates the distribution of the slip on the fault plane. Two changes were introduced, namely a slight reduction in the slip under the zone of the foundation sites, and a moderate increase around about 15 km from the south end of the fault (along the length). The reduction under the crossing was necessary in order to achieve a reasonable approximation to the permanent vertical displacement of 28.8 cm actually measured after the 1908 earthquake near the Calabria foundation (landmark n. 41, Villa San Giovanni, De Natale e Pingue) [11]. In the end, the final simulations reached a value of 30 cm.

The adopted increase of the slip was generally consistent with the results by Pino et al. with an assumed maximum of 3.96 m and a mean value of 1.35 m. This slip distribution yields a final seismic moment of $3.01 \cdot 10^{19}$ Nm (equivalent to magnitude $M_w = 6.95$). This value, slightly lower than that suggested in Pino et al. and Section 5.1.4, strongly depends on the elastic properties (shear modulus G) of the crustal layers crossed by the fault. A further increase in *Mo* would have requested a further significant, but fictitious, modification of the shear modulus, not consistent with the data.

A rise time of 1.6 s of the slip function was selected by comparing the estimates from well-known empirical relations with those obtained in the literature for recent events of comparable magnitude.

The position of the hypocentre on the fault was chosen so as to reach a safety compromise between the duration of the simulated ground motion and the long period response spectral ordinates. All the hypocentres (points of nucleation of rupture) shown in Figure 5.25 were tested and the position "S" was finally selected (in agreement with the discussion of Section 5.1.4).

Figure 5.26 portrays the 5% damped *PSV* spectra computed from one of the simulations, in the most conservative cases, i.e. for the S, 1 and HC hypocentre locations. Design spectra are also shown for comparison. The S location tends to produce the highest *PSV* ordinates, at least in the intermediate period range 1 to 3 s, and the longest significant duration of ground shaking. Note that the resonance periods of the bridge towers are close to 3 s.

Concerning the rupture velocity and the slip distribution in each subfault, a stochastic approach has been used, using the following formula:

$$v_{ri} = 0.75 \cdot V_{Si} \cdot (1 + random\ number_i \cdot 0.15)$$

$$s_i = \hat{s}_i \cdot (1 + random\ number_i \cdot 0.15)$$

where v_{ri} is the rupture velocity, V_{si} the S-wave velocity and s_i the slip in subfault i; $\hat{s}_i$ is the reference (deterministic) slip shown in Figure 5.22.

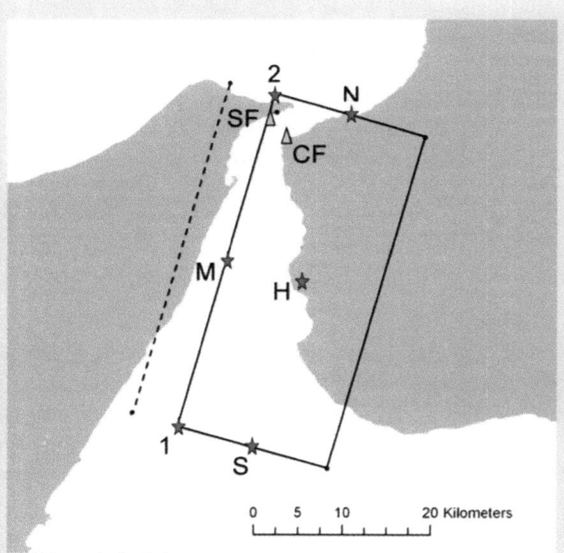

Figure 5.25 ► Plan view of modelled fault with position of the two foundation sites SF, CF (green triangles) and hypocentres (red stars) tested in the analysis.

Random numbers were generated with an internal routine, starting from a "seed". Different seeds generate different sets of random numbers (white noise with zero mean) for each subfault, resulting in different slip distributions and rupture velocities, and, hence, different realisations of the source process and time histories of ground motion. A variability of ±15% with respect to the reference slip was assumed. As a consequence of adding this stochastic component, more realistic signals were generated, because of an enriched high frequency content. The signals exhibit at the same time a slightly lower amplitude in the frequency domain, as energy tends to be distributed across a wider range of frequencies.

Results

Two sets of simulations for each foundation site have been performed, each consisting of 20 different realisations with the 3 ground motion

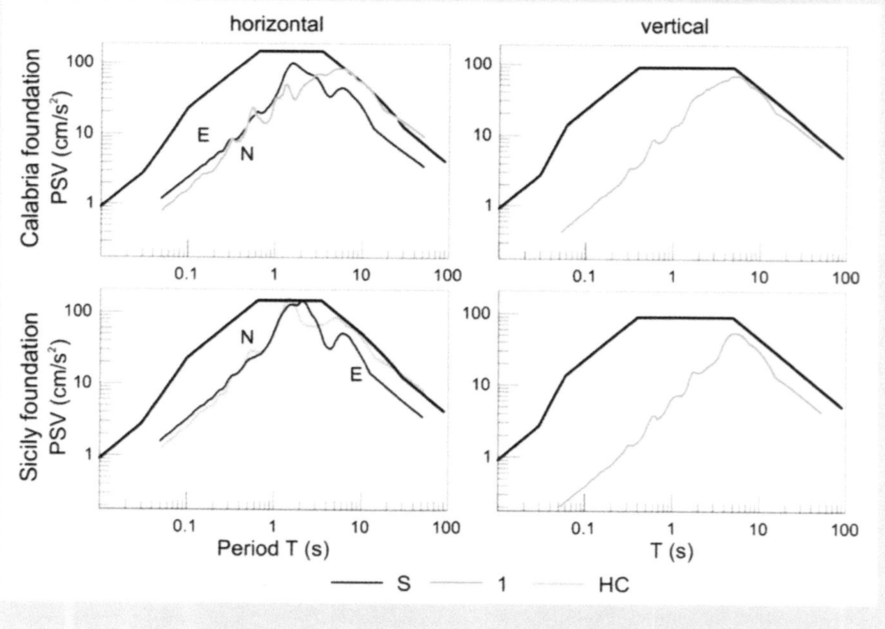

Figure 5.26 ► Example of 5% damped PSV response spectra for Calabria and Sicily foundation sites from deterministic simulations corresponding to one realisation of the source model. Also shown are the design spectra. For comparison of different positions of hypocentre (see Figure 5.21).

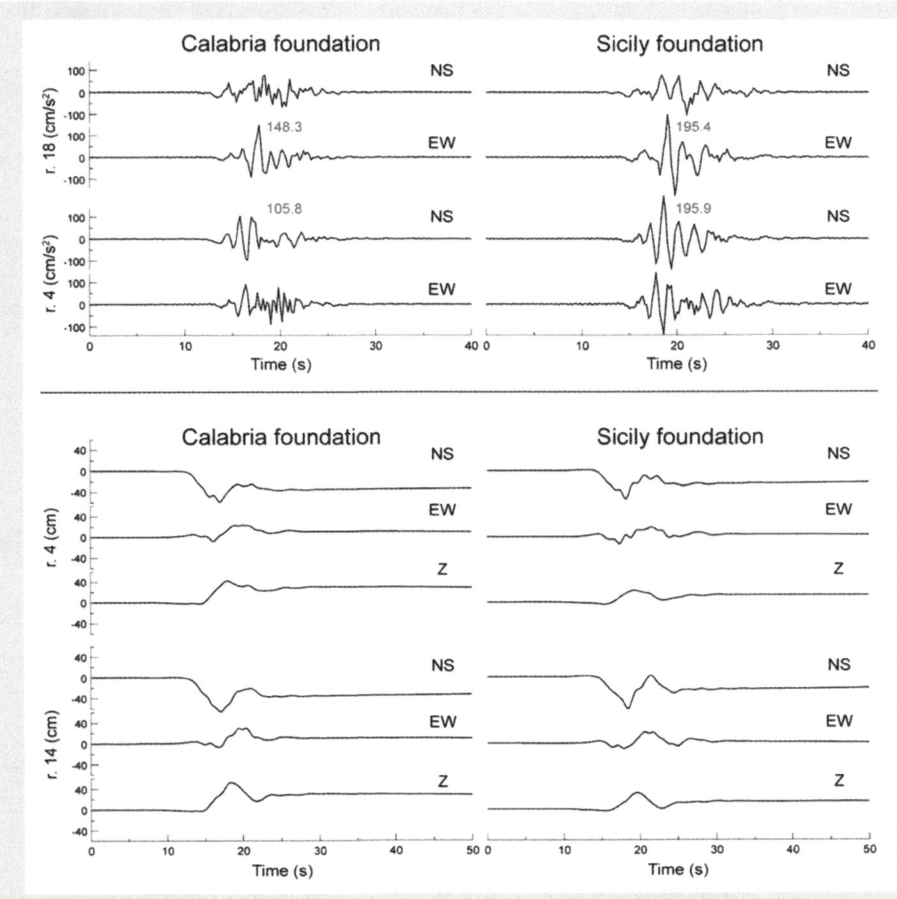

◀ Figure 5.27 Time histories of earthquake ground motion generated at the Calabria and Sicily foundation sites in two different realisations (r.18 and r.4) of the source process. Top: accelerations (peak values marked in red) Bottom: displacements.

components (NS, EW, Z), using the same 'seed' values for corresponding simulations. For each realisation, time histories of acceleration and displacements have been computed, together with *PSV* response spectra. Figure 5.27 displays two different realisations for acceleration and displacement time histories. Note that the low frequency motions are not significantly affected by stochastic variability (i.e. displacement signals are all quite similar). Instead, ground accelerations are more sensitive and display horizontal peak values that range from 205.3 cm/s^2 to 60.61 cm/s^2. As already mentioned, the method computes both static and dynamic Green functions, as clearly seen in the displacement series, where permanent offsets are present.

Generation of high frequency ground motion: f > 1 Hz

The previous ground motion simulations, though realistic, are poor in high frequency content, due to the intrinsic characteristics of the source model (limited in the number of subfaults) and of the method used. To enrich the high frequency content of simulated ground motion (related to interaction between waves and small scale heterogeneities in the Earth's crust) an *ad hoc* approach has been devised. Several methods have been used in the past for the same purpose: among these we relied basically on the generation of modulated white noise. The main steps in this part of the study included:

- generation of white Gaussian noise
- tapering of signals (beginning and end) for 10% of their duration

- multiplication of signals by an appropriate envelope in time domain
- convolution with site specific amplification function
- scaling according to *PSV* values of previous low frequency simulations
- high pass filtering (1 Hz for horizontal motion, 1.5 Hz for vertical motion).

The normalised temporal envelopes used to correct the noise series were obtained from the low frequency realisations and were specific to motion component and foundation site.

The amplification functions used in the convolution accounting for site-specific response were obviously different the two foundation sites, as well as for horizontal and vertical motion (Figure 5.28).

Figure 5.28 ▶ Analytical 1D amplification functions computed for the uppermost 2 km for transverse *(left)* and longitudinal *(right)* waves, applicable to horizontal and vertical motion, respectively.

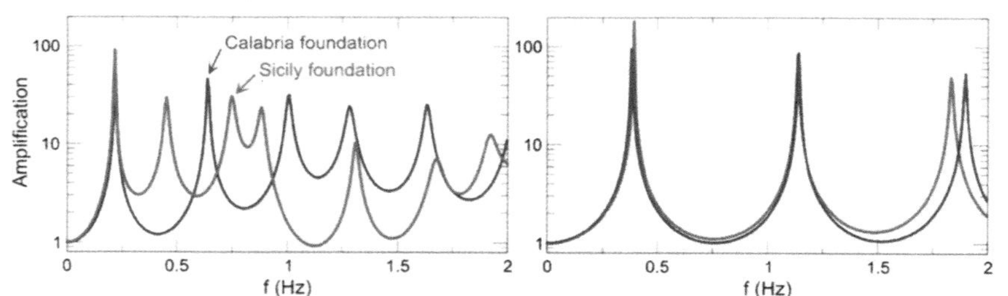

Scaling of the modulated white noise signals was achieved by computing mean *PSV* values between periods of 1 and 2 seconds for each low frequency realisation and for modulated noise, and by using their ratio to multiply the latter. Hence, each high frequency signal was scaled by a different factor. The factors range from 2.23 to 19.00, depending on component and realisation.

With this procedure, 120 (=20 × 3 × 2) uncorrelated time series were originally generated, one for each component (3) of the previous realizations (20), and for each foundation site (2).

Synthesis of ground motion over a wide frequency band

Ground shaking in terms of acceleration signals was generated in this phase by summing the results, in time domain, from source simulation and modulated noise, according to component, foundation site and stochastic realisation.

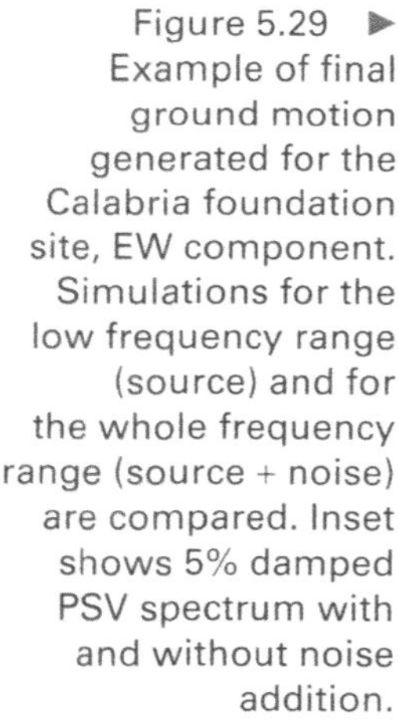

Figure 5.29 ▶ Example of final ground motion generated for the Calabria foundation site, EW component. Simulations for the low frequency range (source) and for the whole frequency range (source + noise) are compared. Inset shows 5% damped PSV spectrum with and without noise addition.

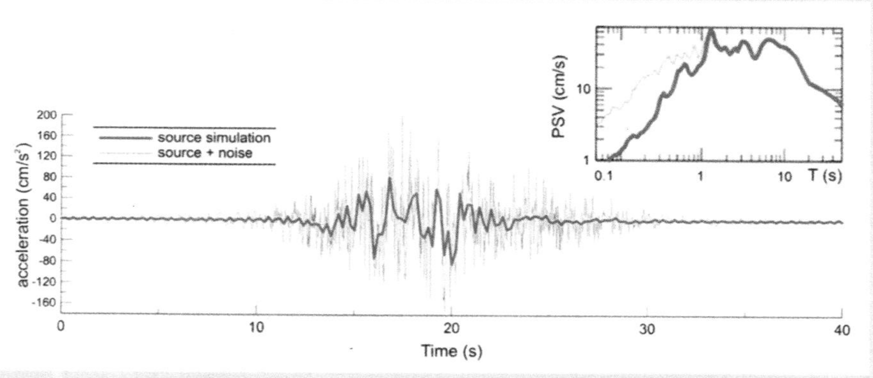

5% damped *PSV* response spectra were then computed for each time series, and mean values subsequently derived for comparison with design spectra. Figure 5.29 shows one example of wide band time series for the EW component, for the Calabria foundation site.

5.3 Geotechnical site characterisation

5.3.1 Introduction

This section deals with the geotechnical site characterisation at the four locations where the bridge interacts with the underlying geomaterials: the two tower foundations and the two anchor blocks.

The chronological sequence of the depositional history of soil and rock formations at the bridge location is summarized in Figure 5.30 which defines the nomenclature used in this section. In-situ and laboratory soil investigations were carried out in the late seventies and eighties were aimed at determining the physical and mechanical properties of the geomaterials which are likely to interact directly with the bridge foundations and anchor blocks under static and earthquake loadings.

CP Coastal Plain Deposits, Holocene
Sand and gravel

MG Messina Gravels, medium Pleistocene
Sand and gravel

VC Vinco Calcarenite, lower Pleistocene/upper Pliocene
Bioclastic calcarenite with lenses of conglomerate

CSC Clay-Sand Complex, Pliocene
Silt and sandy silt with sand and gravel in the lower part

PC Pezzo Conglomerate, Miocene
Weakly cemented conglomerate and sandstone containing weathered rock blocks

BR Cristalline Bedrock
Granite and Gneiss

◄ Figure 5.30 Soil deposits encountered across Messina Strait.

5.3.2 Geotechnical investigations

In addition to the various geological borings (referred to as Series C on the Calabria shore and S on the Sicily shore), which reached depths between 50 and 200 metres, a series of geotechnical borings (BH series) and in situ tests has been carried out on both sides of the Strait to determine the essential information necessary for a safe and optimal foundation design.

The programme of these geotechnical investigations is summarized in Table 5.2, and the locations of the geotechnical borings and in-situ tests carried out along the bridge axis are shown in Figures 5.31 and 5.32, together with some geological borings of C and S series. All of these are crucial for a satisfactory foundation design.

Table 5.2 ► Soil investigation programme.

	CALABRIA		SICILY	
	Foundation	Anchor	Foundation	Anchor
Borings (1)	4	3	3	1
Holes for SPT's (2)	4	4	4	4
Holes for LPT's (2)	4	4	4	4
Shafts for PLT's (3)	None	1	None	1
Pumping test	1	None	1	None
CPT's	None	None	2	3
DMT's	None	None	1	3
CH (4)	1	1	1	1
SASW	1	1	1	1

BH = Boring with sampling
SPT = Standard penetration test
LPT = Large penetration test
PLT = Plate load test
CPT = Static cone penetration test
CH = Seismic wave velocity measurement in cross-hole
SASW = Spectral analysis of surface waves
DMT = Flat Marchetti dilatometer

(1) depth 90 to 100 m
(2) depth 50 m, with rod energy measurements
(3) 2.5 m in diameter cased shaft, depth 18 m, plate diameter 800 mm
(4) 3 holes, depth of each 100 m

Figure 5.31 ► Soil investigation on Calabrian shore.

E556000
UTM 33S

BH *geotechnical boreholes*
C,S *geological borings*
CH *cross hole test*
PLT *plate loading tests in deep shafts*

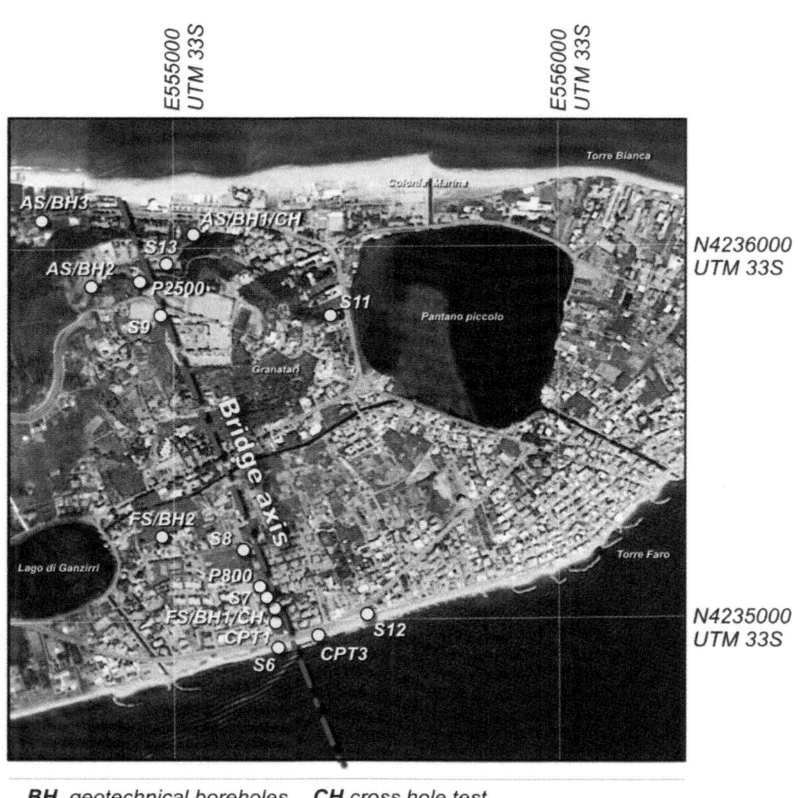

◄ Figure 5.32
Soil investigation on Sicilian shore.

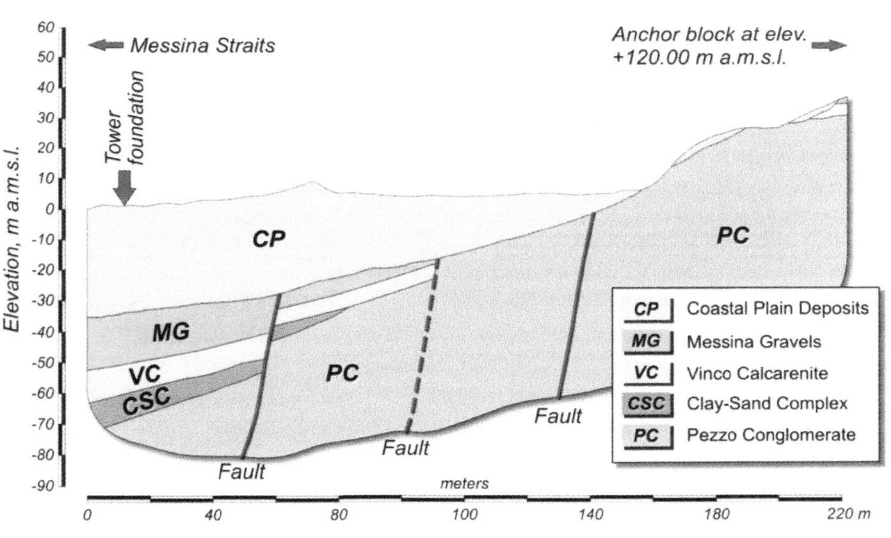

◄ Figure 5.33
Soil profile at Calabrian shore.

The information gained from the BH, C and S borings allowed the identification of the most important soil profiles for the locations of the tower foundations and anchorages, as shown in Figures 5.33 and 5.34.

From Figure 5.34 it can be seen that the Sicily tower is laying on 50 to 70 metres of thick sand and gravel Coastal Plain (CP) deposits, which

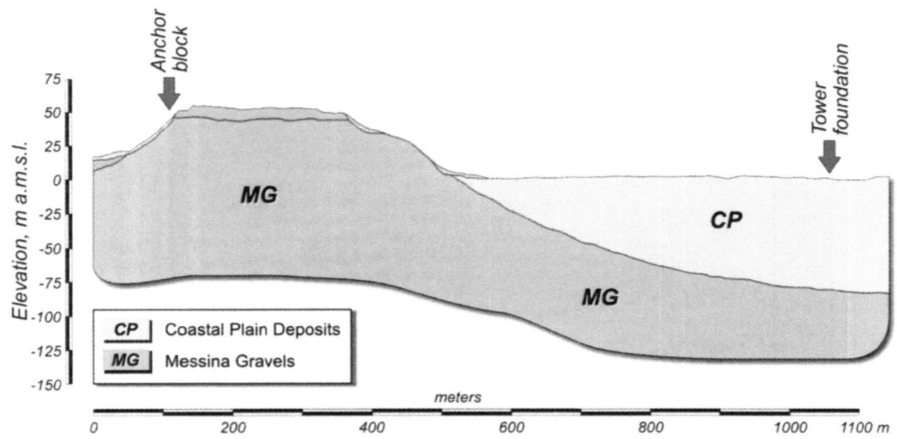

Figure 5.34 ► Soil profile at Sicilian shore.

are of Holocene age, overlaying a deposit of sand and gravel more than 150 metres thick, of mid Pleistocene age, locally named Messina Gravels (MG).

The Calabria tower foundation soil profile (Figure 5.33) appears more complex. Here, the sandy and gravely soils, belonging to the CP and MG formations with a thickness of 35 to 45 metres, overlay the Calcarenite di Vinco (CV) of the upper Pliocene age and the Sandy-Clayey Complex (SCC) of medium to lower Pliocene age. Overall the thickness of such geological units, laying directly on the soft rock formation of large thickness (>150 m) locally named Pezzo Conglomerate (PC), does not exceed about 35 to 50 metres.

As to the soil formation relevant for the anchor block design, the MG and PC materials are found on the Sicilian and the Calabrian sides respectively.

The information obtained from the BH boring series were supported and enhanced by the results of a variety of in situ tests, as listed in Table 5.2. Among them, considering that on both sides of the Strait the foundation designs must deal with difficult-to-sample geomaterials, the following two are worth mentioning:

- The dynamic Large Penetration Test (LPT) which was developed due to concerns that the size of gravely particles can falsify the blow/count obtained from Standard Penetration Tests (SPT).
- The measurements of the shear (V_s) and compression (V_p) wave velocities by means of Cross Hole (CH) tests, which nowadays play a crucial role in geotechnical site characterization. In the examined case, the V_s and V_p measurements permitted the in-situ evaluation of the small strain shear modulus G_0, the porosity n, and an estimate of the bulk density γ.

For the Large Penetration Tests (LPT) the sampler used as the penetration tool had an internal diameter of 110 mm, an outside diameter of 140 mm and housed a plastic liner 5 mm thick. (See figure below) It was driven to penetrate 450 mm below the borehole bottom, using a fall weight (W) of 5298 N and a drop height (H) of 500 mm. Analogously to SPT, the number of blows necessary to penetrate every 300 mm is recorded in LPT tests.

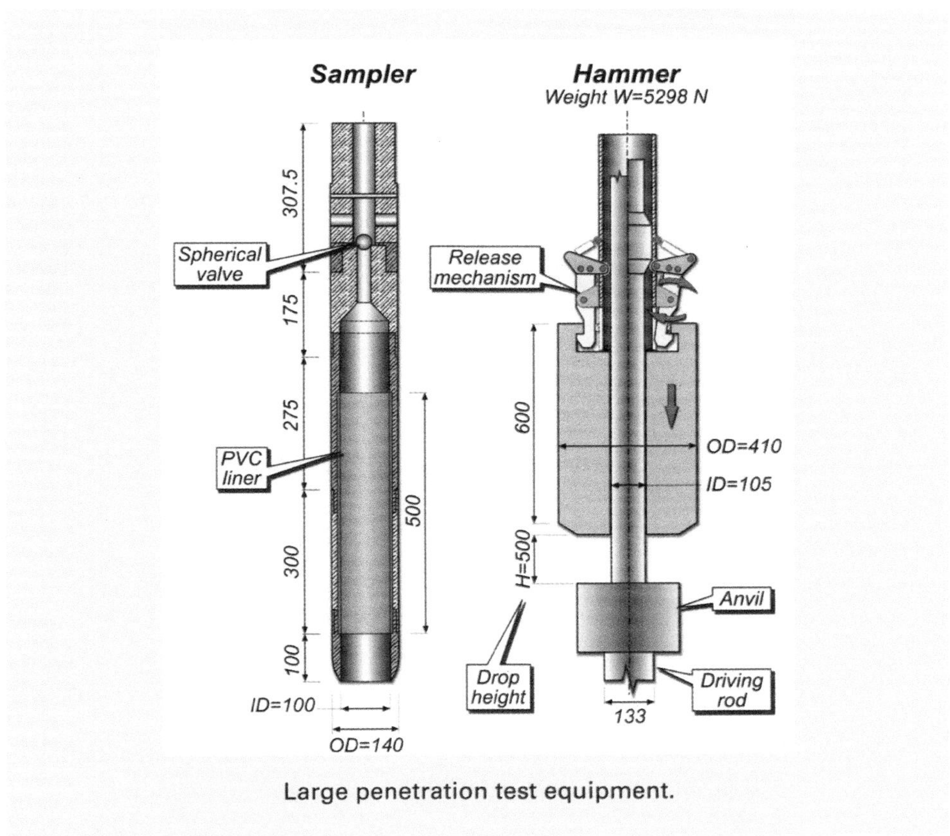

Large penetration test equipment.

To calibrate the LPT and compare its results against SPT, parallel tests with both devices were performed in the well documented uniform Po river sand [46, 47] whose grain size distribution is shown in Figure 5.35. The results of these calibration tests are presented in Figure 5.36.

All SPT's and LPT's, both at the Po river site and the Messina site, were carried out in cased boreholes filled with bentonite slurry and having a diameter of 75 mm and 200 mm respectively. During all the tests, the driving energy

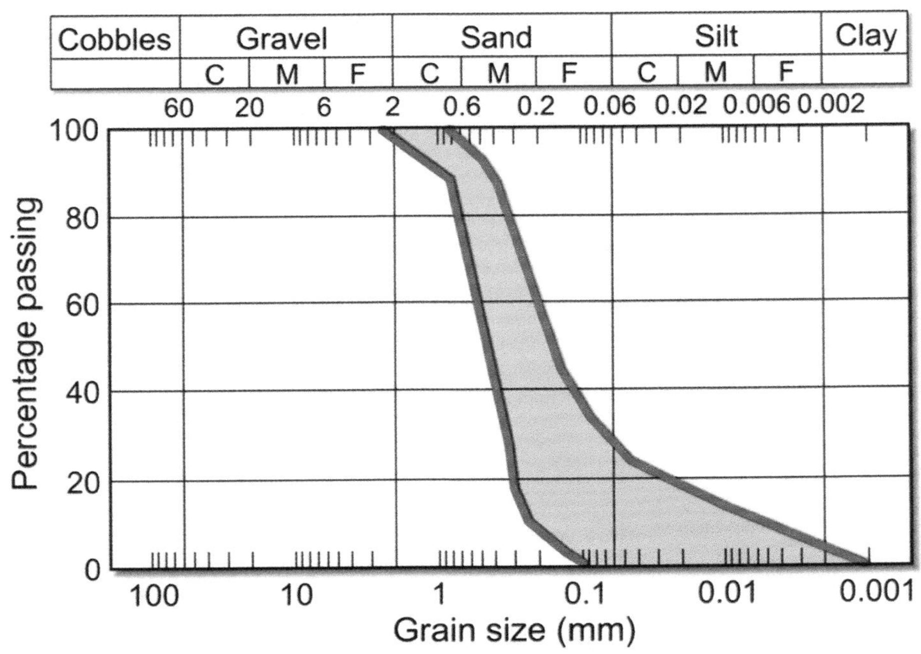

◄ Figure 5.35
Soil grading at Po river site.

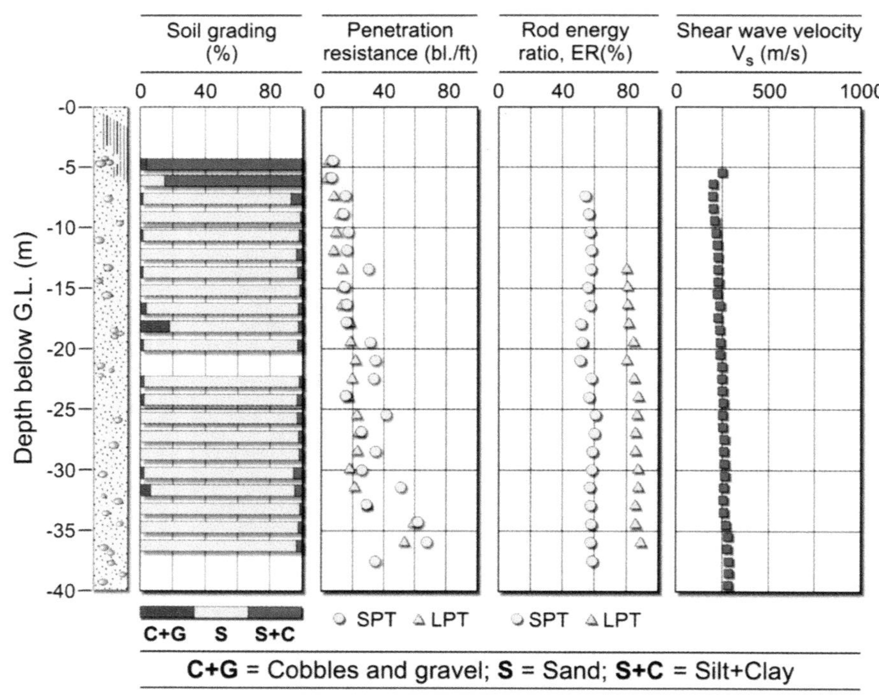

Figure 5.36 ► Soil profile at Po river site.

delivered to the rods was measured using an instrumented 1 metre long rod segment located at the top of the string [48].

Examples of the SPT and LPT results measured at the Messina site are given in Figures 5.37 to 5.40.

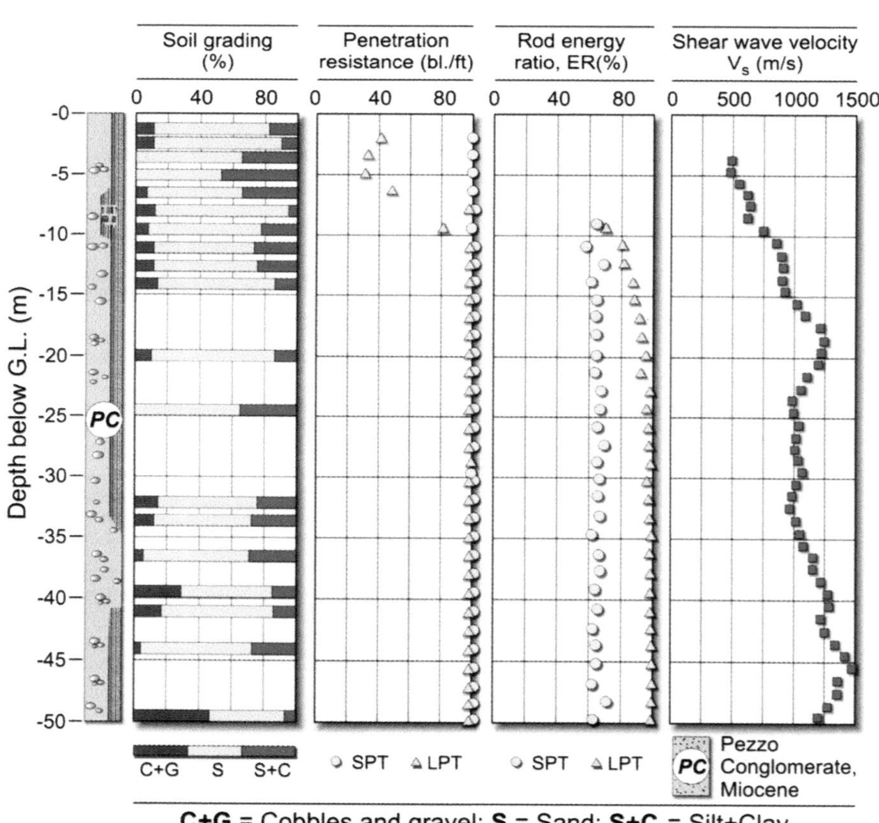

Figure 5.37 ► Soil profile at Calabrian anchor site.

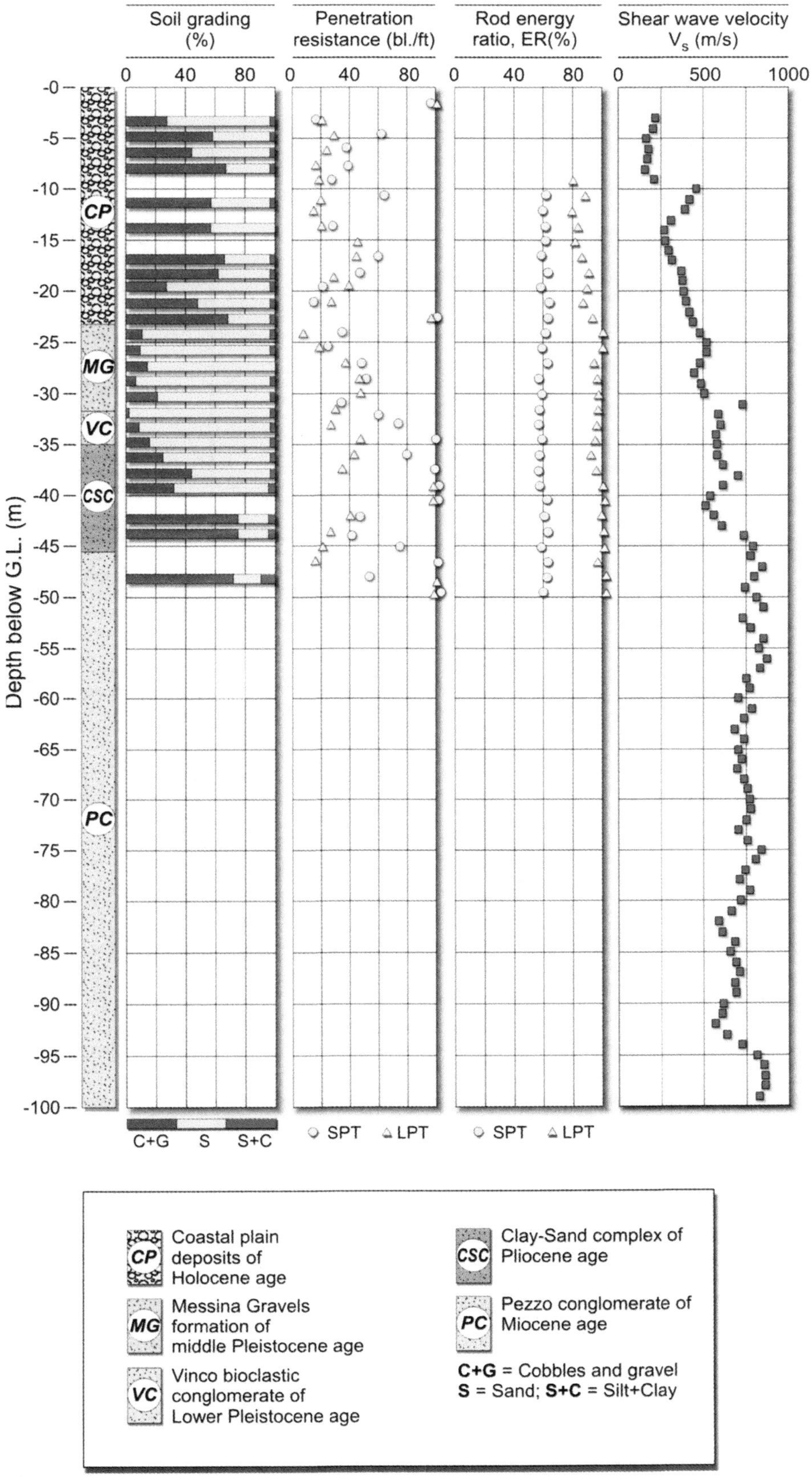

▼ Figure 5.38
Soil profile at Calabrian tower site.

Soil grading (%)
Penetration resistance (bl./ft)
Rod energy ratio, ER(%)
Shear wave velocity V_s (m/s)
Depth below G.L. (m)
CP
MG
VC
CSC
PC
C+G S S+C
SPT LPT
SPT LPT

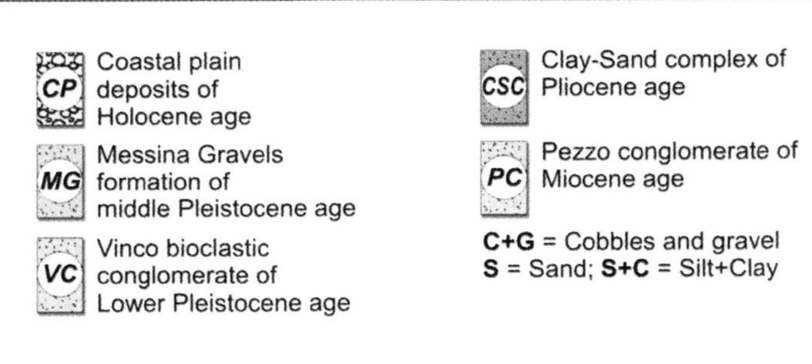

Figure 5.39 ▼
Soil profile at Sicilian anchor site.

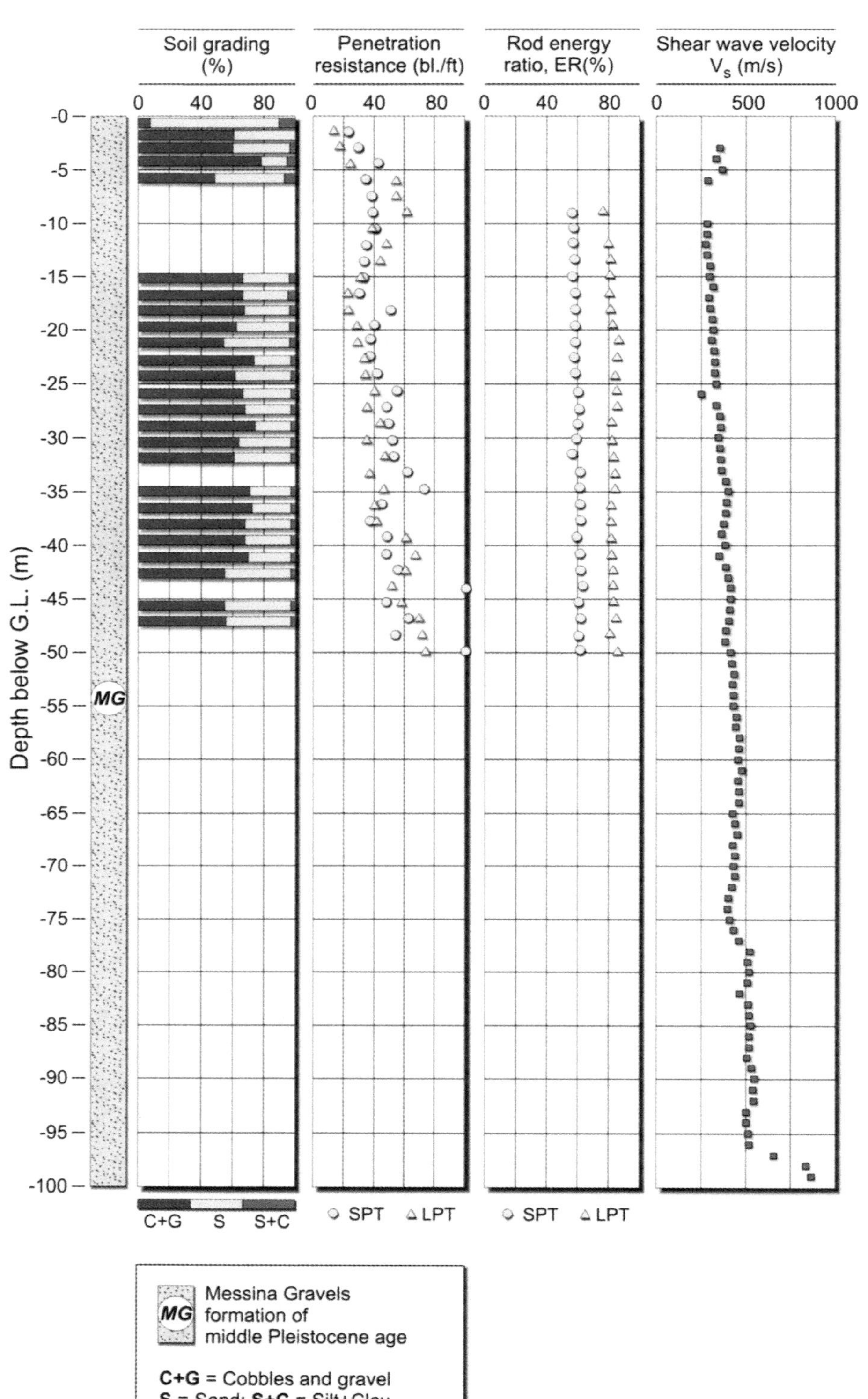

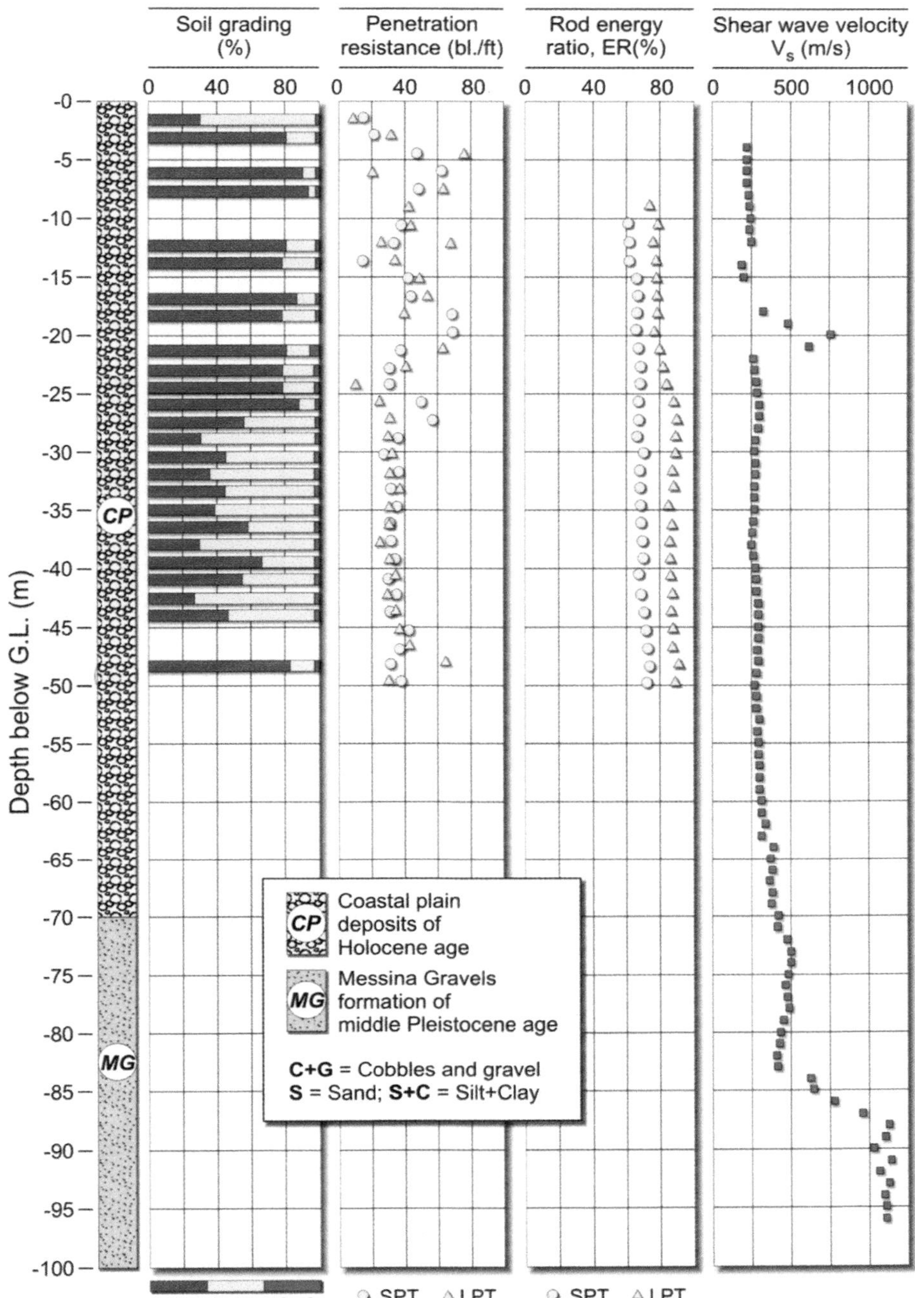

Figure 5.40
Soil profile at Sicilian tower.

Besides the SPT and LPT blow/counts, Figures 5.37 to 5.40 also report the V_s trend with depth together with the energy ratio (ER) for both penetration devices. The ER is defined as the measured driving energy divided by its theoretical value WH. At all the above-mentioned sites, the yielded ER results are quite uniform, ranging between 60 and 65 percent for SPT and 80 to 85 percent for LPT, as shown below.

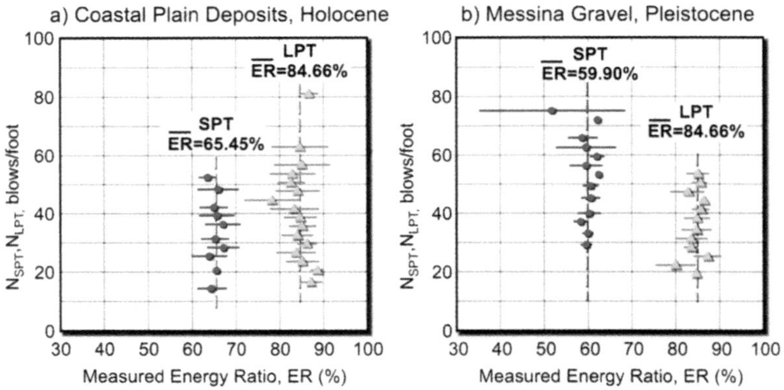

Values of energy ratio measured in Coastal Plain Deposits (a) and in Messina Gravel formation (b).

As to the geophysical tests, they mainly consisted of 100 m deep cross-hole (CH) tests carried out at 1 metre depth intervals. Each test involved three properly equipped boreholes at 5 metre spaces, one housing the source and the others carrying the receivers. This arrangement allowed the V_s and V_p velocities to be computed using a true time interval method which, combined with accurate measurements of the borehole deviations from vertical together with their azimuths, have provided reliable results.

In addition to cross-hole (CH) tests, the Spectral Analysis of Surface Waves (SASW) method [49] was also used to assess the V_s velocity in situ via surface geophysics. The figure below shows the V_s measured in the Messina Gravels at the Sicily anchor site by CH test and using the SASW method. Both yield almost identical values except for the depth between 50 and 55 m where CH tests suggest the presence of a cemented layer not detected by SASW.

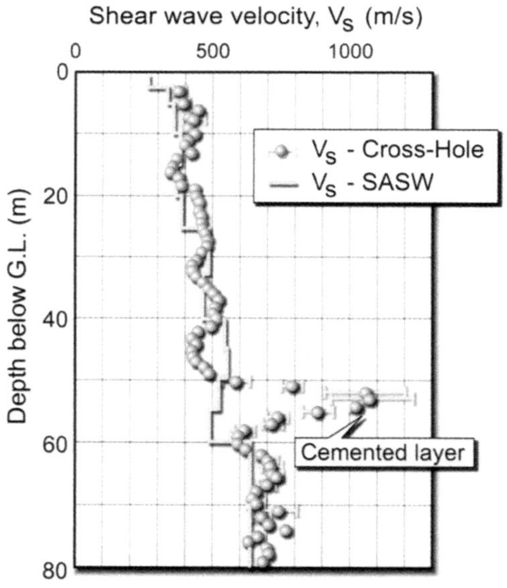

Shear wave velocity measured by two different methods at Sicilian anchor site.

During borings and pumping tests, the ground water level (GWL) conditions at the four locations were determined. At the two tower locations, where the ground level is at an elevation approximately 1 to 3 metres above mean sea level, the GWL is encountered at a depth ranging between about 1 and 2 metres. At the Sicily anchor block location, the GWL is found at about 6 metres below ground level, while on the area of the Calabrian anchor block it is found at around 6 to 10 metres below ground level.

As to laboratory testing, a decision was taken to re-schedule the proposed undisturbed sampling by freezing technique [50, 51, 52, 53] to take place during the final design stage, so only a limited number of laboratory tests have been carried out. Most laboratory tests were aimed at determining the classification and the soil index properties. In addition, some testing was carried out on reconstituted material to determine the sand and gravel stiffness from the CP deposits. The samples of these materials, properly scalped to the maximum grain size of 10 mm, prepared by under-compaction, have been subject to resonant column (RCT) and triaxial compression (TX-C) tests [54, 55, 56].

5.3.3 Engineering properties

The investigation of the engineering properties of the soils encountered across the Messina Strait has been mostly focused on the coarse grained materials belonging to the CP and MG deposits. Their characteristics were used to check the preliminary design of the two tower foundations and the Sicily anchor block, both under static and earthquake generated loadings.

The preliminary design of the Calabria anchor block embedded in the CP formation, where both SPT and LPT always gave refusal and the CH yielded a V_s value higher than 800 m/s, was guided by the outcome of the in-situ tests. A more detailed geotechnical characterization of the PC by seismic and electrical resistivity tomography, as well as through laboratory tests, has been postponed until the final design stage.

Regarding the classification and index property tests, the available information is summarized as follows:

a. The mineralogical composition of CP and MG deposits is similar because both have originated from the disruption and erosion of the same rock formation deposited in a similar, shallow sea and coastal environment. The predominant minerals are feldspar (44%), quartz (25%) followed by mica (2%) and various rock fragments (29%).
b. The grain size characteristics can be inferred from the grading curves shown in Figures 5.41 to 5.43. Moreover, some additional pieces of information on the grading of the CP and MG deposits, as encountered on the Sicily shore, are given in Table 5.3.

   As to PC formation it consists in a succession of horizons, from very weakly to weakly cemented conglomerates and sandstone incorporating rare inclusions of siltstone.

   The cementing matrix of the sand-silt type is rich in filosilicate components.

   The clasts incorporated in the conglomerate and in the sandstone consist of poorly rounded fragments of underlying crystalline bedrock.

   The point load tests carried out on the cores extracted from the borings at the Calabria anchor block yielded values ranging between 0.7 and 1.1 MPa for the conglomerate and between 0.3 and 0.4 MPa for the sandstone.

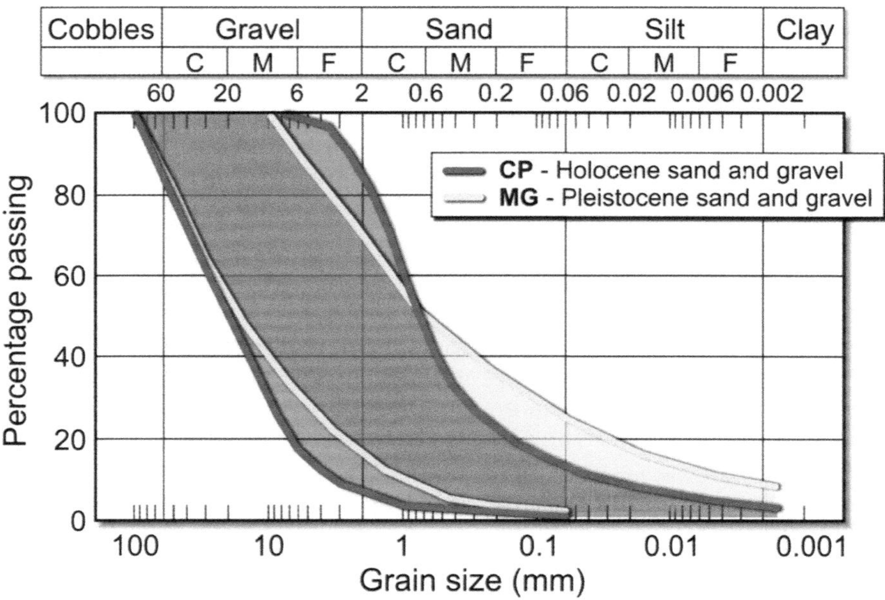

Figure 5.41 ▶ Sicilian tower foundation and anchor block.

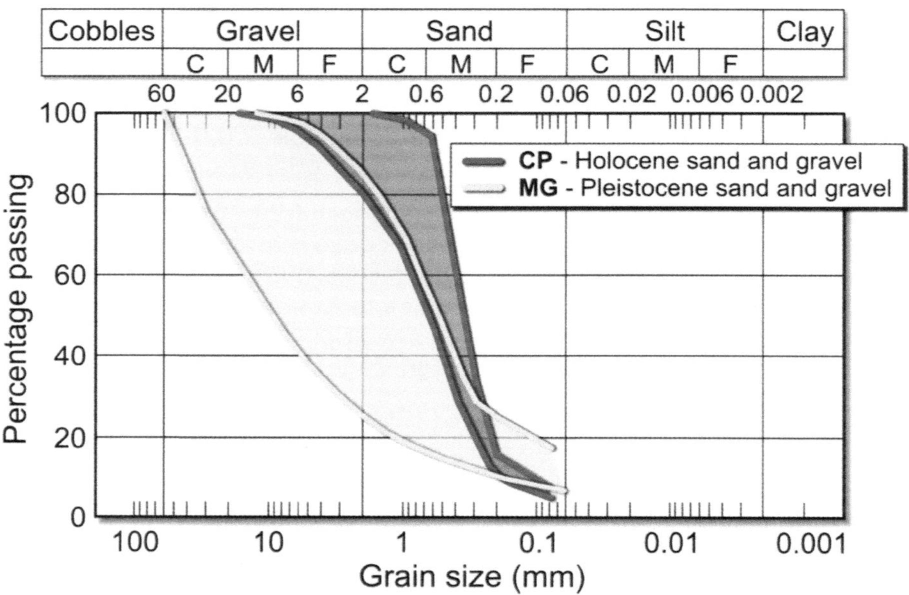

Figure 5.42 ▶ Calabrian tower foundation. Grading of the CP and MG deposits.

c. As to bulk density γ, because of the lack of undisturbed sampling, its determination, at the time of the preliminary design, was limited to measurements carried out in situ during the excavation of the shafts for deep plate loading tests in the MG and PC formations at the Sicily and the Calabria block anchor locations respectively. The obtained values of γ range from 18 to 21 kN/m^3 and from 1.9 to 2.2 kN/m^3 in the MG and PC respectively.

Such values were measured above the phreatic surface in soils whose natural water content W_n varied from 2.5 to 4.5% in the MG and from 4 to 11% in the PC.

More recently, Foti et al. [57] have developed a method allowing the soil porosity (n) to be evaluated in situ based on seismic wave velocity (V_s, V_p) measurements (see box below). The method employs Biot's theory of poroelasticity [58] for low frequency ranges and is applicable to fully saturated geomaterials.

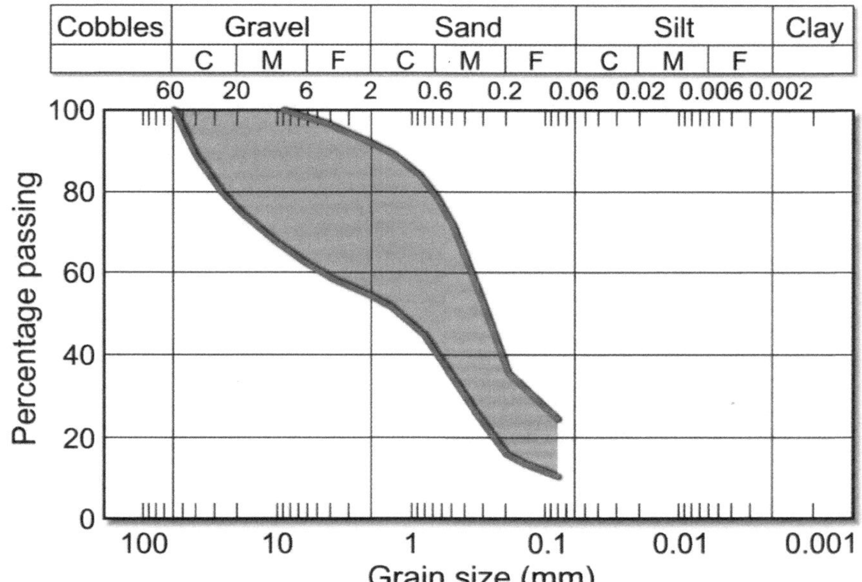

◄ Figure 5.43 Calabrian anchor block. Grading of the disaggregated PC.

▼ Table 5.3 CP and MG sands and gravels; salient features of their grading.

			Gravel (%)		Sand (%)		Fines (%)		D_{50}(mm)		D_{max}(mm)		Tests
			mean	st.dev.	mean	st.dev.	mean	st.dev.	mean	st.dev.	mean	st.dev.	N.
Tower foundation	Holocene CPD	BH	58.0	22.1	34.4	18.4	8.6	6.7	0.53	9.60	47.0	17.1	46
		LPT	65.5	22.5	31.7	21.6	2.8	2.2	0.60	21.30	57.9	24.1	73
		SPT	59.6	17.4	36.4	16.7	4.0	2.1	0.60	15.00	30.1	8.1	103
	Pleistocene MGF	BH	57.1	16.3	36.3	16.4	6.6	3.3	0.92	44.00	56.3	22.2	35
		LPT	85.7	13.2	12.7	12.2	1.7	1.2	7.00	20.00	73.3	26.6	6
		SPT	56.3	11.9	39.0	12.6	4.7	1.0	1.00	11.00	38.3	7.5	6
Anchor block	Pleistocene MGF	BH	56.9	14.9	33.5	11.9	9.6	6.7	0.20	12.00	39.9	12.4	74
		LPT	67.0	11.5	30.4	6.6	4.0	2.8	0.54	36.00	57.0	22.1	115
		SPT	57.1	9.8	36.5	9.3	6.4	2.1	0.95	7.50	31.3	8.2	117

The formula to compute soil porosity n is as follows:

$$n = \frac{\rho_s - \left[\rho_s^2 - \dfrac{4(\rho_s - \rho_f) K_f}{V_p^2 - 2\left(\dfrac{1-\nu_s}{1-2\nu_s}\right) V_s^2} \right]^{0.5}}{2(\rho_s - \rho_f)} \tag{5}$$

where
n = connected porosity
ρ_s = soil particles (mass density)
ρ_f = pore fluid (mass density)
K_f = bulk modulus of pore fluid
ν_s = Poisson ratio of soil skeleton

The above equation was validated in a number of fine grained saturated deposits where the computed values of porosity or void ratio (e) were compared against values measured in the laboratory on high quality undisturbed samples [59, 60].

Table 5.4 gives typical values of the void ratio (e) evaluated below the phreatic surface for the Strait deposits based on Cross Hole (CH) tests results. The table also provides an estimate of γ assuming full saturation and adopting $\rho_s = 2.67$.

Table 5.4 ► Void ratio and bulk density from seismic tests[(1)].

Formation	Void ratio[(2)] e (-)	Bulk density[(3)] γ (kN/m³)	Depth range (m)
Coastal plain deposits	0.35 to 0.42	21.0 to 21.8	5 to 60
Messina gravels	0.28 to 0.40	21.0 to 22.1	20 to 100
Pezzo conglomerate	0.15 to 0.20	22.5 to 23.8	45 to 100[(4)]

Notes: (1) Cross hole tests;
(2) By means of equation (1), Foti et al. [57];
(3) From computed e, assuming full saturation and adopting $G_s = 2.67$;
(4) At Calabria tower location.

d. During the preliminary design stage, in the early eighties, the estimate of the relative density (D_r) in situ for the sandy and gravely soils was inferred from SPT results, based on the empirical correlation developed for sands and given by Skempton [61]. In the late nineties, Cubrinovski and Ishihara [62] developed a specific empirical correlation for gravely geomaterials which was calibrated in Japan against the relative density of undisturbed samples retrieved by means of in-situ soil freezing. The correlation assumes that the Energy Ratio (ER) is equal to 78%, typical in the Japanese practice. Consequently, the N_{SPT} and N_{LPT} values used to compute D_r were corrected by multiplying the blow/count by the measured Energy Ratio divided by 78%.

The relationships allowing the estimate of D_r, suggested by Cubrinovski and Ishihara [62] are the following:

$$D_r^2 = \frac{(N_1)_{78}}{C_d} \quad (6a)$$

$$C_d = \frac{9}{e_{max} - e_{min}} \quad (6b)$$

$$e_{max} - e_{min} = 0.23 + \frac{0.06}{D_{50}} \quad (6c)$$

$$N_1 = \left(\frac{p_a}{\sigma'_{vo}}\right)^{0.5} N_{SPT} \quad (6d)$$

$$(N_1)_{78} = \frac{ER}{78} N_1 \quad (6e)$$

where:

N_1 = SPT and LPT blow/count normalized according to Liao and Whitman (1986) with respect to effective overburden stress σ'_{vo}
p_a = reference stress equal to 98.1 kPa
ER = measured energy ratio in percent
e_{max} = maximum void ratio
e_{min} = minimum void ratio
D_{50} = mean grain size in mm

Figures 5.44 and 5.45 show the comparison between values of D_r evaluated by means of formulae 6a to 6e. The results suggest that the approach by Skempton [61] tends to overestimate D_r for sands in gravely soils. However, the issue is whether this overestimate is as pronounced as it seems from Figures 5.44 and 5.45. We will know the answer to this question when undisturbed samples retrieved from the Strait sites by means of the freezing technique become available.

For the time being, the values of the relative density, to be considered as actual for the geomaterials in question, should be those obtained using the Cubrinovski and Ishihara [62] approach, giving values of D_r between 40 and 60% for both CP and MG deposits.

Regarding the assessment of the stress-strain, strength and hydraulic conductivity characteristics for the preliminary design, considering the coarse grained nature of geomaterials encountered, it was mostly based on the results of in-situ tests with some limited contribution from the laboratory tests on reconstituted specimens. The available information can be summarized as follows:

a) The small strain shear modulus (G_0) has been obtained from the value of V_s measured during CH tests by means of the following formula:

$$G_0 = V_s^2 \rho \quad (7)$$

The value of shear modulus yielded by Equation 7 holds for the shear strain level $\gamma < 1 \times 10^{-5}$ below which the stress-strain relationship can be assumed to be linear [56, 63, 65, 66, 68].

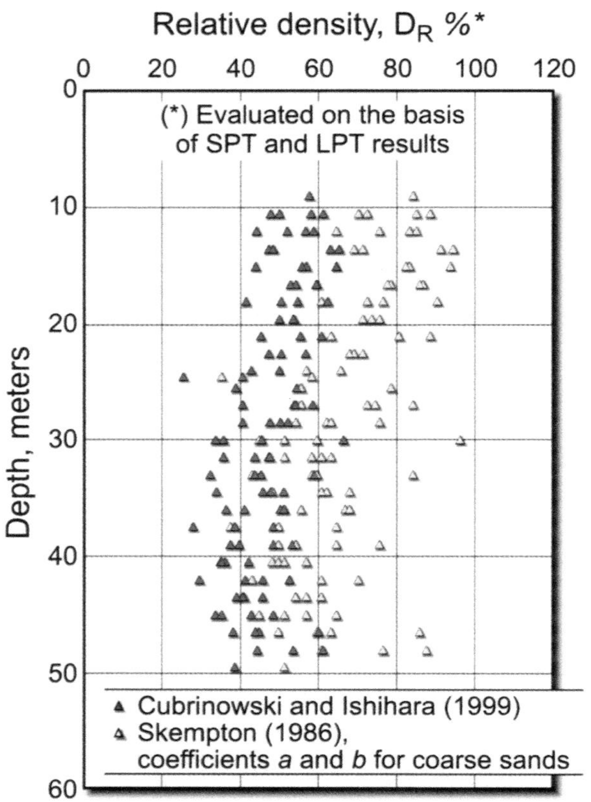

Figure 5.44 ▶ Relative density of Coastal Plain deposit.

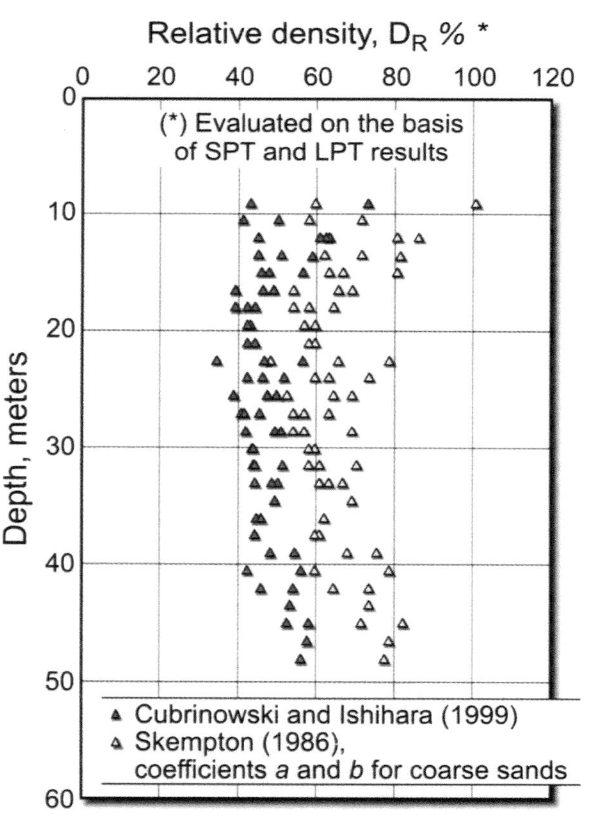

Figure 5.45 ▶ Relative density of Messina Gravel formation.

At larger strains, both under monotonic and cyclic loading, the shear modulus is subject to degradation reflecting the non-linearity of the soil behaviour [65, 66, 67, 68].

The relationship $G = f(\gamma)$ has been investigated by means of the RC and TX-C tests on reconstituted specimens of the CP deposits having values of D_r ranging between 50 and 60% [54, 55]. Figure 5.46 shows the variations of shear modulus G normalized with respect to G_0 (being a value of shear modulus at $\gamma < 1 \times 10^{-5}$) and shear damping ratio D in relation to shear strain γ, as obtained from the above mentioned laboratory studies.

Unfortunately, the shear modulus values obtained on reconstituted specimens, although apparently tested under similar stress and density conditions that exist in-situ, differ from those in an undisturbed state because they do not reflect the influence of depositional and post-depositional phenomena, such as structure, ageing, or even slight cementation that all natural deposits to a lesser or larger extent exhibit [69, 70].

An evident proof is represented by the comparison of V_s measured in CP and MG deposits of Holocene and medium Pleistocene age respectively. As already mentioned, these two soils were deposited in a similar environment and exhibit similar mineralogical composition, grading and penetration resistance (see Figure 5.47), but they differ significantly in age by about 600,000 years.

The reasons for the above differences in V_s, and hence in G_0, can be explained by the amount of ageing experienced by the MG as compared to the CP deposits, and by the presence of several forms of very light

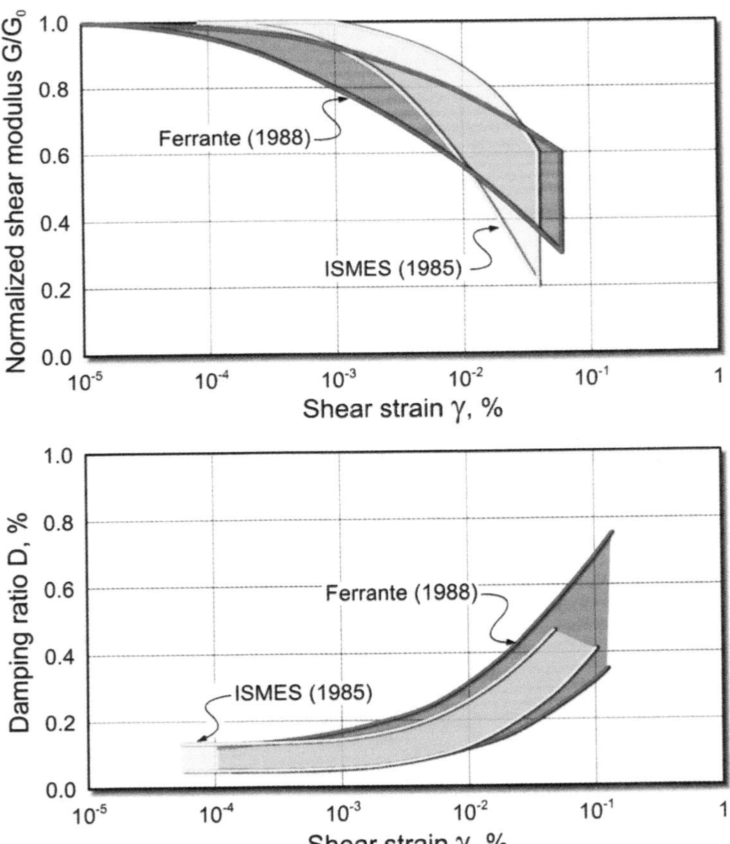

◀ Figure 5.46
Results of RC tests on reconstituted sand and gravel.

Figure 5.47 ▶ Comparison of CP versus MG deposits.

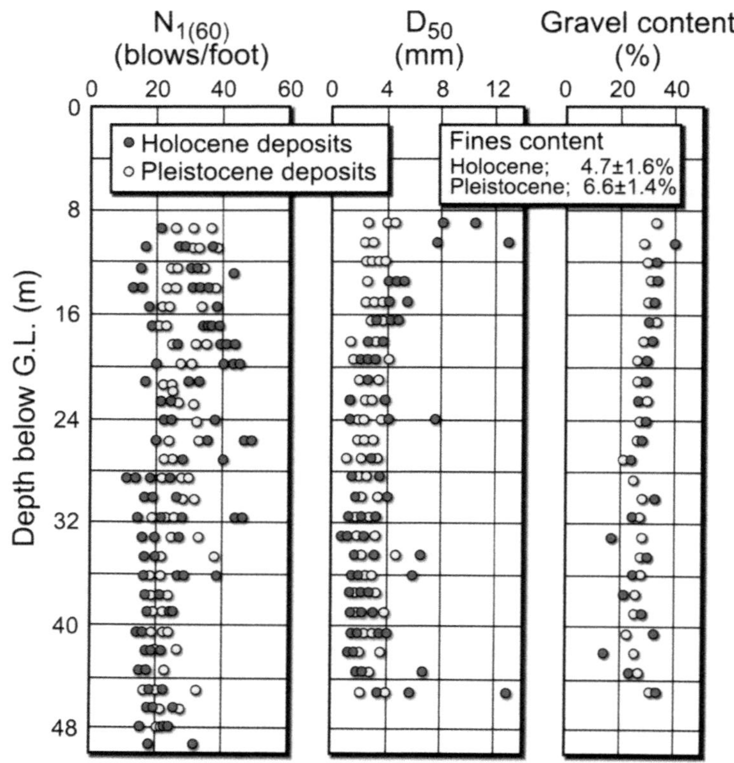

cementation (see Figure 5.48) discovered in outcrops of the MG formation but completely absent in the CP deposits [71, 72].

In view of the above, the degradation curve of strain dependent shear modulus in the field can be estimated for the time being as suggested by Ishihara [73], benefiting from the plentiful experimental evidence [73, 74, 75, 76, 77, 78] that the G/G_0 vs, $\log_{10}\gamma$ curves of undisturbed coarse grained soils are similar in shape to those of the same material in reconstituted state (see for example Figure 5.49). This means that the G_0 vs. $\log_{10}\gamma$ curve determined in the laboratory can be taken as the field degradation curve, normalizing the G value with respect to the field G_0 value inferred from the V_s measured in-situ.

Figure 5.48 ▶ Types of light cementation encountered in MG formation.

Adapted from Bosi, C. (1990) for Stretto di Messina S.p.A.

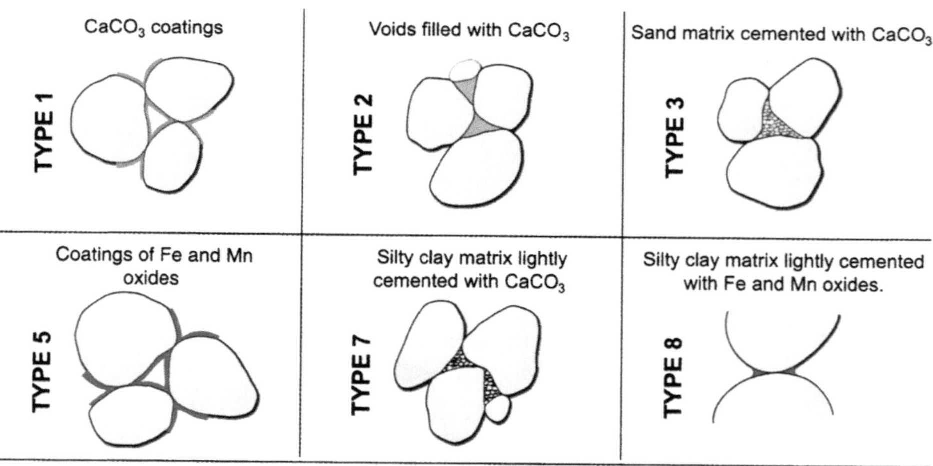

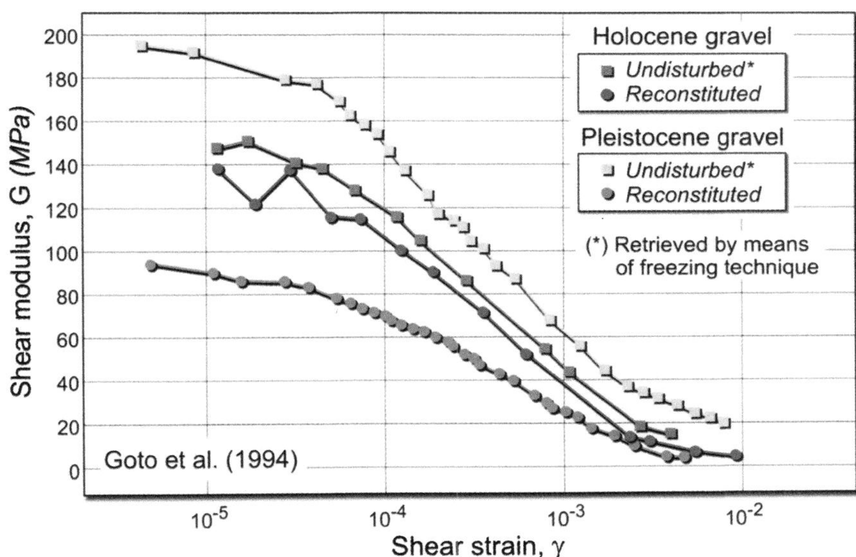

◀ Figure 5.49 Strain dependent shear modulus; undisturbed vs. reconstituted.

Some additional information about the non-linearity of the shear modulus can be deduced from the results of loading tests performed on an 800 mm diameter plate at the bottom of 2.5 metre diameter shafts which were carried out at the anchor block locations. Although the interpretation of such tests is far from being straightforward, especially, because of the large difference between shaft and plate diameter, it is worth reporting the ratios of unload-reload modulus G_{ur} inferred form the plate loading tests normalized with respect to G_0 obtained at the same depth as the V_s. The ratio of G_{ur}/G_0 falls in the following range:

Calabria anchor; $0.40 < G_{ur}/G_0 < 0.48$, for $0.05\% < s/D < 0.20\%$
Sicily anchor; $0.72 < G_{ur}/G_0 < 0.93$, for $0.18\% < s/D < 0.24\%$
where:
s = plate settlement
D = plate diameter.

b) The drained shear strength has been estimated by referring to empirical correlations between large dynamic penetration tests (LPT) and the peak angle of shearing resistance ϕ', for example as proposed by Hatanaka and Uchida [79].

According to Hatanaka and Uchida [79], the ϕ' value of gravely soil is given by the following relationship:

$$\phi = C + (20\ N_1)^{0.5}{}_{LPT} \qquad (8)$$

where:

N_1 = LPT blow-count normalized with respect to the vertical effective stress σ'_{vo},
C = a constant ranging between 17 and 23.

According to these authors Equation 8 holds also for $(N_1)_{SPT}$ values.

However, it is necessary to recall the readers' attention that Equation 8 has been set up according to Japanese practice. Therefore, the N_{LPT} and N_{SPT} values used in connection with the correlation in question should be referred to a value of ER = 78%. Using the results of LPT and SPT obtained in the CP and MG deposits, Jamiolkowski et al. [56] found $37° < \phi' < 47°$.

An alternative approach to evaluating ϕ' for coarse grained soils is contained in Bolton's work [80] that proposed the following equation from which ϕ can be estimated:

$$\phi' = \phi_{cv} = m\,[D_R(Q - \ln p'_f)] - 1 \qquad (9)$$

where:

ϕ'_{cv} = constant volume angle of friction in degrees
m = dimensionless coefficient equal to 3 or 5 for axisymmetric and plane stress conditions
Q = particle strength parameter
D_r = relative density
p_f' = mean effective stress at failure in kPa.

Keeping in mind the mineralogical composition of the CP and MG deposits, the parameters that can be used in association with Equation 9 are:

Q = 9 to 10,
ϕ'_{cv} = 35°,
D_r = 60%, and

p'_f in the range of practical interest, between 250 and 500 kPa.

With this assumption Equation 9 yields $39° < \phi' < 42°$.

Unlike Equation 8, Equation 9 incorporates the framework of the Rowe's stress-dilatancy theory [81], rendering the ϕ' value dependent on the stress level at failure, thus implicitly incorporating the features of a curvilinear shear strength envelope.

c) The hydraulic conductivity of the CP deposits was evaluated at the two tower locations by pumping from wells and observing the ground water level drawdown in the surrounding piezometers. Pumping wells were used which penetrated the CP deposits to depths of 25 m and 35 m on the Calabria and Sicily shores respectively and had filters along their entire length, with piezometers located at 10 m and 20 m from the wells.

The hydrological conditions considered in the interpretation of the tests were an unconfined aquifer with partially penetrating wells and the phreatic surface coincident with the mean sea level.

Pumping was carried out until the attainment of a flow rate of 0.08 m^3/s in steady state conditions corresponding to the drawdown in the wells of 0.7 metres. In these conditions, pumping was carried on for 72 hours, monitoring changes in sea level and ground water level during this time.

The test results were interpreted in accordance with Todd [82], disregarding, on account of the modest drawdown, the horizontal flow components. The information assumed from the pumping tests is shown in Table 5.5.

Table 5.5 ▶ Hydraulic conductivity from pumping tests.

Location	$T \cdot 10^{-3}$ (m^2/s)	$k_h \cdot 10^{-3}$ (m^2/s)	$S \cdot 10^{-3}$ (–)	Gravel + cobbles (%)	Fines (%)
Sicily shore	65	5.0	1.7	44 to 79	3 to 8
Calabria shore	65	2.6	1.7	12 to 80	1 to 16

T = Transmittivity; S = Storage coefficient.

The information given in this section enables the reader to become acquainted with some of the key issues regarding the foundation soils, as well as the techniques used for their characterization so that the geotechnical parameters the for Messina Strait Bridge foundation design could be acquired.

5.4 Wind

5.4.1 Introduction

The evaluation of the design wind speed for the Messina Strait Bridge is a complex aspect which is critical to the design, construction, operation and safety of the work. This consideration has given rise to many activities over the years, the first of which was the design and construction of a meteorological station at Punta Faro [83] at the northern end of the Strait. Using the data provided by this and other neighbouring stations, together with numerical simulations of the surrounding area, the Politecnico di Milano and the University of Genoa developed in 1991 the first derivation of the design wind speed for use in the 1992 preliminary design [84].

In 2003, with reference to technological developments in the intervening decade and new available data, a further study was carried out by the University of Genoa in order to improve the earlier derivation. The results of this study are discussed below under seven headings: analysis of the data measured at the meteorological station of Torre Faro; probabilistic analysis of data measured at meteorological stations close to the bridge; implementation of a digital model of the Strait area; evaluation of the scenarios of the mean wind speed, representative of the local climatology; probabilistic analysis of the mean wind speed at the bridge; evaluation of the turbulence fields at the bridge; simulation of space-time variation of the wind speed at the bridge.

5.4.2 Analysis of the data measured by Torre Faro station

The meteorological station at Torre Faro (Figure 5.50) was operational on both the sides of the Messina Strait between 1985 and 1994. Since 1994, only the sensors on the Sicilian side have been operational.

◄ Figure 5.50 Location of Sicilian meteorological station of Torre Faro.

The present study has been carried out examining the data recorded by six anemometers: three Gill anemometers are put at 64, 78, and 92 m above ground level (agl), on a tower previously used by ENEL (the Italian national company responsible for electricity production) to operate the overhead electrical link between Sicily and the mainland, later dismissed by ENEL and used by the Meteo Centre; and three cup anemometers, one on the same tower at 128 m and two on an independent mast at 7 and 15 m above ground level. The placement of these anemometers has both advantages and disadvantages. Considering the four anemometers on the ENEL tower, on one hand they are at altitudes which are representative of the bridge, but on the other hand they are each affected, depending on the wind direction, by shielding effects caused by the structure of the tower itself and its inspection ladder. Thus, they only provide reliable data for certain unshielded wind directions. Similarly, the anemometer on the independent mast at 7 m suffers the shielding effect of a boundary wall for a certain wind directions. Therefore, apart from the anemometer at 15 m on the mast, the data obtained from the other five anemometers is reliable only for certain wind directions, and this means that not all the data are suitable for complete and reliable probabilistic analysis.

Therefore, a detailed analysis of the data, including a thorough comparison of wind speeds and directions at different positions, played a crucial role in the identification of data points possibly affected by measurement, transduction, and/or storage errors. These errors were firstly identified and then removed through a process of selection and control. At the same time, correlation and regression laws were developed to simulate not measured values on the basis of the recorded data. This led to the production of a corrected database that is new, richer, and free from error.

Figure 5.51 shows the joint frequencies of the horizontal components of the mean wind speed for two Gill anemometers at 64 and 92 m above ground level. They demonstrate that the prevailing wind comes from directions between 180° and 210°. Such directions are oblique with respect to the bridge axis (also marked on the figure) so do not give rise to maximum wind loading effects which are related to a direction normal to the bridge. Nevertheless, the design has assumed the maximum wind speed to apply normal to the bridge, providing, in effect, an additional safety margin.

The analysis of the atmospheric turbulence was developed with reference to the wind speeds recorded by the Gill anemometers. First, stationary records corresponding to unshielded directions were selected. Then, longitudinal (horizontal and parallel to the wind direction), lateral (horizontal and normal

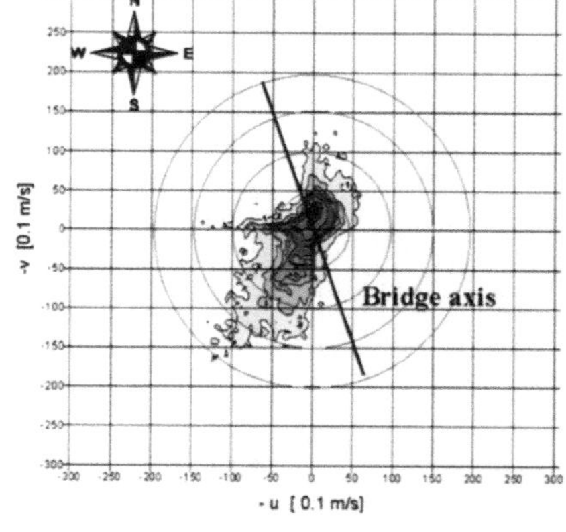

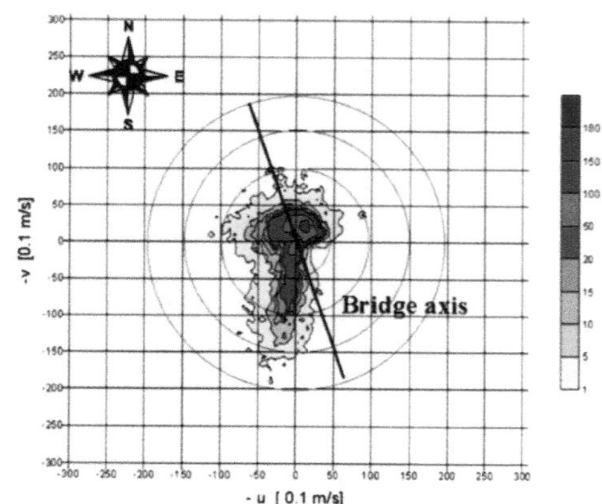

Figure 5.51 ▶ Joint frequencies of the horizontal components of the mean wind speed recorded by Gill anemometers at 92 m (a) and 64 m (b) above ground level.

to the wind direction) and vertical turbulence histories were extracted. The selected turbulence histories were grouped into homogeneous families, each related to wind directional sectors characterized by uniform terrain roughness. The turbulence histories in each family were analysed to extract turbulence intensities, power spectral densities and integral length scales. Finally, the numerical models best suited to represent these quantities were selected and calibrated. Figure 5.52 compares the theoretical power spectral densities proposed in [85] with a few experimental results.

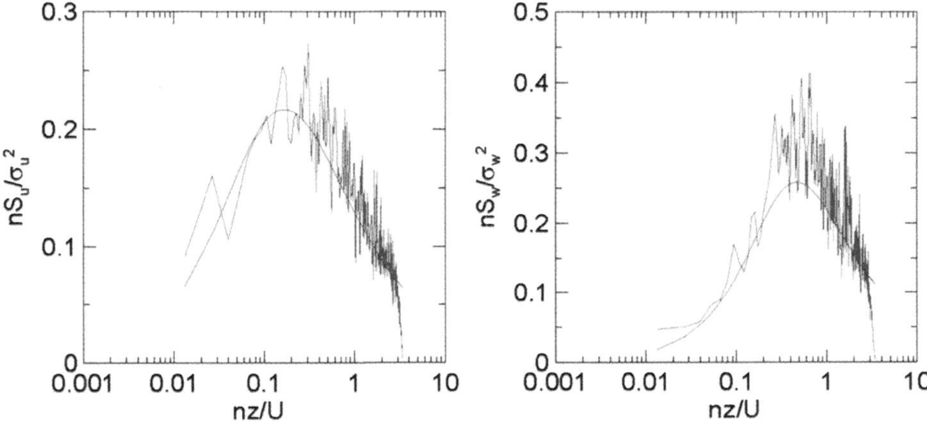

◄ Figure 5.52 Theoretical and measured power spectral densities, wind from 180° – 220° direction sector, Gill anemometer at 92 m above ground level: longitudinal left and vertical right turbulence components.

5.4.3 Probabilistic analysis of data measured by stations close to the bridge

As highlighted by the previous section, the data recorded by the meteorological station of Torre Faro is not enough to fully describe the wind climatology of the Messina Strait in a robust and exhaustive way, especially in relation to extreme winds. In particular, the data obtained from the sole reliable sensor from a probabilistic viewpoint covers a period which is relatively short for the aims of this study. Therefore, the available data was increased through measurements coming from neighbouring stations. Figure 5.53 shows the anemometers in the Strait area (white stars). The stations of Messina, Reggio Calabria and Catania (red stars) are the only ones with a long historical record.

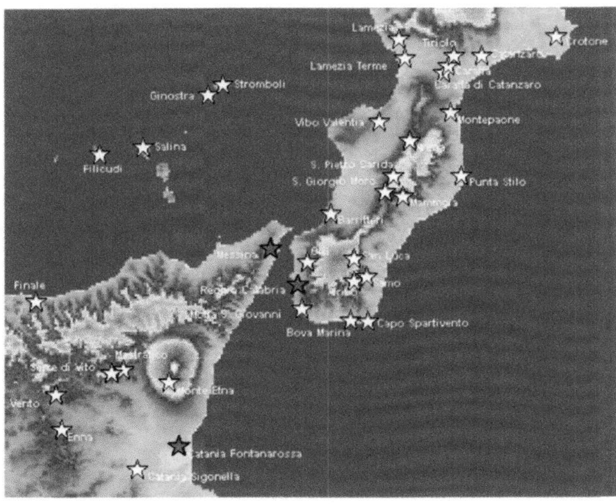

◄ Figure 5.53 Anemometers located in the Messina Strait area.

Unfortunately, these stations are also not fully reliable. The anemometer mast of the Messina station stands on a building roof in the town centre, the Reggio Calabria anemometer does not record any data during most night-times, and although the anemometer of the Catania station is ideal it is very far from the bridge. Thus, there was a need to perform probabilistic analysis on all the data, keeping an essential margin for a critical judgement.

Based on this principle, the probabilistic analysis of the mean wind speed was carried out considering eight instruments: five of the six anemometers at Torre Faro (the 7 m anemometer on the mast was neglected) and the three anemometers at Messina, Reggio Calabria and Catania. For each of these instruments the probabilistic analysis of the current and extreme values was performed using the methods described in [86, 87].

The current values of the mean wind speed and direction were subjected to directional and non-directional analyses using a Weibull model. Figure 5.54 shows the exceedance probability of the mean wind speed, using different regression criteria.

The yearly maxima of the mean speed were subjected to non-directional analyses by using the I-type asymptotic distribution, the process analysis, and the Pareto analysis. The I-type or Gumbel distribution is used in most codes, and usually leads to overestimates that are higher and higher as the return period grows [88]. The Pareto distribution was recently debated in a critical way [89], and often provides reduced estimates for high return periods. The process analysis produces results that are usually reliable and intermediate between the I-type and the Pareto distributions, especially for high return periods. Figure 5.55 shows that this behaviour is exaggerated at the Torre Faro station because of the limited duration of the available data.

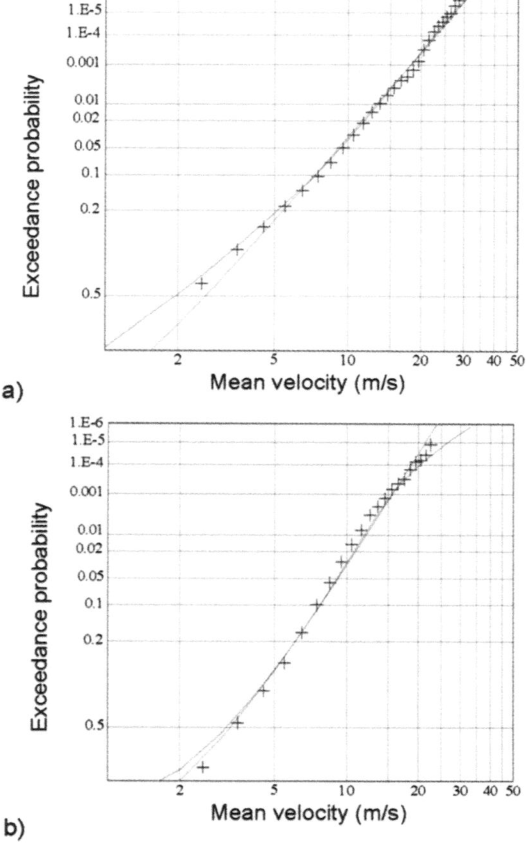

Figure 5.54 ▶ Exceedance probability of the mean wind speed: (a) Met One on the Torre Faro mast at 15 m above ground level; (b) Reggio Calabria station.

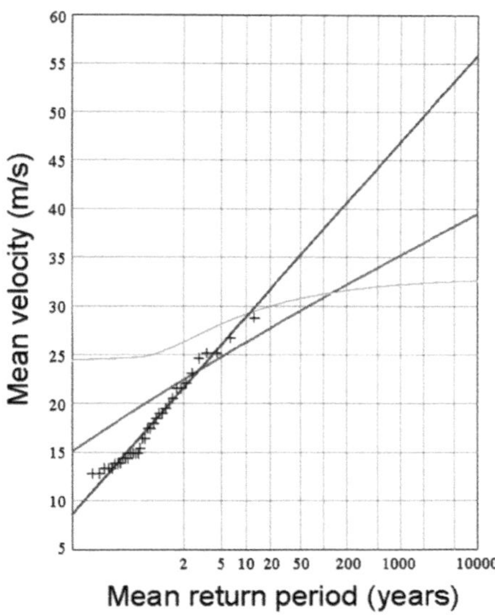

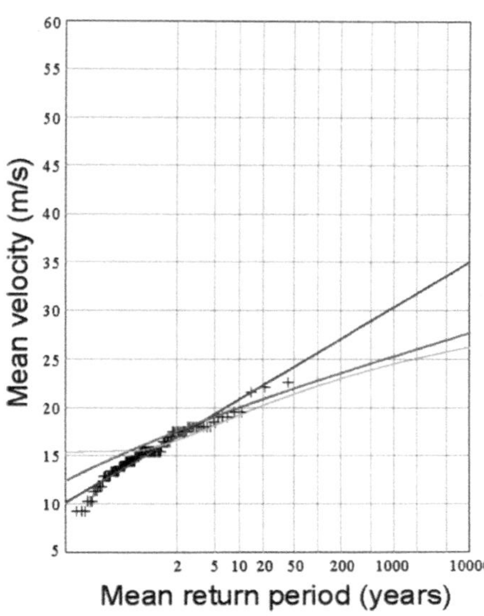

◀ Figure 5.55 Distribution of the yearly maximum mean wind speed: (left) Met One of the Torre Faro mast at 15 m above ground level; (right) Reggio Calabria station.

Although measurements from the Catania station cannot be considered representative of the Strait area, they were used to investigate the reliability of the Reggio Calabria station which does not record measurements during most of the night-time. This check showed that the Reggio Calabria data can be used correctly, if it is suitably interpreted.

5.4.4 Implementation of a digital terrain model

The assessment of the wind climate at the bridge site requires the definition of suitable mathematical methods to relate the results of the probabilistic analyses from the anemometer sites to the bridge site. This required a numerical model of the surrounding terrain to be built, encompassing the anemometer sites, the bridge area and all the topographic features affecting the wind at the bridge site.

Two digital models of the topography and the terrain roughness were created at two different spatial scales relative to two different simulation domains referred to as "macro-area" and "micro-area" (Figure 5.56).

The macro-area (shown dark in Figure 5.56) contains the main features affecting the wind speed and direction in the bridge area, therefore including a large portion of the sea, Mount Etna and Aspromonte. This area extends approximately 173 km in the West-East direction and 165 km in the North-South direction. The grid step is about 963 m in longitude and 920 m in latitude.

The micro-area (shown white) is limited to a small area around the bridge, including the considered meteorological stations (with the exception of Catania). It extends approximately 38.5 km in the West-East direction, and 50.6 km in the North-South direction. The grid step is about 243 m in longitude and 230 m in latitude.

The terrain topography in Figure 5.56 was derived from computerized tables compiled by the Italian Military Geographic Institute. Ground roughness was deduced from the digitized CORINE Land Cover database, through suitable conversion tables. Figure 5.57 shows the roughness scale of the terrain in the micro-area.

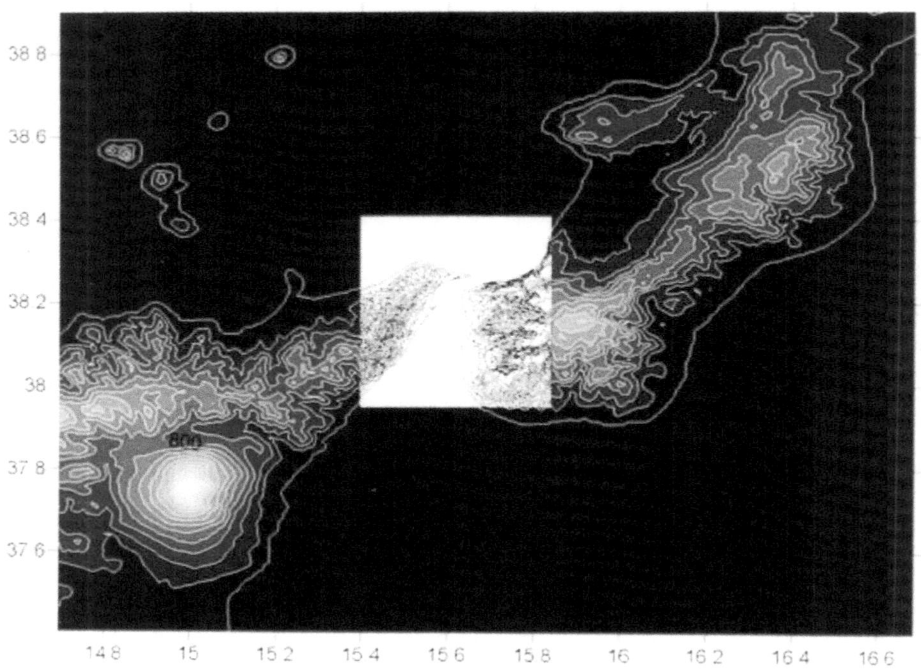

Figure 5.56 ▶ Digital models of the macro-area (dark) and micro-area (white).

To include the topography features in the model areas, the vertical size of the calculation volume was set equal to 5,000 m above sea level. This volume was discretised through surfaces that become denser close to the ground. The lowest of these follow the terrain surface, while they become flatter and flatter towards the top of the domain.

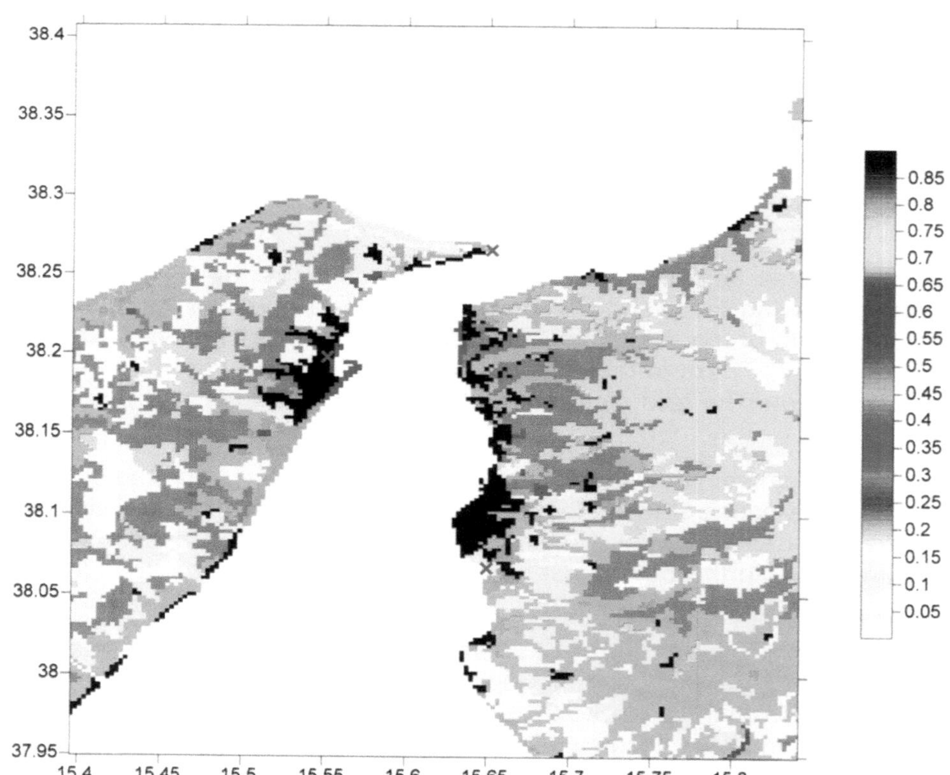

Figure 5.57 ▶ Roughness length of the terrain (m) in the micro-area.

A post-processor was used to extract the mean wind speed and direction for a set of nodes in the micro-area corresponding to the bridge deck, the towers, the cables and the considered anemometers.

5.4.5 Simulation of mean wind velocity scenarios

The simulation of mean wind speed scenarios in the Messina Strait area was performed by the mass-consistent numerical code WINDS [90], schematising the main characteristics of the terrain by the digital models described above.

The simulations were carried out by assigning uniform values of wind speed and direction at the top of the calculation volume. Six values of the wind speed (between 5.3 and 85 m/s) and 16 values of the wind direction (every 20°) were selected. All these configurations were studied under the hypothesis of a neutral atmospheric stratification, which is typical of intense wind speeds. Furthermore, cases of moderate mean wind speeds were studied assuming stable and unstable atmospheric regimes on the land. Altogether, 160 simulations were carried out. Figure 5.58 shows a typical scenario at 10 m above ground level over the micro-area.

Figure 5.59 shows the derived profiles of the neutral mean wind speed along the deck axis, normalized with respect to the mid-span for different wind directions. The origin of the abscissa corresponds to the axis of the Sicilian tower. The mean wind speed is greater in the main span where friction between air and sea is minimal, compared to the side-spans where the flow is slowed down due to terrain roughness and topographic effects, especially on the Calabrian side.

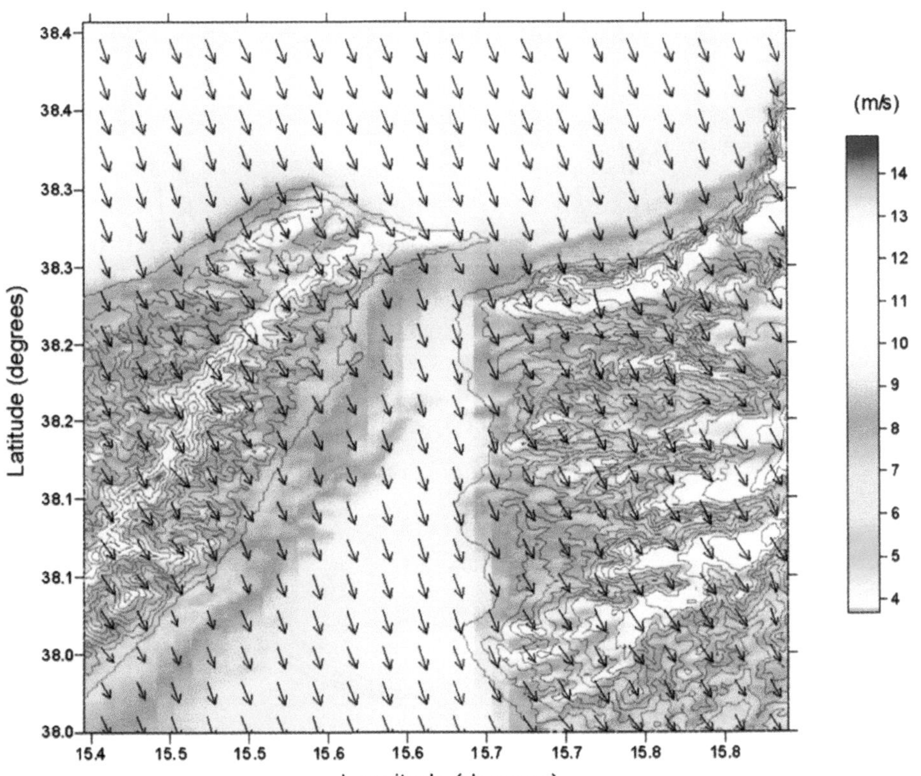

◄ Figure 5.58 Neutral mean wind speed at 10 m in the micro-area (at the top of the calculation volume the speed is 11 m/s and the direction is 348.75°).

Figure 5.60 shows the profiles of the mean wind speed over the height of the Sicilian and Calabrian towers, normalized to the same mean speed at mid-span for different wind directions as in Figure 5.59. Regardless of the dependence on wind direction, it is apparent that representation of the vertical profile of mean wind speed by simple logarithmic laws is not possible; such laws are only valid, at best, in the first 200 m above ground or sea level.

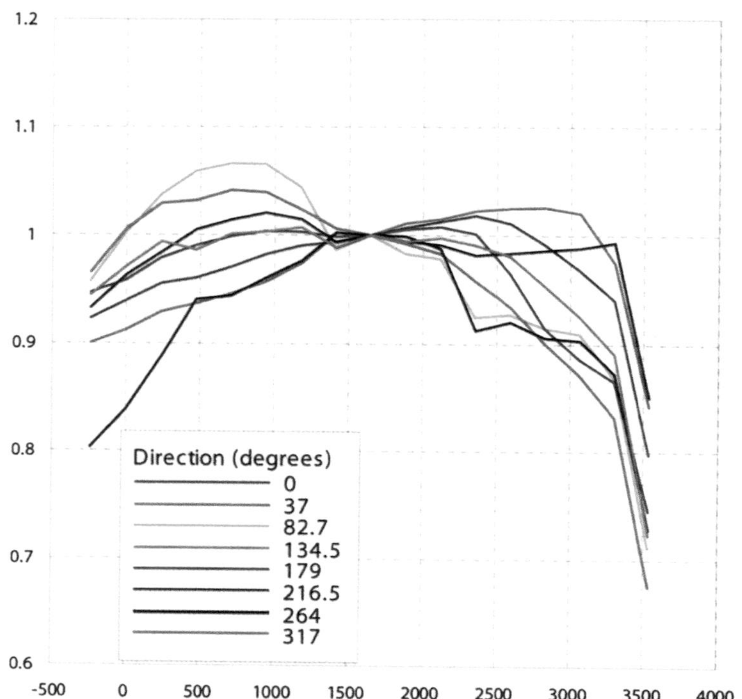

Figure 5.59 ► Profiles of the normalized mean wind speed along the deck axis (in m).

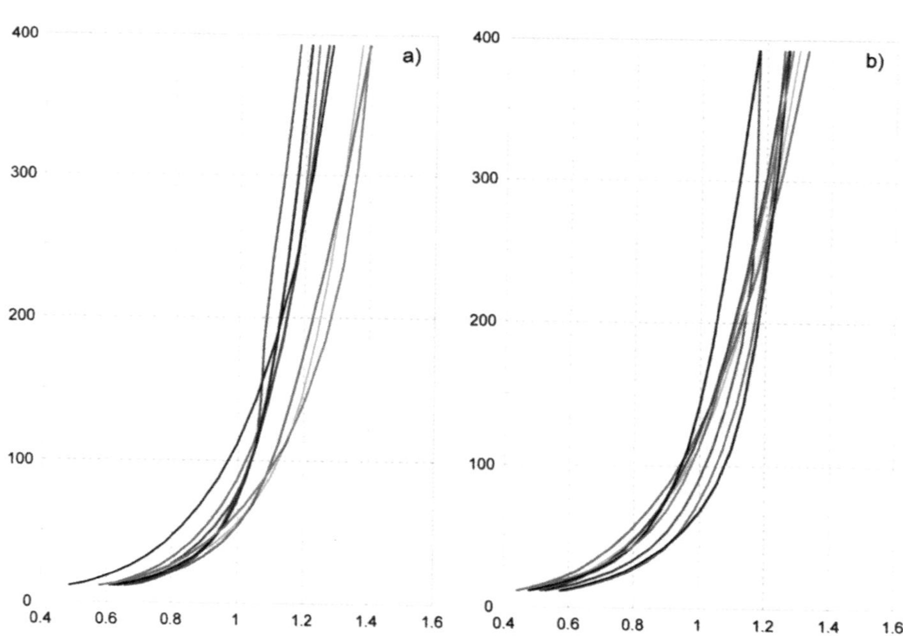

Figure 5.60 ► Profiles of normalized mean wind speed over the height (in m) of the Sicilian (a) and Calabrian (b) towers.

5.4.6 Probabilistic analysis of the mean speed at the bridge

Using the scenarios of the mean wind speeds as described in the previous section, a procedure was developed to transfer the data from the anemometer stations to the bridge. Such data was subjected to new probabilistic evaluations (analogous to those developed in Section 5.4.3), in order to obtain the design values of mean wind speed at the bridge.

Figure 5.61 shows the mean speed at the mid-span, as a function of the return period, comparing: a) the results achieved from the examined stations; b) the analysis estimates derived during the preliminary design period: black lines corresponding to mean values (solid) and mean plus or minus one/two standard deviations (dashed/dash-dot); c) the values finally adopted for the Preliminary Design; and d) the values supplied by the Italian code [91].

The results highlight that the three Gill anemometers and the Met One anemometer at 128 m provide estimates much lower than all the others. Therefore, remembering the problems described in Sections 5.4.3 and 5.4.4, they were considered as unreliable from a probabilistic viewpoint. On the contrary, the Met One anemometer at 15 m and the Messina and Reggio Calabria stations lead to more reasonable, homogeneous and prudent estimates.

Considering these results, the inherent uncertainties, the values selected for the Preliminary Design and those supplied by the Italian code, it was judged

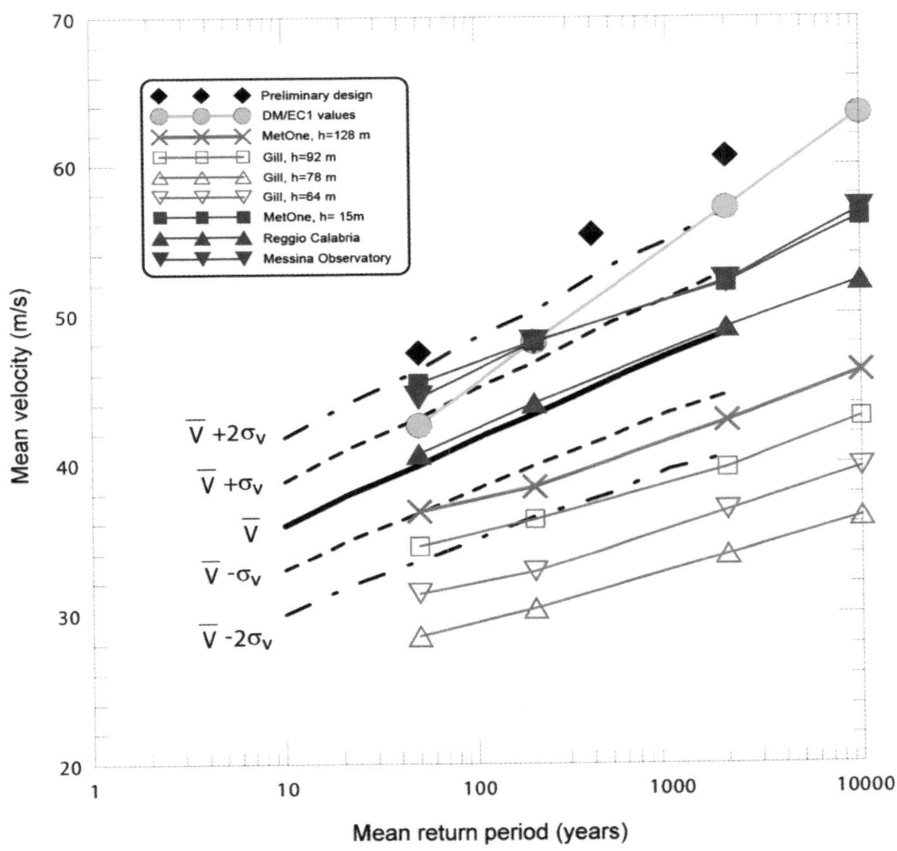

◀ Figure 5.61 Distribution of yearly maximum mean wind speed at mid-span of the bridge, as provided by different analysis criteria.

appropriate that the design of the bridge should be based on values not lower than those given in Table 5.5. It is also worth noting that such values derive from non-directional probabilistic analyses so they do not take into account the crucial aspect that dominant winds are not orthogonal to the bridge axis as noted in Section 5.4.2.

Table 5.5 ▶ Mean wind speeds at bridge mid-span, for different return periods.

Return period (years)	50	200	2000	10000
Mean velocity (m/s)	45	48	52	57

In parallel to this study at the mid-span, the probabilistic analysis of the mean wind speed at the bridge towers was developed in four positions: the top of the Sicilian and Calabrian towers (h = 382.60 m) and the intersection of the bridge deck with the Sicilian (h = 55.40 m) and Calabrian (h = 64.68 m) towers. Table 5.6 shows the main results.

Table 5.6 ▶ Mean wind speeds (m/s) for different return periods R (years).

Location	R = 50	R = 200	R = 2000	R = 10000
Deck level at Sicilian tower	41	44	48	52
Deck level at Calabrian tower	43	46	50	54
Top of Sicilian tower	53–58	58–61	65–66	66–71
Top of Calabrian tower	50–54	55–56	59–62	62–67

A comparison between the Tables confirms that the mean wind speed along the deck is larger at mid-span and smaller next to the towers. In addition, because of the lower level of the deck, it is smaller on the Sicilian side compared to the Calabrian side. On the other hand, due to the lower ground roughness and the less complex topography, the mean wind speed at the top of the Sicilian tower is greater than at the top of the Calabrian tower. Note that these results also do not consider the potential benefits due to the direction of the oncoming flow being on the safe side.

The parametric mean speed profiles and the data reported in Tables 5.5 and 5.6 were also used to derive simplified laws for the mean wind speed. Such laws represent a project alternative to more refined numerical solutions.

5.4.7 Simulation of atmospheric turbulence scenarios

The simulation of turbulence scenarios at the bridge was carried out by a three step procedure [92]. First, the vertical mean wind speed profiles were extracted from the mean wind scenarios depicted in Section 5.4.5. Second, these profiles were schematised through local logarithmic profiles characterised by "effective parameters". Third, using such parameters, the intensities and spectral densities of the longitudinal, lateral and vertical turbulence components were estimated [85].

Figure 5.62 shows the neutral profiles of the longitudinal turbulence intensity along the deck axis for the same set of wind directions given in Figure 5.59. The results demonstrate that the turbulence intensity is lower at mid-span, where the sea roughness is minimum, compared to the side-spans where the flow is disturbed by the ground roughness and topographic effects, especially on the Calabrian side.

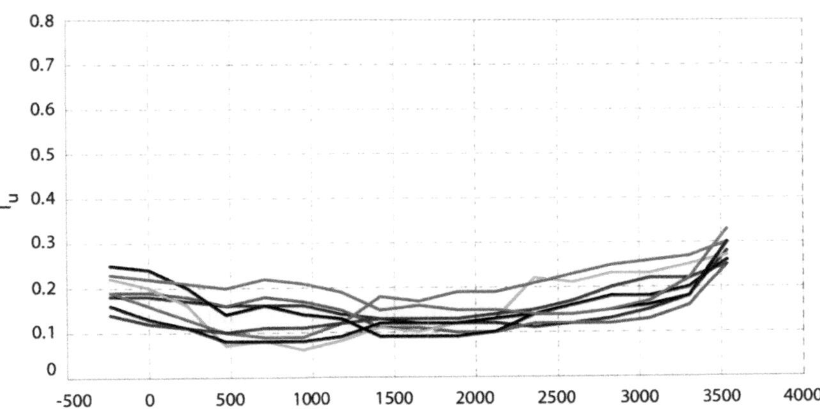

◄ Figure 5.62 Longitudinal turbulence intensity profiles along the deck axis (length in meters).

Figure 5.63 shows the neutral profiles of longitudinal turbulence intensity over the height of the towers for the same set of wind directions. A comparison between the simulations and the measured data at Torre Faro (Section 5.4.2) reveals satisfactory agreement.

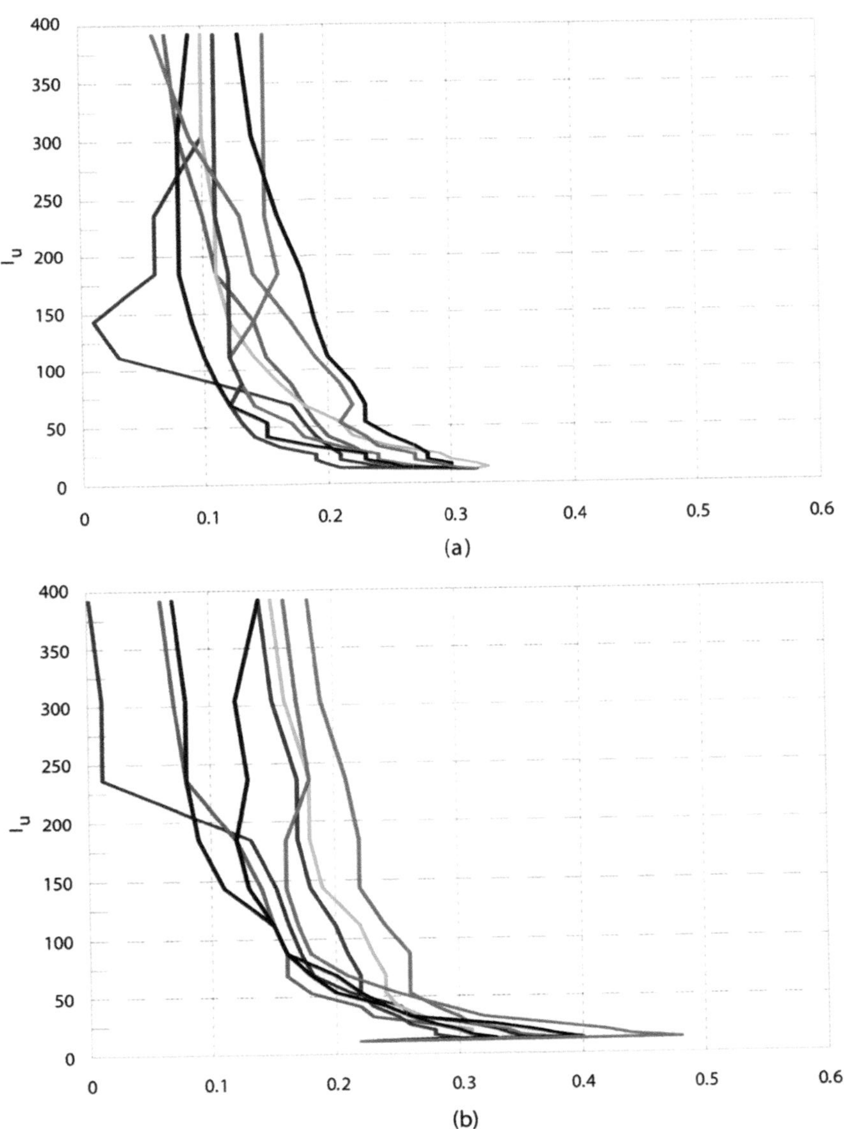

◄ Figure 5.63 Longitudinal turbulence intensity profiles along the axes of the Sicilian (a) and Calabrian (b) towers (Height in meters).

The situation is much more critical with respect to the representation of turbulence coherence, a crucial parameter for the design of a structure of the scale of the Messina Strait Bridge. Unfortunately, both the measurements at Punta Faro and the simulations do not provide enough data to model this parameter. Thus, it was necessary to apply the models proposed in literature, filtered by a critical judgement inspired by the peculiarities of the structure in question.

The coherence models available in literature can be divided into two main classes. The first [93] expresses coherence by a classical exponential decay involving a unit value when the frequency or the distance between the considered points is null. The second [94] derives coherence through refined expressions leading to values less than one when the frequency is null, but the distance between the considered points is greater than zero.

A comparison between these models reveals substantial differences. The former is well-known and widely applied; it is approximate but easy to use, supported by many measurements, and usually provides solutions on the safe side. The latter is less well-known and more difficult to use; it represents the physical phenomenon and the theory, but unfortunately the available data is not enough to calibrate the model parameters.

When dealing with small structures, the differences in using these models are small. However, the situation is more critical for the Messina Bridge, a structure whose size magnifies such differences. Analysis has shown that use of the second model would allow relevant reductions in wind loading compared to the first. However, in spite of this, without enough data to justify its application, the use of the first model was considered as more reasonable and prudent in this case.

5.4.8 Simulation of velocity time histories

The Monte Carlo simulation of the time-space variation of wind speed along the bridge deck, towers and cables represents the final stage of the study and a basic tool for the bridge design. It involves considerable complexity of a computational nature, due to the exceptional size of the problem. Indeed, the structural modelling of the bridge was made using a large 3-D Finite Element model, so the simulations had to be made for a large number of nodes distributed in a 3-D space, taking into account three turbulence components.

In order to carry out this operation, two novel approaches were developed specifically in relation to this bridge project. The first [95] involves advanced tools (e.g. spectral representation, Fast Fourier Transforms, proper orthogonal decomposition and parallel calculus) aimed at improving the computational efficiency of the simulation. The second [96] introduces the concept of effective turbulence in order to clarify the physical aspects of the phenomenon and to minimize the time and space steps of the simulations.

Thanks to such approaches, many families of multi-correlated time histories of the wind speed were generated at the nodes of the Finite Element model. These families involve, as input data, the profiles of the mean wind speed (Section 5.4.5) and the turbulence properties (Sections 5.4.2 and 5.4.7). Such quantities depend on the return period, a focal parameter to select and calibrate with reference to the different verifications to apply. Figure 5.64 shows an example of these simulations, and Figure 5.65 illustrates the precision with which the simulated power spectral densities and the density functions reproduce the target models.

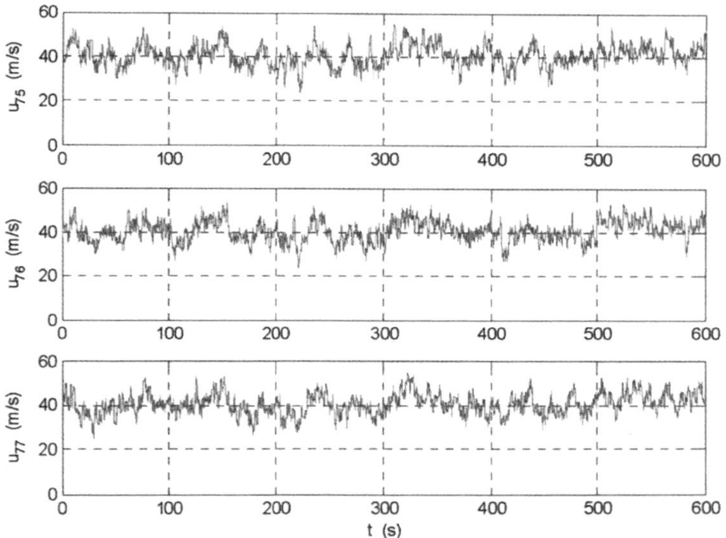

◄ Figure 5.64
Examples of simulated histories of wind speed. Neutral mean wind speed at 10 m in the micro-area (at the top of the calculation volume the speed is 11 m/s and the direction is 348.75°).

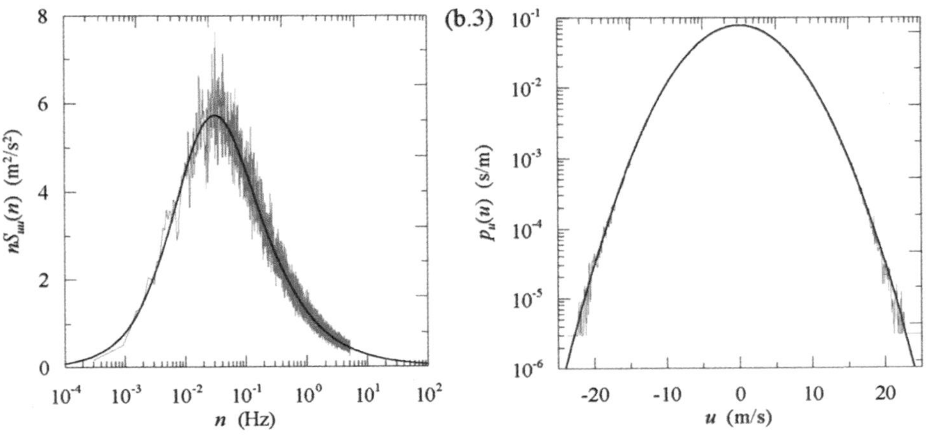

◄ Figure 5.65
Comparison between the power spectral densities (left) and the density functions (right) of the simulated time-histories and their target models.

5.4.9 Conclusions

This Section has described the methods applied and the results obtained with reference to the determination of the time-space configuration of the design wind speed for the Messina Strait Bridge. An articulated framework has emerged. On the one hand, the available data is heterogeneous and rather different from the ideal data to use, if it were available. On the other hand, the complexity and the scale of the problem are so large as to magnify the typical difficulties of these analyses.

Being aware of this situation, the University of Genoa has studied the available data under different perspectives, often comparing heterogeneous values with a critical viewpoint aimed at finding relatively prudent solutions. Furthermore, methods and models inspired by the best state-of-the-art techniques and developed through a broad band of alternatives have been selected so as to provide a wide framework within which to evaluate and interpret the obtained results. In those cases where the available knowledge was not sufficient, research activities have been performed, focussing on developing innovative procedures.

Altogether, the developed evaluations have led to results that seem to be homogeneous and well balanced, and physically and quantitatively in keeping with the scope and purpose of the study.

5.4.10 Acknowledgements

This study has been the result of a wide team activity involving G. Piccardo, L. Carassale, L.C. Pagnini, M.P. Repetto, F. Tubino (Department of Civil, Environmental and Architectural Engineering, DICAT, University of Genoa), C.F. Ratto, M. Antonelli, M. Burlando, R. Festa, E. Georgieva, L. Villa (Department of Physics, DIFI, University of Genoa), and F. Castino (ARPAL, Ligurian Regional Agency for the Environmental Protection). Special thanks are due to Professor Ratto who coordinated the work carried out at DIFI.

References and Further Readings

Section 5.1 References

[1] Baratta M. (1910). La catastrofe sismica calabro-messinese (28 dicembre 1908). Soc. Geogr. It., Roma, pp. 496.

[2] Omori, F. (1909). Preliminary report on the Messina-Reggio earthquake of Dec. 28, 1908. *Bull. Imperial Earth. Invest. Comm.*, 3-2, 37-46.

[3] Pino, N. A., D. Giardini and E. Boschi (2000).The December 28, 1908, Messina Straits, southern Italy, earthquake: waveform modeling of regional seismograms. *J. Geophys. Res.*, vol. 105, n. 25,pp. 473-25, 492.

[4] Loperfido, A. (1909). Livellazione geometrica di precisione eseguita dall'I.G.M. sulla costa orientale della Sicilia, da Messina a Catania, a Gesso ed a Faro Peloro e sulla costa occidentale della Calabria da Gioia Tauro a Melito di Porto Salvo. Relaz. Comm. Reale Acc. Naz. Lincei, 35 pp.

[5] Boschi, E., E. Guidoboni, G. Ferrari, D. Mariotti, G. Valensise and P. Gasperini (eds) (2000). Catalogue of strong italian earthquakes from 461 B.C. to 1997. Ann. Geophys. 43(4), 259 pp., with CD-Rom.

[6] Schick, R. (1977). Eine seismotektonische Bearbeitung des Erdbebens von Messina im Jahre 1908. *Geol. Jahrb.*, R.E., H., 11, 74 pp.

[7] Bottari, A., E. Carapezza, M. Carapezza, P. Carveni, F. Cefali, E. Lo Giudice, and C. Pandolfo (1986).The 1908 Messina Strait earthquake in the regional geostructural framework. Journal of Geodynamics, 5, 275-302.

[8] Mulargia, F., and E. Boschi (1983). The 1908 Messina earthquake and related seismicity. In: H. Kanamori and E. Boschi (eds), Proc. Int. School Phys. "E. Fermi" on "Earthquakes: observation, theory and interpretation", North Holland Publ. Co., 493-518.

[9] Capuano, P., G. De Natale, P. Gasparini, F. Pingue and R. Scarpa (1988). A model for the 1908 Messina Straits (Italy) earthquake by inversion of levelling data. Bull. Seism. Soc. Am., 78, 1930-1947.

[10] Boschi, E., D. Pantosti and G. Valensise (1989). Modello di sorgente per il terremoto di Messina del 1908 ed evoluzione recente dell'area dello Stretto. Proc. 8° Meeting G.N.G.T.S., Rome 1989, 245-258.

[11] De Natale, G., and F. Pingue (1991). A variable slip fault model for the 1908 Messina Straits (Italy) earthquake, by inversion of levelling data. *Geophys. J. Int.*, 104, 73-84.

[12] Ghisetti, F. (1984). Recent deformations and the seismogenic source in the Messina Strait (Southern Italy).Tectonophysics, 109, 191-208.

[13] Montenat, C., P. Barrier and P. Ott d'Estevou (1991). Some aspects of the recent tectonics in the Strait of Messina, Italy.Tectonophysics, 194, 203-215.

[14] Monaco, C., and L.Tortorici (2000). Active faulting in the Calabrian arc and eastern Sicily, J. Geodynamics, 29, 407-424, doi:10.1016/S0264–3707(99)00052-6.

[15] Catalano, S., G. De Guidi, C. Monaco, G.Tortorici and L.Tortorici (2003). Long-term behaviour of the late Quaternary normal faults in the Straits of Messina area (Calabrian arc): Structural and morphological constraints, Quat. Int., 101-102, 81-91, doi:10.1016/S1040–6182(02)00091-5.

[16] Amoruso, A., L. Crescentini and R. Scarpa (2002). Source parameters of the 1908 Messina Straits, Italy, earthquake from geodetic and sesmic data. J. Geophys. Res., 107, B4, 2080, doi:10.1029/2001JB000434.

[17] Valensise, G., and D. Pantosti (1992). A 125 kyr-long geological record of seismic source repeatability: the Messina Straits (southern Italy) and the 1908 earthquake (Ms 71/2).Terra Nova, 4, 472-483.

[18] McGuire, J.J., Li Zhao, andT.H. Jordan (2002). Predominance of unilateral rupture for a global catalog of large earthquakes. Bull. Seism. Soc. Am. 92 (8), 3309–3317, doi: 10.1785/0120010293.

[19] Michelini, A., A. Lomax, A. Nardi, A. Rossi, B. Palombo and A. Bono (2005). A modern re-examination of the locations of the 1905 Calabria and the 1908 Messina Straits earthquakes, in European Geosciences Union, 2005 General Assembly Abstracts.

[20] Neri, G., G. Barberi, G. Oliva and B. Orecchio (2004).Tectonic stress and seismogenic faulting in the area of the 1908 Messina earthquake, south Italy. Geophys. Res. Lett., 31, 10, doi 10.1029/2004GL019742.

[21] Caputo, M., G. Folloni, L. Pieri and M. Unguendoli (1974). Geodimetric Control across the Straits of Messina. Geophys. J. R. astr. Soc., 38, 1-8.

[22] Anzidei, M., P. Baldi, C. Bonini, G. Casula, S. Gandolfi and F. Riguzzi (1998). Geodetic surveys across the Messina Straits (southern Italy) seismogenetic area. J. Geodynamics, 25 (2), 85-97.

[23] D'Agostino, N., and Selvaggi, G., 2004. Crustal motion along the Eurasia-Nubia plate boundary in the Calabrian Arc and Sicily and active extension in the Messina Straits from GPS measurements. J. Geophys. Res., 109, B11 402, doi:10.1029/2004JB002998.

[24] Piatanesi, A., S. Tinti and E. Bortolucci (1999). Finite-element simulations of the 28 December 1908 Messina Straits (southern Italy) tsunami. Phys. Chem. Earth, 24, 145-150.

[25] Azzaro, R., F. Bernardini, R. Camassi and V. Castelli (2007). The 1780 seismic sequence in NE Sicily (Italy). Shifting an underestimated and mislocated earthquake to a seismically low rate zone. Nat. Hazards, 42, 149-167.

[26] DISS Working Group (2007). Database of Individual Seismogenic Sources (DISS),Version 3.0.4: A compilation of potential sources for earthquakes larger than M 5.5 in Italy and surrounding areas. http://www.ingv.it/DISS/, © INGV 2007 – All rights reserved.

[27] Valensise, G., and D. Pantosti (eds.) (2001). Database of Potential Sources for Earthquakes Larger than M 5.5 in Italy. Ann. Geophys. 44, Suppl. 1, with CD-Rom.

[28] Basili, R., G.Valensise, P.Vannoli, P. Burrato, U. Fracassi, S. Mariano, M.M.Tiberti and E. Boschi (2008).The Database of Individual Seismogenic Sources (DISS), version 3: summarizing 20 years of research on Italy's earthquake geology, Tectonophysics, 452, doi:10.1016/j.tecto.2007.04.014.

[29] Bordoni, P., and G. Valensise (1998). Deformation of the 125 ka marine terrace in Italy: tectonic implications. in: I. Stewart e C. Vita-Finzi (eds), CoastalTectonics, Geol. Soc. London Spec. Pub., 146, 71-110.

[30] Cernobori, L., H. Hirn, J.H. McBride, R. Nicolich, M. Petronio, M. Romanelli and Streamers Profile Working Group (1996). Crustal image of the Ionian basin and its Calabrian margins. Tectonophysics, 264, 175-189.

[31] Monaco, C., L. Tortorici, R. Nicolich, L. Cernobori and M. Costa (1996). From collisional to rifted basins: an example from the southern Calabrian arc (Italy). Tectonophysics, 266, 233-249.

[32] Guidoboni, E., A. Muggia and G. Valensise (2000). Aims and methods in Territorial Archaeology: possible clues to a strong IV century A.D. earthquake in the Straits of Messina (southern Italy) in: B. McGuire, D. Griffiths e I. Stewart (eds), The Archaeology of geological catastrophes, Geol. Soc. London Spec. Pub., 171, 45-70.

Section 5.2 References

[33] Norme Tecniche per le Costruzioni (2008), Decree 140108 (January 14, 2008), Ministry of Infrastructures and Transportation, Rome.

[34] Gruppo di Lavoro per la redazione della mappa di pericolosità sismica (2004), INGV, Rapporto Conclusivo, http://zonesismiche.mi.ingv.it/.

[35] Newmark N. M., J. Blume K. Kapur (1973). Seismic Design Spectra for Nuclear Power Plants. *J. Power Div., Proc. ASCE*, vol. 99(PO2), pp. 287-303.

[36] Stretto di Messina (2004), INGEGNERIA – PROGETTAZIONE DEFINITIVA ED ESECUTIVA, SPECIFICHE TECNICHE DI PROGETTAZIONE, Requisiti e linee guida per lo sviluppo della progettazione, Integrazione e aggiornamento del quadro geosismotettonico, Rev. 1b, Ref. GC.F.05.03.

[37] Hisada, Y. (1994). An efficient method for computing Green's functions for a layered half-space with sources and receivers at colse depths, I. *Bulletin Seismological Society of America*, vol. 84, pp. 1456-1472.

[38] Hisada, Y. (1995). An efficient method for computing Green's functions for a layered half-space with sources and receivers at colse depths, II. *Bulletin Seismological Society of America*, vol. 85, pp. 1080-1093.

[39] Hisada, Y., e J. Bielak (2003). A Theoretical method for computing near-fault ground motions in layered half-spaces considering static offset due to surface faulting, with a physical interpretation of fling step and rupture directivity. *Bulletin Seismological Society of America*, vol. 93, n. 3, pp. 1154-1168.

[40] Bottari, A., E. Lo Giudice, D. Schiavone (1979). Geophysical study of a crustal section across the Straits of Messina. *Annals of Geophysics*, vol. 32, pp. 241-261.

[41] Faccioli, E. (1994). Seismic ground amplification, stability analyses and 3-dimensional SSI studies for the 3300 m one-span suspension bridge across the Messina Straits. *10th European Conf. on Earthquake Eng.*, August 28th – September 2nd, Vienna, Austria. G. Duma (ed.), Balkema, Rotterdam, ISBN 90 5410 528 3.

[42] Neri, G., G. Barberi, B. Orecchio e M. Aloisi (2002). Seismotomography of the crust in the transitino zone between Tyrrhenian and Sicilian tectonic domains. *Geophysycal Research Letters*, vol. 29, n. 23, 2135, doi: 10.1029/2002GL015562.

[43] Stretto di Messina (1991). Studi di amplificazione locale per determinare gli spettri di sito mediante analisi bidimensionali e monodimensionali. Metodologia e analisi per un modello di faglia sismogenetica. Ismes SpA, Bergamo, Studio Geotecnico Italiano srl, R.4326/6-FAC/PAO/rc, 20.02.91, Milano.

[44] Ismes (1989a). Indagine geotecnica delle fondazioni del ponte sullo stretto di Messina. Prima fase. Fondazione Calabria. Indagini cross-hole. Progr. ASP-3691 – Doc. RAT-DGF-042.

[45] Ismes (1989b). Indagine geotecnica delle fondazioni del ponte sullo stretto di Messina. Prima fase. Fondazione Sicilia. Indagini cross-hole. Progr. ASP-3691 – Doc. RAT-DGF-062.

Section 5.3 References

[46] Baldi, G., Bruzzi, D., Superbo, S., Battaglio, M. and M., Jamiolkowski (1988). "Seismic Cone in Po River Sand". Proc. ISOPT-1, Orlando, Fla., Vol.2, pp.643-650.

[47] Baldi, G., Jamiolkowski, M., Lo Presti, D.C.F., Manfredini, G. and G.J., Rix (1989). "Italian Experience in Assessing Shear Wave Velocity from CPT and SPT". Earthquake Geotechnical Engineering, Proc. of Discussion Session on Influence of Local Conditions on Seismic Response. XII ICSMFE, Rio De Janeiro, Brasil, Vol.4, pp.157-168.

[48] Schmertmann, J.H. and Palacios, A. (1979). "Energy Dynamics of SPT". Journal of GED, ASCE. Vol.105, No.GT8, pp.909-926.

[49] Nazarian, S. and Stokoe, K.H. II (1983). "Use of the Spectral Analysis of Surface Waves for Determination of Moduli and Thickness of Pavement Systems". Transportation Research Record, No.954, TRB, Transportation Research Board, Washington, D.C.

[50] Yoshimi, Y., Hatanaka, M. and Oh-oka, H. (1977). "A Simple Method for Undisturbed Sampling by Freezing". Specialty Session 2 on Soil Sampling 9th Int. Conf. on Soil Mech. and Found. Engng., Tokyo, pp.23-28.

[51] Yoshimi, Y., Hatanaka, M. and Oh-oka, H. (1978). "Undisturbed Sampling of Saturated Sands by Freezing". Soils and Foundations, Vol.18, No.3, pp.59-73.

[52] Yoshimi, Y., Tokimatsu, K., Kaneko, O. and Makihara, Y. (1984). "Undrained Cyclic Shear Strength of a Dense Niigata Sand". Soils and Foundations, Vol.24, No.4, pp.131-145.

[53] Kokusho, T. and Tanaka, Y. (1994). "Dynamic Properties of Gravel Layers Investigated by In-situ Freezing Sampling". Ground Failures under Seismic Conditions, GSP 44, ASCE, pp.121-140.

[54] ISMES (1985). "Results of Laboratory Tests Carried out on Reconstituted Specimens of the Coastal Plain Deposits". R-2685 and R-2690.

[55] Ferrante, G. (1988). "Comportamento in colonna risonante di sabbie frantumabili e della ghiaia di Messina". Politecnico di Torino, Department of Structural and Geotechnical Engineering.

[56] Jamiolkowski, M. and Lo Presti, D.C.F. (2002). "Geotechnical Characterisation of Holocene and Pleistocene Messina Sand and Gravel Deposits". Int. Workshop on Characterisation and Engineering Properties of Natural Soils (Tan et al., eds.), Singapore, Swets and Zeitlinger, pp.1087-1119.

[57] Foti, S., Lai, C.G. and Lancellotta, R. (2002). "Porosity of Fluid Saturated Porous Media from Measured Seismic Wave Velocities". Geotechnique, Vol.52, No.5, pp.359-373.

[58] Biot, M.A. (1956). "Theory of Propagation of Elastic Waves in Fluid-Saturated Porous Solid, I Lower Frequency Range". Journal of Acoustical Society of America, No.2, pp.168-178.

[59] Foti, S. and Lancellotta, R. (2004). "Soil Porosity from Seismic Velocities". Geotechnique, Vol.54, No.8, pp.551-554.

[60] Arroyo, M., Ferreira, C. and Sukolrat, J. (2006). "Dynamic Measurements and Porosity in Saturated Triaxial Specimens". Symposium on Soil Stress-Strain Behaviour: Measuring Modelling and Analysis, Rome, Eds.: H.I., Ling, L., Callisto, D., Leshchinsky and J., Koseki, Springer, pp.537-546.

[61] Skempton, A.W. (1986). "Standard PenetrationTests Procedures and the Effects in Sands of Overburden Pressure, Relative Density, Particle Size, Ageing and Overconsolidation". Géotechnique, Vol.36, No.3, pp.425-447.

[62] Cubrinovski, M. and Ishihara, K. (1999). "Empirical Correlation between SPT-N Value and Relative Density for Sandy Soils". Soils and Foundations, No.5, pp.61-71.

[63] Jardine, R.J. (1985). "Investigations of Pile-Soil behaviour with Special Reference to the Foundations of offshore Structures". Ph.D. Dissertation, University of London, London.

[64] Jardine, R.J. (1992). "Some Observations on the Kinematic Nature of Soil Stiffness". Soils and Foundations, Vol.32, No.2, pp.111-124.

[65] Tatsuoka, F., Jardine, R.J., Lo Presti, D.C.F., Di Benedetto, H. and Kokaka, T. (1997). "Characterising the Pre-failure Deformation Properties of Geomaterials".Theme Lecture for Plenary Session No.1, 14th Int. Conf. on Soil Mech. and Found. Engng., Hamburg, Vol.4, pp.2129-2164.

[66] Hight, D.W. and Leroueil, S. (2002). "Characterisation of Soils for Engineering Purposes". Int. Workshop on Characterisation and Engineering Properties of Natural Soils (Tan et al. eds.), Singapore, Swets and Zeitlinger, pp.255-360.

[67] Derendeli, M.B. (2001). "Development of New Family of Normalised Modulus Reduction and Material Damping Reduction Curves". Ph.D. Dissertation, University ofTexas at Austin.

[68] Jardine, R.J., Kuwano, R., Zdravkovic, L. andThornton, C. (2001). "Some Fundamental Aspects of the Pre-Failure behaviour of Granular Soils". 2nd Int. Symp. on Pre-Failure Deformation Characteristics of Geomaterials, IS Torino '99 Jamiolkowski et al., eds., Balkema, Vol.2, pp.1077-1111.

[69] Barton, M.E., Mockett, L.D. and Palmer, S.N. (1993). "An Engineering Geological Classification of the Soil-Rock Borderline Materials between Sands and Sandstone". 26th Annual Conf. on Engineering Geology of Weak Rock (Cripps et al. eds.), Leeds, UK, Balkema, pp.125-138.

[70] Cresswell, A.W. (1999). "Sampling and StrengthTesting of Unbounded Locked Sand". Ph.D. Dissertation, Southampton University, U.K.

[71] Bosi, C. (1990). "Studio dello stato di aggregazione delle Ghiaie di Messina". Report by Sitec S.r.l., Roma.

[72] Crova, R., Jamiolkowski, M., Lancellotta, R. and Lo Presti, D.C.F. (1992). "Geotechnical Characterization of Gravelly Soils at Messina Site, Selected Topics". Proceedings of the Wroth Memorial Symposium, Thomas Telford, London, pp.199-218.

[73] Ishihara, K. (1996). "Soil Behaviour in Earthquake Geotechnics". Oxford Science Publications.

[74] Goto, S., Suzuki,Y., Nishio, S. and Oh-oka, H. (1992). "Mechanical Properties of UndisturbedTone-river Gravel Obtained by in Situ Freezing Method". Soils and Foundations, Vol.32, No.3, pp.15-25.

[75] Goto, S., Nishio, S. andYoshimi,Y. (1994). "Dynamic Properties of Gravels Sampled by Ground Freezing". Ground Failures under Seismic Conditions. GSP 44, ASCE, pp.141-157.

[76] Yasuda, N., Otha, N. and Nakamura, A. (1994). "Deformation Characteristics of Undisturbed Riverbed Gravel by In-Situ Freezing Sampling Method". 1st Int. Symp. on Pre-Failure Deformation Characteristics of Geomaterials, IS Hokkaido '94 Sapporo, Japan (Shibuya et al. eds.), Balkema, Vol.1, pp.41-46.

[77] Hatanaka, M. and Uchida, A. (1995). "Effects of Test Methods on the Cyclic Deformation Characteristics of High Quality Undisturbed Gravel Samples". GSP 56, pp.136-151.

[78] Lo Presti, D., Pallara, O., Froio, F., Rinolfi, A. and Jamiolkowski, M. (2006). "Stress-Strain Strength behaviour of Undisturbed and Reconstituted Gravelly Soils Samples". Rivista Italiana di Geotecnica, Vol.40, No.1, pp.9-27.

[79] Hatanaka, M. and Uchida, A. (1996). "Empirical Correlation Between Penetration Resistance and Internal Friction Angle of Sandy Soils". Soils and Foundations, Vol.36, No.4, pp.1-10.

[80] Bolton, M.D. (1986). "The Strength and Dilatancy of Sands". Géotechnique, Vol.36, No.1, pp.65-78.

[81] Rowe, P.W. (1962). "The Stress-Dilatancy Relation for Static Equilibrium of an Assembly of Particles in Contact". Proc. Royal Society, 269A, pp.500-527.

[82] Todd, D.K. (1980). "Groundwater Hydrology". Wiley.

Section 5.4 References

[83] Gasparetto, M., Bocciolone, M. (1992). Wind measurements on Messina Strait. *J. Wind Engng. Ind. Aerod.*, 41, 393-404.

[84] Bocciolone, M., Gasparetto, M., Lagomarsino, S., Piccardo, G., Ratto, C.F., Solari, G. (1993). Statistical analysis of extreme wind speeds in the Strait of Messina. *J. Wind Engng. Ind. Aerod.*, 48, 359-377.

[85] Solari, G., Piccardo, G. (2001). Probabilistic 3-D turbulence modeling for gust buffeting of structures, *Prob. Engng. Mech.*, 16, 73-86.

[86] Solari, G. (1996). Wind speed statistics, in *Modelling of the Atmospheric Flow Fields*, D.P. Lalas and C.F. Ratto Editors, World Scientific Publ., Singapore, 637-657.

[87] Solari, G. (1996). Statistical analysis of extreme wind speeds, in *Modelling of the Atmospheric Flow Fields*, D.P. Lalas and C.F. Ratto Eds, World Scientific Publ., Singapore, 659-678.

[88] Lagomarsino, S., Piccardo, G., Solari, G. (1992). Statistical analysis of high return period wind speeds. *J. Wind Engng. Ind. Aerod.*, 41, 485-496.

[89] Harris, I. (2006). Errors in GEV analysis of wind epoch maxima from Weibull parents. *Wind Struct.*, 9, 179-191.

[90] Ratto, C.F. (1996). The AIOLOS and WINDS codes. In *Modelling of the Atmospheric Flow Fields*, D.P. Lalas and C.F. Ratto Editors, World Scientific Publ., Singapore.

[91] *Istruzioni per l'applicazione delle "Norme tecniche relative ai criteri generali per la verifica di sicurezza delle costruzioni e dei carichi e sovraccarichi" di cui al decreto ministeriale 16.1.1996*, Ministero LL. PP., Roma, Circolare 4 luglio 1996, n. 156AA.GG/STC, 1996.

[92] Burlando, M., Carassale, L., Georgieva, E., Ratto, C.F., Solari, G. (2007). A simple and efficient procedure for the numerical simulation of wind fields in complex terrain, *Bound Lay Meteorol*, 125, 417-439.

[93] Pielke, R.A., Panofsky, H.A. (1970). Turbulence characteristics along several towers. *Bound. Lay. Meteorol.*, 1, 115-130.

[94] Mann, J. (1994). The spatial structure of neutral atmospheric surface-layer turbulence. *J. Fluid Mech.*, 273, 141-168.

[95] Carassale, L., Solari, G. (2006). Monte Carlo simulation of wind velocity fields on complex structures, *J. Wind Engng. Ind. Aerod.*, 94, 323-339.

[96] Tubino, F., Solari, G. (2007). Double modal transformation and effective turbulence for the gust buffeting of long span bridges, *Eng. Struct.*, 29, 1698-1707.

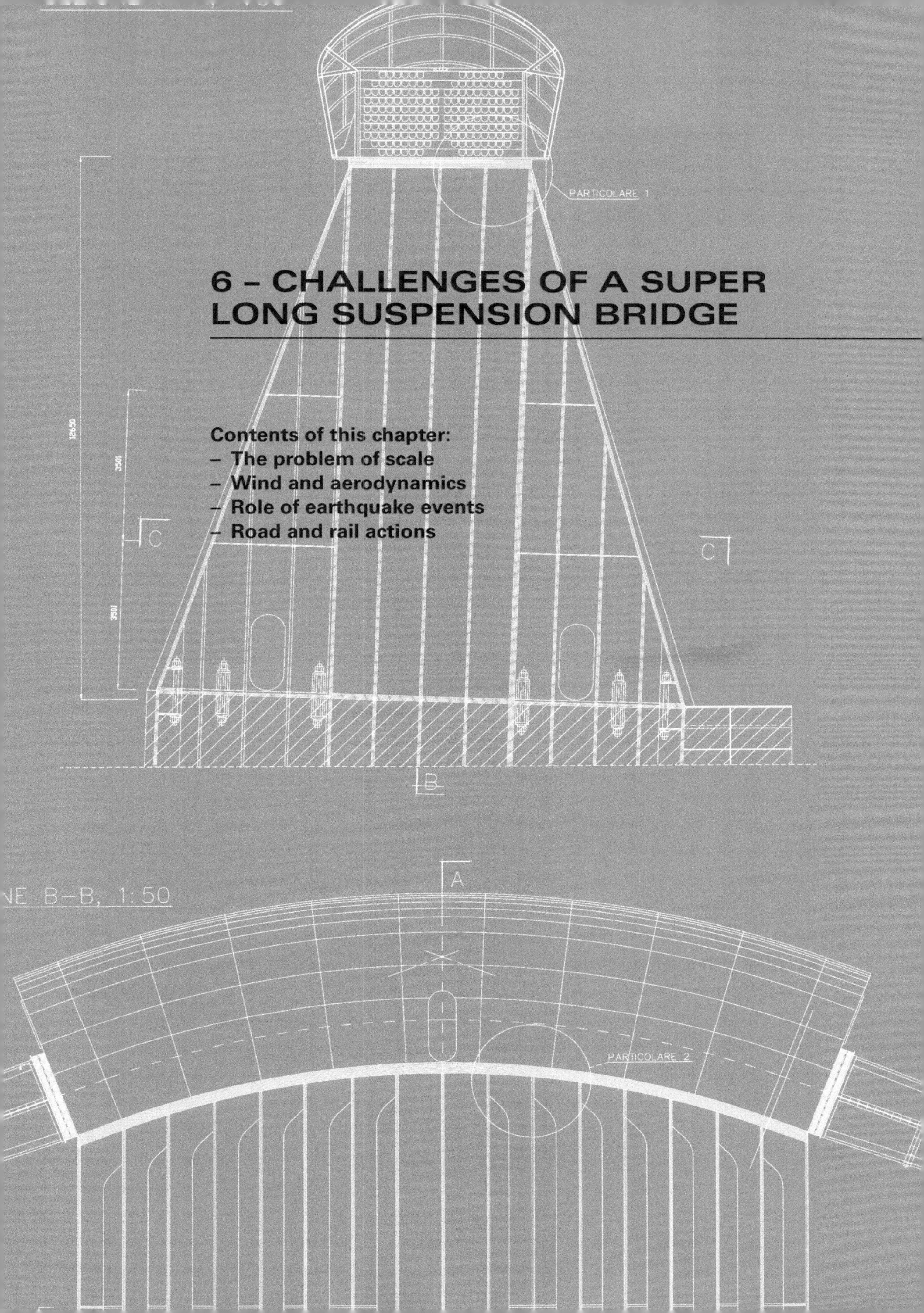

6 – CHALLENGES OF A SUPER LONG SUSPENSION BRIDGE

Contents of this chapter:

- **The problem of scale**
- **Wind and aerodynamics**
- **Role of earthquake events**
- **Road and rail actions**

This chapter discusses some of the dominant factors affecting the design of a very long suspension bridge. Extensive research and design development has been undertaken over many years, as described herein, to advance the appreciation of how the very long span influences the behaviour, properties and response of a suspension bridge. The current design accommodates the lessons learnt from this far reaching activity, and the implications for this and other suspension bridges are discussed.

6.1 The problem of scale

A suspension bridge is a fundamentally simple structure, with a "serial type" of static behaviour in carrying loads. In simple terms, the loads are applied to the deck, carried to the main cable through the hangers and subsequently brought to ground mainly through the towers for the vertical component and through the anchor blocks for the horizontal one. (Figure 6.1)

When subject to live loads for example, as in any redundant system, a suspension bridge undergoes an interaction in which forces are carried through the structure depending on the relative stiffness of the different elements involved. The primary interaction is that between the deck, whose stiffness is mainly related to flexure, and the main cables, whose stiffness is instead mainly geometric and in turn dependent upon the present tensile load. Since the tensile force for suspension bridges is mainly connected to weight, such geometric cable stiffness is often referred to as "gravity stiffness".

The change in relative stiffness between the deck (in bending) and the main cable (due to tension and geometry) with span is the main parameter dominating the behaviour of suspension bridges for many aspects. For example, with a constant tensile load, the cable stiffness decreases as span length increases, while for constant cross-section the deck stiffness also decreases, but in proportion to a higher power of the span.

For comparatively short spans, such as those of 19th century suspension bridges, the two stiffnesses are comparable, and both the deck and the cables carry significant shares of the applied loads. This led to the idea of the deck as the "stiffening girder", collaborating with the main cable to carry, through its own stiffness, a significant share of the applied loads.

Figure 6.1 ▶ Suspension bridge scheme with force flow.

For increasing spans, the decrease in cable stiffness due to the greater length is partly mitigated by the increase in axial load with span and the consequent increase in cable size. On the contrary, as the deck size does not depend directly on span, the relative deck stiffness decreases rapidly and any significant stiffening role by the deck for global loads is lost. The deck acts merely as the element collecting live loads and distributing them between the hangers, and does not contribute globally to carry large applied loads that are virtually all transferred by the main cables.

This trend is illustrated in Figure 6.2 which shows the proportion of the total applied load carried by the deck as a function of span length for two different bridge deck girders.

The above result depends upon assumptions about deck type, sag to span ratio, side span configurations etc., but in any case it is evident that the main cable is the main element conferring stiffness to the bridge; it is the real backbone of the structure. This means that the static and dynamic behaviour of a large suspension bridge is strongly connected to that of a catenary.

There is nothing new in this; the simple concept presented above has been well understood since at least the early 20th century. For example, when presenting his design for the Washington Bridge (1931), with its record span of 1030 m, Othmar Amman observed that the key to providing adequate service performance capabilities to the bridge was "... its enormous gravity stiffness ... ".

Less apparent, but also easily understandable, is that the same is true for any load type – not only vertical live loads. Important among these other load types are lateral loads, mainly due to wind, and eccentric live loads producing twisting moment about the bridge axis, i.e. "torsion" in the deck.

For large suspension bridge spans, wind loads on the deck are mainly transmitted to the supporting towers not through deck lateral flexural stiffness but via the main cables due to lateral inclination, or "pendulum" effect, of the hangers. Wind loads are thus largely carried to the tower tops and hence typically dominate the tower transversal design, and are decisive in determining the type of elements to provide for absorbing them, such as cross framed elements and bracings. Eccentric live loads inducing "torsional" moments about the bridge axis are also absorbed primarily by the

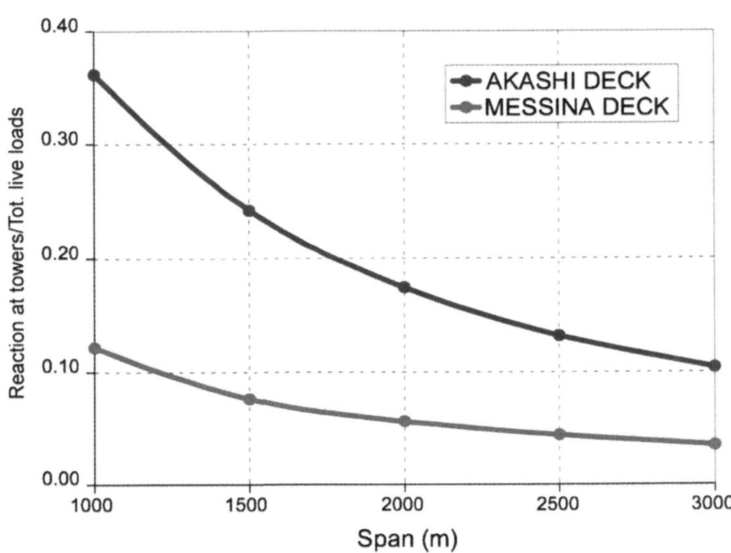

◄ Figure 6.2
The proportion of load carried by the deck girder as a function of span for two different girder types.

cables. In this case, it is apparent how a decisive parameter in determining the bridge "torsional", or to say better "rotational", stiffness is the distance between the main cable suspension planes, as this provides the lever arm to transfer the deck rotation into vertical cable forces and stiffness.

But although the transfer of stiffness towards the main cables is the most obvious consequence of span increase it is not the only one. Other aspects emerge which form the basis of what shall be referred to herein as "the problem of scale", underlining how the size of the bridge changes its behaviour not only quantitatively but also qualitatively. Such scale aspects are manifold, with a few of them becoming dominant for feasibility and design. For simplicity, these have been collected into the following two issues which are different in nature although derived from the same source; the first is concerned with overall sustainability and financial feasibility, and the second with very fundamental structural safety:

- Scale issue one: cable steel self weight and quantity.
- Scale issue two: dynamic properties and aeroelastic stability.

6.1.1 Scale issue one: cable steel self weight and quantity

The question is straightforward: it has already been shown that the size of the cable increases with span. Figures 6.3(a) to (d) show the variation in cable tension with span due to different loading types; road and rail live loads, deck weight and cable self weight. The four Figures 6.3a to 6.3d cover four different deck types and deck self-weight. The plots are normalised for the different load types, which do not have the same distribution, so as to allow a direct comparison.

It can be seen that for spans around one thousand metres the deck weight is the largest component of cable tension, with significant contributions from rail loads and less from the cable self weight. For growing spans the increase in the deck contribution is, as expected, proportional to the span. The road and rail live load contribution is less than proportional, as for very large spans average live load intensities decrease, due to standard probability considerations.

By contrast, the cable self weight contribution increases more than proportionally with increasing span. For spans between 1500 and 2000 metres,

Figure 6.3a ▶ Road only: box deck.

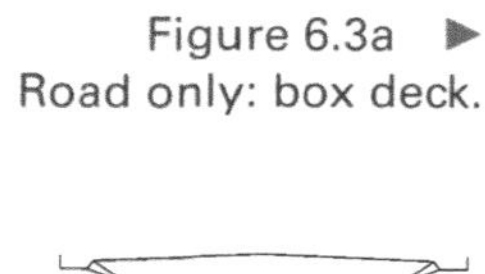

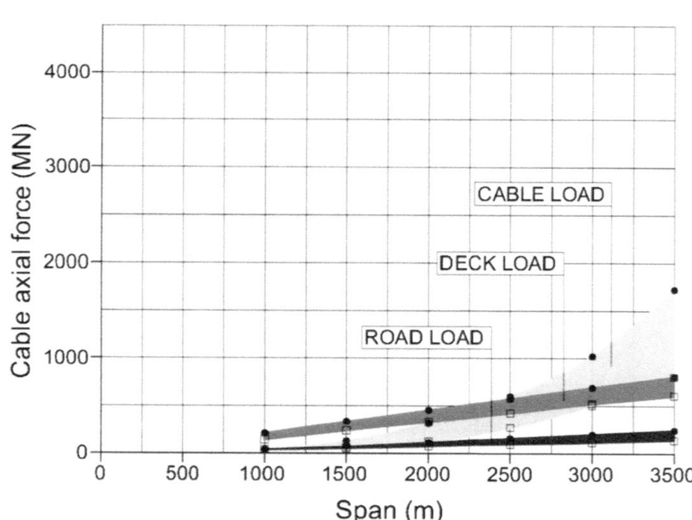

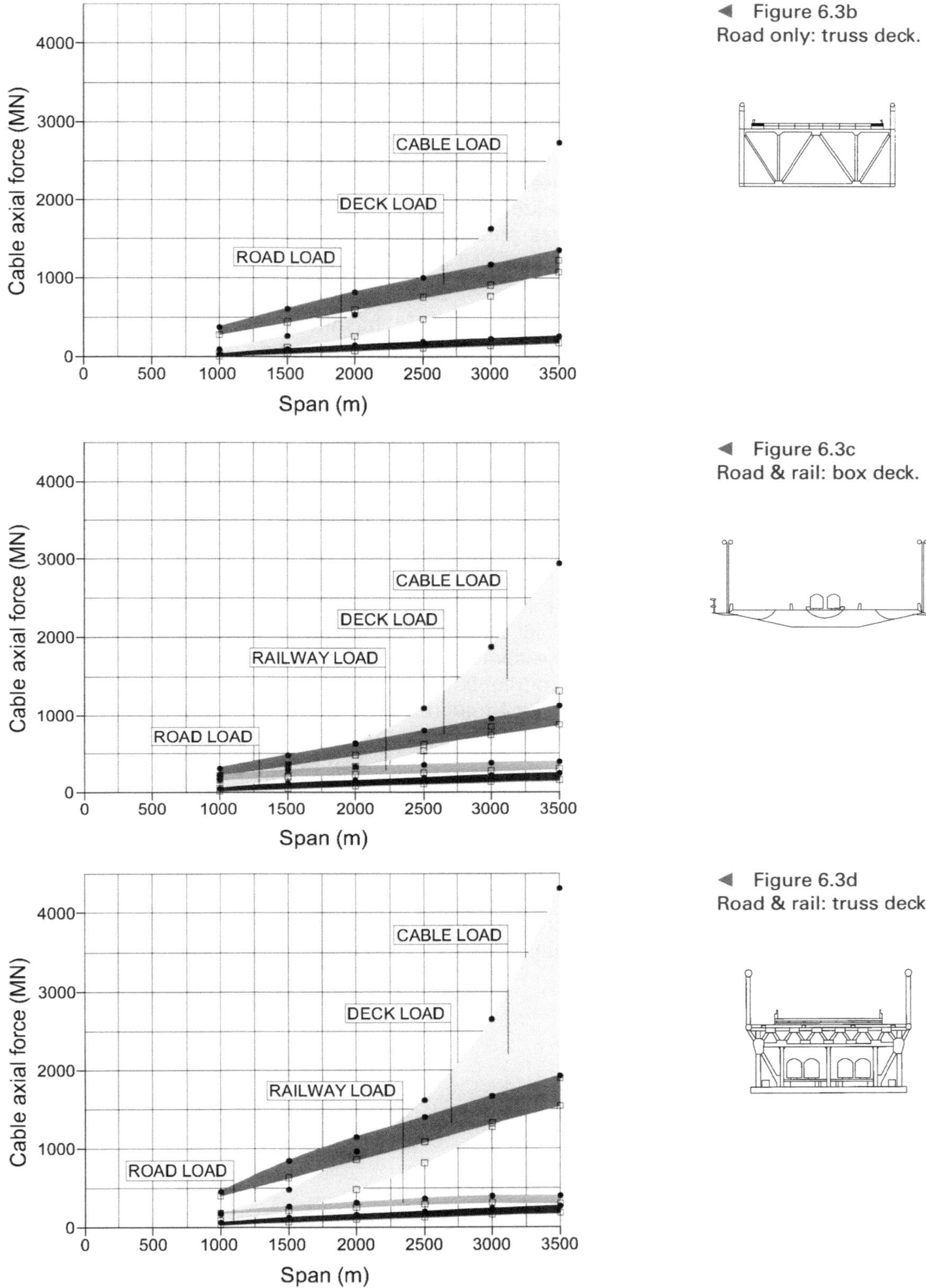

◄ Figure 6.3b
Road only: truss deck.

◄ Figure 6.3c
Road & rail: box deck.

◄ Figure 6.3d
Road & rail: truss deck.

the contribution of cable self weight to cable tension surpasses the effect of railway loads. Between 2000 and 2500 metres it equals the contribution of the deck, and it becomes clearly the largest contribution for spans over 3000 metres. In fact, the cable size variation with span follows the characteristic elbow shaped theoretical curve which approaches asymptotically the maximum span achievable for given cable steel material properties and deck weight plus live load. One such curve, derived from the Messina project assumptions, is given in Figure 6.4, and shows a maximum theoretical span of about 7000 metres.

This means that for spans as large as the one needed for Messina, the cable becomes the heaviest and most expensive component of the superstructure, including the towers, and the large cable size in turn results in higher sizes and costs for the towers, foundations and anchor blocks because these are all elements that support or restrain the cable weight and forces.

Limiting the cable weight is thus the most fundamental design target to be achieved to deal with the first large scale issue, so as to allow the overall sustainability and financial feasibility of a very large span bridge.

As the cable weight stems from the loads it must carry and from the maximum allowed working stress in the steel wire, every effort must be devoted to:

- Selecting deck configurations which are as lightweight as possible, adopting high strength steels in order to reduce weight where appropriate.
- Keeping all deck fittings, surfacing and equipment at their lowest weight consistent with suitable performance.
- Adopting for the main cables steel wires of the highest material strength compatible with other required performance characteristics.
- Careful selection of partial safety factors and working stress levels suitable for the specific case of a structure with very high self weight and dead load percentages.
- Adopting sag to span ratios as high as possible, consistent with other necessary performance requirements.

The final design for the Messina Bridge has a structural deck weight of about 18 t/m, with an average cable weight of 32 t/m. In simple terms, this means that one extra kilogram of deck weight implies more than one and a half extra kilograms in the main cable. In existing suspension bridges of lesser span this parameter is far less dominant. For the current world record span of the Akashi Kaikyo Bridge, the deck weight of about 23 t/m and cable weight of about 12 t/m, mean that one extra kilogram in the deck results in only about half an extra kilogram in the main cables.

Figure 6.4 ▶ Variation in cable size with span.

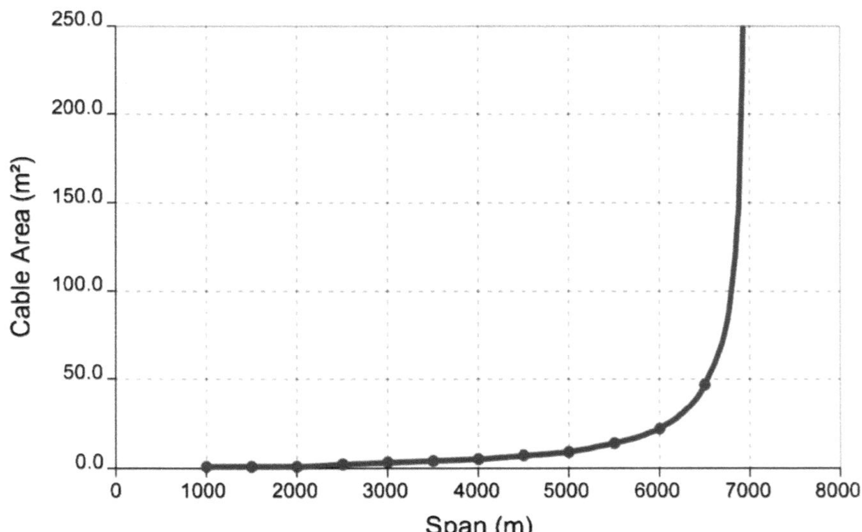

6.1.2 Scale issue two: dynamic properties and aeroelastic stability

Having seen how cable stiffness varies with span and how it becomes progressively dominant for static behaviour, it is also clear that the same becomes progressively dominant for the dynamic behaviour of the bridge, in relation to its inertial properties, namely the quantity and distribution of mass. Figures 6.5(a) to (f) show the typical first mode frequencies of a suspension bridge, based on the Messina analysis, which do not necessarily occur in the order listed:

- The first two lateral modes, a) symmetric and b) antisymmetric, are both associated with a lateral "pendulum" motion of the deck and cables, and are dominated by geometric stiffness. The symmetric lateral mode is usually the first absolute structural mode of a long suspension bridge; i.e. the mode with the lowest natural frequency.
- The first two vertical modes are c) symmetric and d) antisymmetric. Antisymmetric modes are dominated by geometric stiffness, while symmetric ones involve a higher participation of axial strain in the cables and hence of its axial stiffness. Depending on cable configuration, either mode can have the lower frequency, usually becoming the second absolute bridge mode. These modes are often referred to as "flexural" or "bending" modes because of the deflected shape of the deck. However, such terms can be somewhat misleading, since it is the cable stiffness and not the deck stiffness that is dominant, so the term "vertical" modes is preferred herein.
- The first two rotational modes are e) symmetric and f) antisymmetric. As for the vertical ones, the antisymmetric modes are dominated by geometric stiffness, while the symmetric ones involve a higher participation of axial strain in the cables. Again, either mode can have the lower frequency depending on the span. These modes, higher than the corresponding lateral or vertical modes, are often indicated as "torsional" due to the form of deck deformation, but again it is considered that such terms can be somewhat misleading, as the cable stiffness and not the deck stiffness is dominant. Therefore, the term "rotational" is preferred herein.

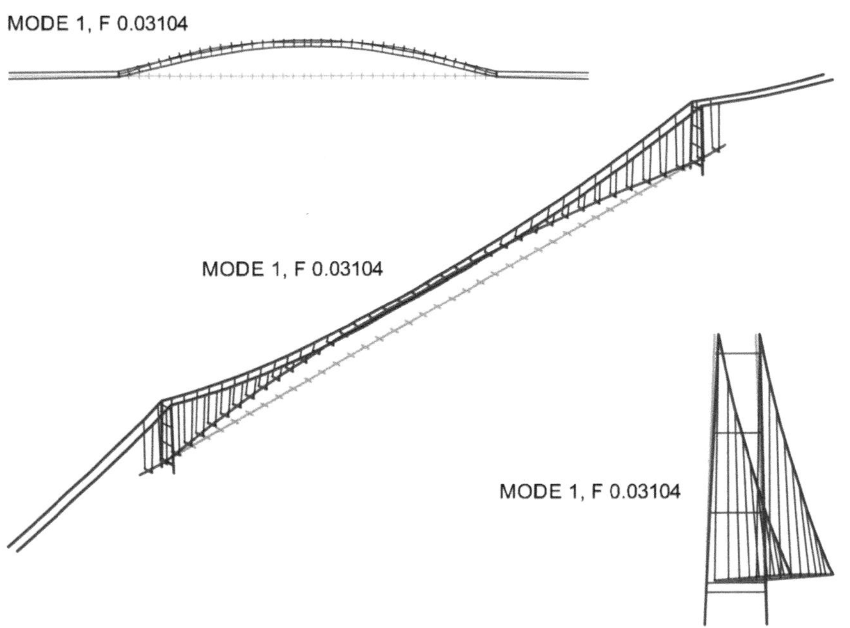

◄ Figure 6.5a Symmetric lateral modes. Period for the Messina Bridge ≈32 sec.

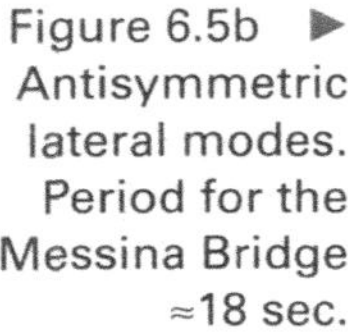

Figure 6.5b ▶ Antisymmetric lateral modes. Period for the Messina Bridge ≈18 sec.

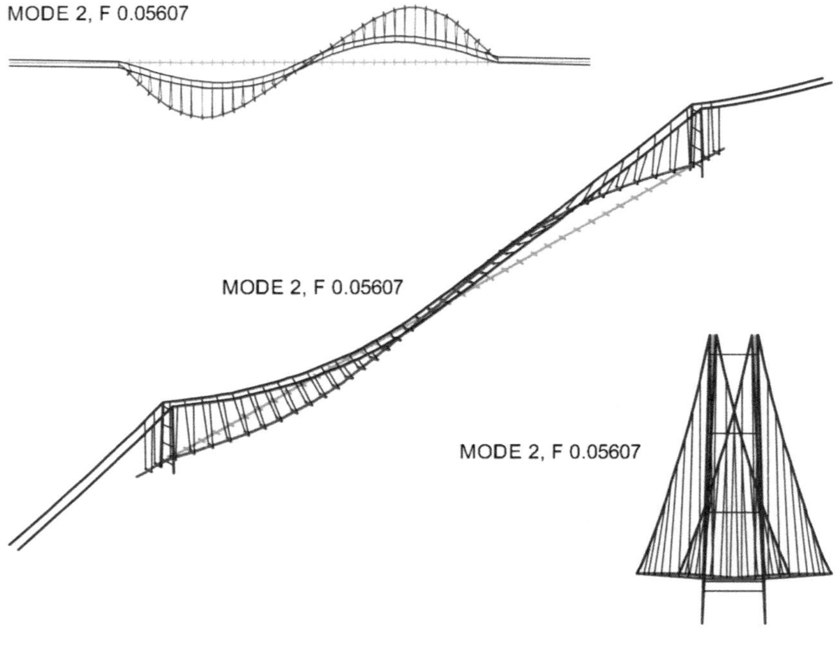

Figure 6.5c ▶ Symmetric vertical modes. Period for the Messina Bridge ≈12 sec.

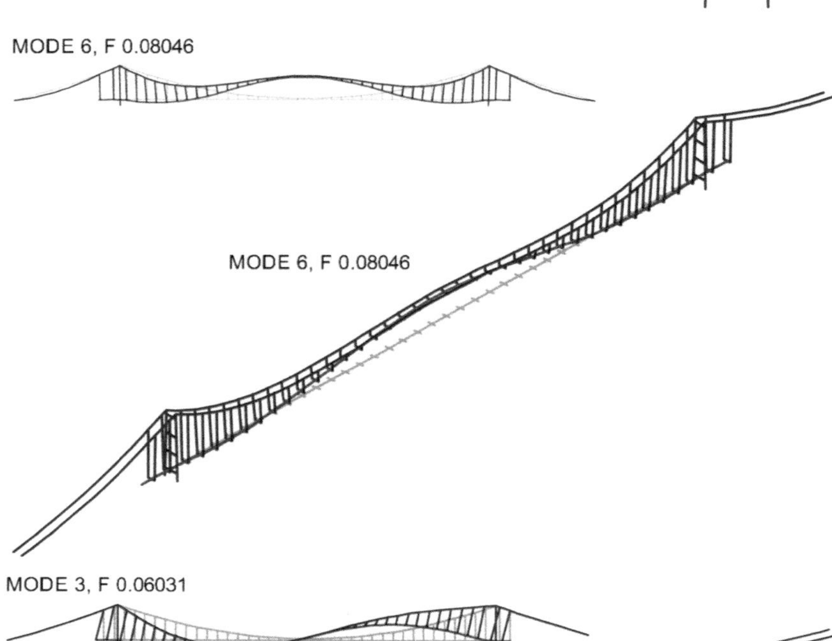

Figure 6.5d ▶ Antisymmetric vertical modes. Period for the Messina Bridge ≈17 sec.

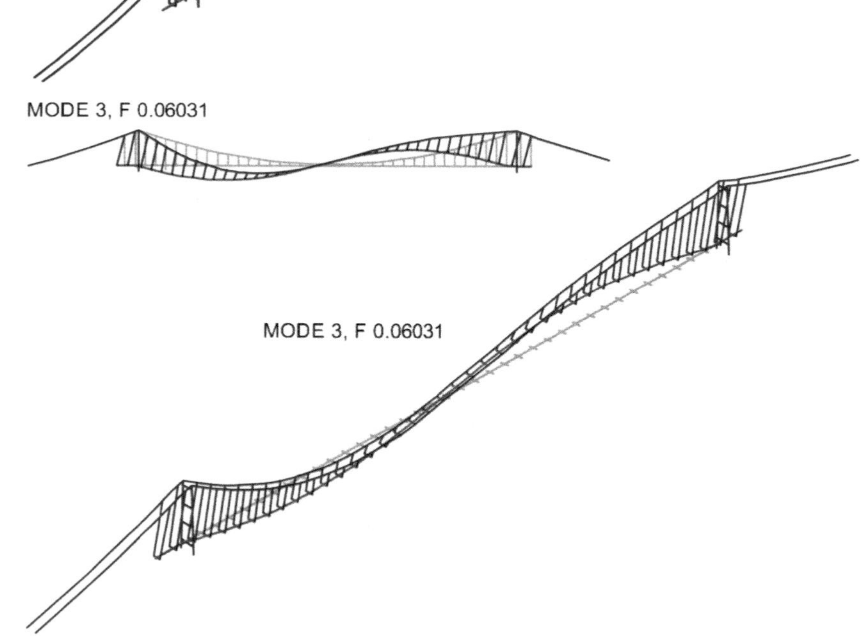

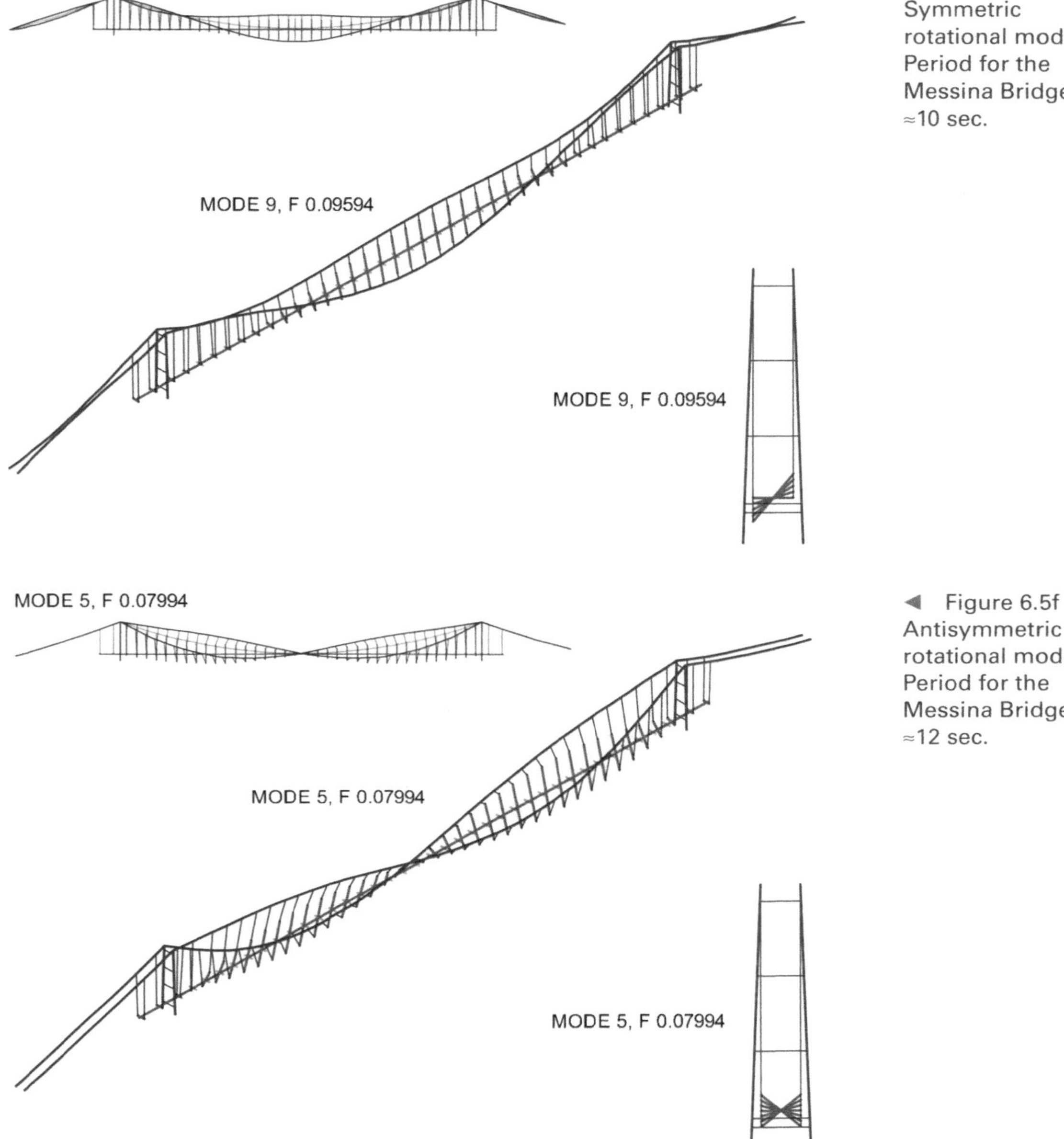

◄ Figure 6.5e Symmetric rotational modes. Period for the Messina Bridge ≈10 sec.

◄ Figure 6.5f Antisymmetric rotational modes. Period for the Messina Bridge ≈12 sec.

The main cable stiffness is dominant in defining the dynamic properties for any suspension bridge, as has also been shown for static behaviour. What changes the picture for very long spans is the fact that the main cables become the largest element for weight and mass, so that they also dominate the inertia distribution. In other words, the mode shapes and frequencies of a very large span suspension bridge become progressively more similar to those of stand-alone cables, while the effect of the other bridge elements becomes progressively smaller. This is of particular relevance for the vertical and rotational modes, whose frequencies become closer for increasing spans. At the limit, if the two main cables were in a stand-alone, perfectly restrained condition, the two modes would have the same frequency, corresponding to in-phase and out-phase oscillations of the two identical perfect cables.

For a given configuration and sag to span ratio, a number of factors, of different significance, contribute to maintaining a certain frequency separation between the two modes, the main ones being:

- The ratio between deck torsional and flexural stiffness, with high values of this parameter increasing the rotational frequencies. This parameter can be tuned in the design, although with more difficulty and less effectiveness for increasing spans.
- The different inertia distribution of deck and main cables. The cable mass is concentrated in the suspension cable planes, i.e. generally at the edges of the structure, while the deck mass is roughly evenly distributed between the cable planes. For vertical modes this means that the inertia contribution does not vary with location. The opposite applies for rotational modes, since the deck contribution is less significant than the cables, leading to increased rotational frequencies with respect to vertical ones with increasing span. The smaller the inertia radius with respect to half the horizontal distance between the cable planes, the larger is the increase in rotational frequency with respect to the vertical.
- The tower properties and specifically the ratio between its overall longitudinal flexural stiffness and its stiffness for rotations about its vertical axis. The former participates in vertical bending modes of the bridge, and the second, which is usually higher due to the connections between the tower legs, participates instead in bridge rotational modes.
- The longitudinal stiffness of the connection between the main cables and the deck. If vertical hangers only are present such stiffness is low, and in some bridges the main cables are connected directly to the deck at midspan via triangulated steel struts or ropes. This has the effect of modifying the relative values of vertical and rotational stiffness in the bridge.

This trend to decrease frequency ratio with span is illustrated in Table 6.1, which shows the rotational/vertical frequency ratios for some existing bridges and the proposed Messina Bridge. While the existing bridges have a ratio well over 2, with a minimum of 2.35 for the largest existing span of the Akashi Bridge, the value for Messina reduces to about 1.36.

The attention given here to the above parameter is because of the significance of the second fundamental aspect that dominates the design of very large span bridges. The rotational/vertical frequency ratio, together with the deck aerodynamic properties, determines the wind response properties of the bridge for the most dangerous possible form of aeroelastic instability, i.e. classic flutter, which arises due to coupling of rotational and vertical modes. This will be considered in the next section, but suffice it here to say that the closer the modal frequencies, or in other words the lower the above frequency ratio, the lower is the wind speed at which instability arises. The wind speed at which instability arises increases as the frequency ratio increases, or as the aerodynamic lift and moment on the deck decrease. This subject will be dealt with in Section 6.2.

Table 6.1 ▶ Rotational/vertical frequency, for first modes with a corresponding number of half waves in the mode shape.

Bridge	Span (m)	Deck type	Frequency ratio
Severn	988	Box	2.65
Humber	1410	Box	2.80
Storebaelt	1624	Box	2.79
Akashi	1991	Truss	2.35
Messina	3300	Multi-box	1.36

Existing bridges tend to exhibit critical wind speeds for flutter of the order of 60–70 m/s. Reaching similar values for a bridge with a frequency ratio as low as 1.36 is extremely challenging.

This then defines the second main scale issue which dominates the design of very long spans, namely achieving adequate flutter stability for a bridge characterised by intrinsically close rotational and vertical mode frequencies and shapes.

While for medium to large spans it is possible, up to a certain extent, to improve stability by making changes to structural parameters, e.g. modifying the deck stiffness to increase frequency ratios, for very large spans the main solution must be to improve the deck aerodynamic properties. That is to leave aside for the moment other possible countermeasures, such as introducing active control devices, which still require further research, particularly into providing adequate robustness and reliability, before they can be used on such a major piece of infrastructure. It is firmly believed that at present such devices are to be considered only for temporary construction conditions or for the control of phenomena which do not endanger bridge safety, e.g. repeated small amplitude oscillations and fatigue. Global stability must still be achieved through intrinsic, passive design measures, since only these can provide the required reliability.

Other less significant, but nevertheless cost-important, objectives connected with the static wind behaviour include the achievement of low aerodynamic drag so as to minimise lateral wind forces, which are the main influence on lateral tower behaviour (in addition to seismic effects), and of low aerodynamic moment so as to minimise the cross-fall slope of the road and rail platforms.

Such objectives must of course be obtained together with achieving minimum weight, in order to satisfy the first scale issue as well. Needless to say, several of these objectives can be contradictory. Before entering the description of how this has been handled for the Messina case, it is useful to comment on how the matter has been tackled for some existing bridges, for which the same considerations are valid although to a lesser extent.

As is well known, the problem of suspension bridge aerodynamic instability was dramatically brought to light by the 1940 Tacoma Narrows Bridge collapse. It is worth noting that the Tacoma designers had indeed understood that the deck is a secondary element for global static behaviour, seeking low weight solutions. The Tacoma deck abandoned the classic truss configuration of the 19th and early 20th century bridges and was simple, statically effective and light, comprising only two shallow I-beams in the suspension planes, plus small transversal I-beams, to carry the road platform, with an overall inverted U shape. While statically effective, such a geometry exhibits extremely undesirable aerodynamic properties, and these allowed the build up of large amplitude oscillations, leading to the collapse under a wind speed lower than 20 m/s.

The immediate reaction to the problem was the return to large truss decks providing high torsional stiffness. Very many bridges with such decks were built in the second half of the 20th century, often also introducing minor aerodynamic measures into intrinsically safe configurations. However, although truss decks can be made aerodynamically stable, they do not meet all the requirements stated by a long way. In the first place they are extremely heavy, and secondly (and by no means negligibly) they exhibit high drag due to their large depth. Figure 6.6 shows an example to represent them all: the deck of the Minami Bisan Seto Bridge in Japan. This is part of the Honshu-Shikoku project, and it is chosen as it comprises a road and rail platform such as is required for the Messina crossing.

Figure 6.6 ▶ Minami Bisan Seto deck (1988) Depth ~ 15 m Average self-weight ~ 1.5 t/m².

Courtesy of Honshu-Shikoku Bridge Expressway Company Limited – Japan.

A significant step forward (for certain span ranges) was achieved with the adoption of an orthotropic stiffened plate streamlined closed box deck. Such a deck, used for the first time in 1966 for the Severn bridge in Great Britain (Figure 6.7), is very light-weight and exhibits good torsional stiffness thanks to the closed box shape, a low flexural stiffness due to the shallow depth, together with a very low wind drag. Such a configuration implies overall costs definitely lower than truss decks for similar spans, and has been extremely successful in the last forty years, having been adopted for a large number of bridges worldwide.

This type of decks meets all the requirements for large spans except for the one that has been said to be of paramount importance.The flat wing shape and large solid lower surface results in high lift forces and thus hampers aerodynamic performance. In other terms, its intrinsic stability properties are good but not excellent; it is well known in aeronautics that a perfect wing suffers from two degree-of-freedom flutter instability. Orthotropic plate "wing" box decks are therefore the winning solution for spans up to about 1500 m, but progressively lose their best properties for longer spans. Already for two of the existing longest span bridges with box decks, the Humber Bridge (1410 m, 1981) and the Storebaelt Bridge (1624 m, 1998) it was necessary to increase the deck depth from the typical 3 m to about 4.50 m, to increase the torsional stiffness and achieve a not exceptional critical flutter speed of about 60 m/sec. Increasing the depth also results in a modest increase in drag and self weight.

Figure 6.7 ▶ Severn Bridge deck (1966) Depth ~3 m Average self-weight ~0.30 t/m².

It is believed that, while not impossible, a 2000 m span with such a deck would become comparatively complex, requiring a very large box and ultimately losing several of the positive aspects of this type of solution. This is confirmed by the solution adopted for the 1991 m span Akashi Bridge design. After carefully comparing different box deck schemes (Figure 6.9), the designers decided to adopt a deep, torsionally stiff truss scheme (Figure 6.10). It must however be borne in mind that the selection of the Akashi deck section was made in the late seventies, twenty years earlier than its completion, before the progress in understanding that was to be made in the following years. The weight issue was already well understood and important enough to require the use of high strength steel for the truss structure, not so much to save on the deck itself but to minimise the cable size as much as possible. The resulting design achieved an average self weight of about 0.85 t/m^2, which is excellent for a large truss but much higher than about 0.30 t/m^2 which would be typical of a Severn type steel box deck.

The authors believe that, on the grounds discussed, the two current longest spans in the world, Storebaelt and Akashi, represent bridge decks that are close to the limit of effectiveness in their own class. While precise limits do not exist in structural design, it is considered that either deck type would become unwieldy for spans over 2000 m and result in very high costs to achieve adequate performance. Neither is a possible solution for spans over 3000 m. The chronicle of how concepts and solutions evolved to meet such a challenging goal for the 3300 m span of the Messina bridge span is presented in the following section.

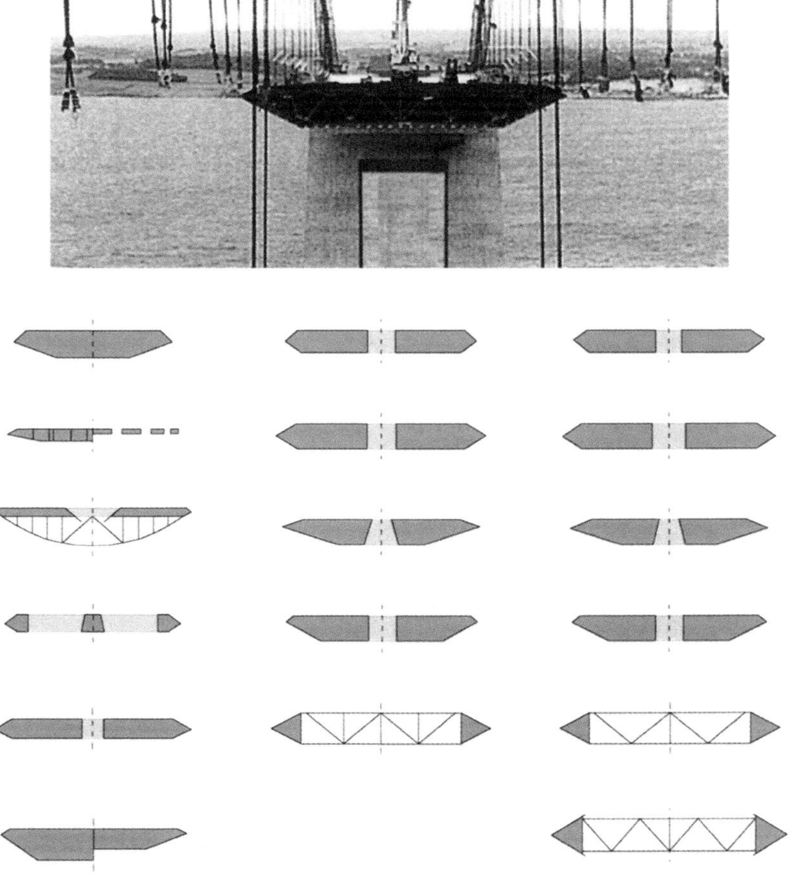

◄ Figure 6.8
Storebaelt Bridge deck (1998) Depth ~4.50 m Average self-weight ~0.32 t/m^2.

◄ Figure 6.9
Box alternatives considered for the Akashi Bridge.

Figure 6.10 ▶ Akashi Bridge deck (1998) Depth ~14 m Average self-weight ~0.85 t/m².

Courtesy of Honshu-Shikoku Bridge Expressway Company Limited – Japan.

6.1.3 Evolution of low-weight, low-lift highly stable decks for the Messina Bridge, 1977–2005

This section discusses how the final deck solution for the Messina deck was reached, in the period from the early seventies to the early nineties (see Chapter 1) to be subsequently improved and optimised.

Already in the seventies it had been appreciated that working on the structural side of the deck properties, e.g. by providing sufficient torsional stiffness to achieve aerodynamic stability by adopting a truss type deck, was a dead end towards feasibility for a three kilometre span. That Severn type box decks could not achieve sufficient stability at such a span was also well understood and proven in wind tunnel tests carried out in 1976–77.

It was therefore understood to be necessary, on the contrary, to work on the aerodynamic properties of the deck, reducing aerodynamic forces at birth, and a concept was already available. Immediately after the Tacoma collapse in 1940, the idea of inserting voids or gaps into bridge decks to reduce aerodynamic forces arose in the suspension bridge world. Such measures were adopted, for example, for the Tagus Bridge in 1966, among others, adopting grid strips within the road platform. (Figure 6.11)

While the idea in itself was not new, the early applications of this concept were rather simplistic and were only proposed to the limited extent of improving the behaviour of truss deck, which already exhibited adequate stability on its own at the spans in question and therefore possibly did not strictly need such measures.

Figure 6.11 ▶ Tagus Bridge deck (1966) Depth ~12 m Average self-weight ~1.2 t/m².

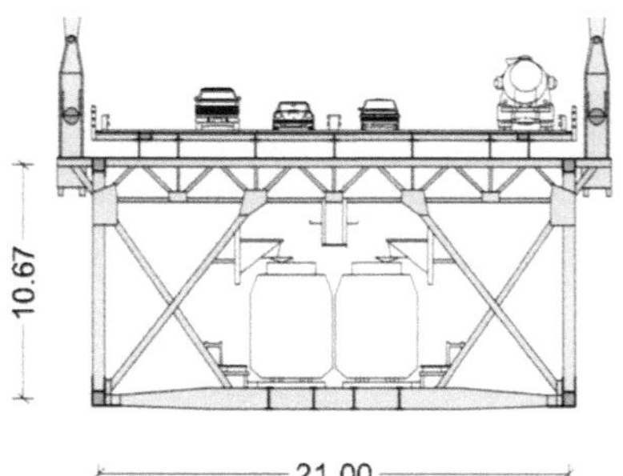

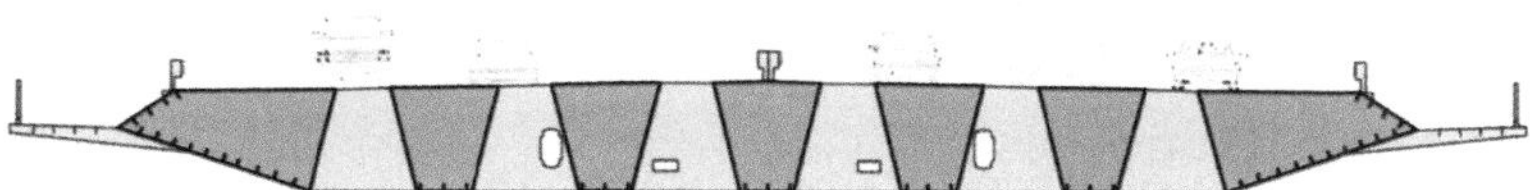

◄ Figure 6.12 The "vented deck concept" of the seventies.

A genuine evolution of the concept was proposed in the seventies by W.C. Brown who proposed the idea of combining low weight Severn type box decks with voids: the so called "vented deck" concept. This was not specifically developed for Messina but was proposed as a general way forward towards very large span suspension bridges. The scheme was based on the adoption of several small boxes with intermediate voids, closed by grids.

When W.C. Brown started his collaboration with the Messina design group in the late seventies, the vented deck concept was adapted to the road and rail bridge challenge, resulting in the configuration shown in Figure 6.13: a "double-decker", with the rail platform at the lower level, connected via a hammer strut to the upper road platform, formed by numerous small steel boxes with intermediate voids closed by grids, plus inclined ropes to connect the two levels and stiffen the whole.

The GPM design was a fundamental step forward, as for the first time it was demonstrated, also through wind tunnel testing, that ensuring adequate stability on a three kilometre span was indeed possible. Nevertheless, its configuration possessed a number of drawbacks and non-optimal solutions:

a. First, most relevant and intrinsic in the early vented deck concept, the running surface comprised alternating longitudinal strips from solid/asphalt to grids/steel, with transverse solid strips at cross girder locations. This is a clear hindrance to comfort and safety of vehicular traffic, and definitely serious in wet weather conditions. If not unthinkable in the seventies, such a surface would definitely be unacceptable for modern best practice.
b. Maintenance would have been difficult and costly, due to the large number of comparatively small elements present, all prone to corrosion and some not easily inspected.
c. Ensuring the correct balance of the stiffening rope tension with a relatively flexible strut was questionable.
d. From an aerodynamic standpoint, although effective, the void distribution was uniform across the deck width and hence intrinsically non optimal. Low lift force and twisting moment were achieved simply via a large total proportion of empty spaces, evenly distributed, without a specific rationale.
e. The distance between the cable planes had to be increased with respect to the road platform width by cantilevering the cross girders, in order to improve the critical mode frequency ratios.
f. The double deck configuration exhibited a moderately high aerodynamic drag.

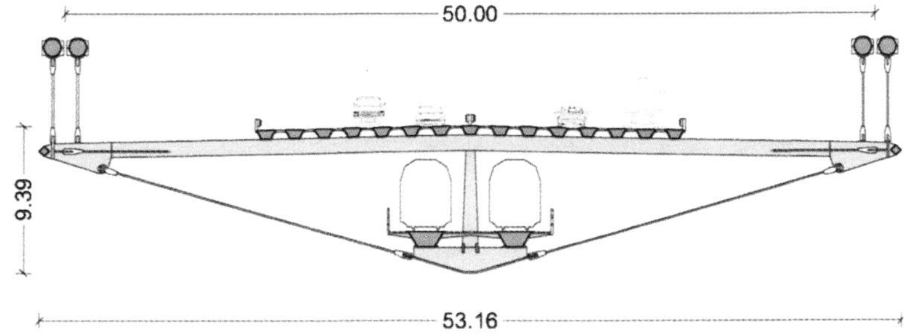

◄ Figure 6.13 GPM single span bridge deck, depth ~10 m, average self-weight ~0.45 t/m^2, (1977).

Hence, when activity was resumed by SdM in the early eighties, immediate attention was devoted to improving the scheme. The main new concepts were:

a. "Combining" the numerous small boxes and gaps into fewer larger ones.
b. As a considerable distance between cable planes was independently needed, the road and rail decks were located at the same level, using the lateral space available.
c. The lower hammer and rope structure was replaced by a 3-D truss girder that carried the road and rail boxes.

This resulted in the scheme shown in Figure 6.14 that was set at the basis of subsequent SdM feasibility studies and was the reference scheme for the feasibility report of 1986.

This configuration, although viable and adequate to demonstrate feasibility of the 3300 m single span suspension bridge, was nevertheless overshadowed by certain concerns:

a. The weight was considerable, due to the presence of a large truss stiffening structure below the road and rail decks.
b. The shape of the box was improved for structural efficiency and maintenance, but was rather simplistic from an aerodynamic standpoint. In consequence, the aeroelastic performance and stability limits were adequate but not outstanding.
c. Drag was non-optimal, due to the large depth, even with the open lower structure.
d. Maintenance was non-optimal, due to the large number of complex surfaces present.

Nevertheless, the work carried out to improve the aerodynamic behaviour was considerable and introduced the idea of adding to the deck passive aerodynamic profiles, examining different locations, sizes and shapes.The final decision was to adopt large wing-shaped aerofoils, aimed at both achieving the required smoothness and slope of the aerodynamic moment curve and acting as additional passive aerodynamic dampers. Seen as sizeable elements in Figure 6.14, such elements were subsequently optimised, to perform an extremely important role.

While the feasibility studies were being completed, consideration was given to further improving the configuration in a more basic manner, seeking all possible further improvements for both aerodynamic performance and low weight. Fundamental to this process was the confirmation that significant vertical deck stiffness was indeed not needed and even harmful in certain

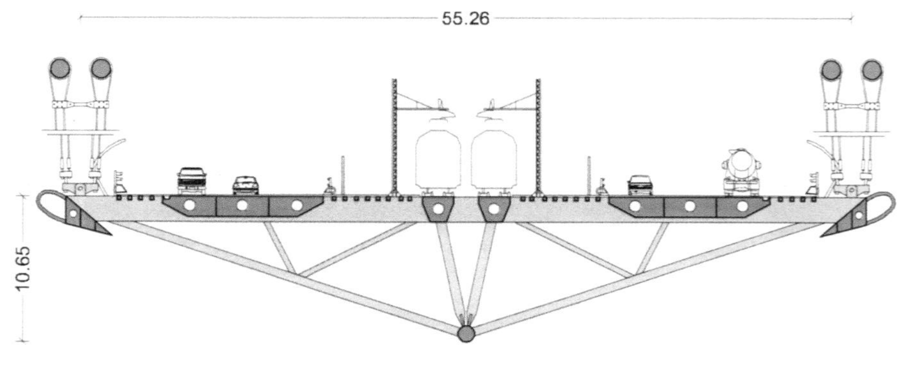

Figure 6.14 ▶ SdM reference scheme for feasibility analyses (1983) Depth ~11 m Average self-weight ~0.60 t/m².

circumstances: the design could do without the stiffening truss addition altogether as demonstrated above. Deck concepts were therefore sought to:

a. Drastically simplify the structure, reducing the number of elements to a minimum.
b. Introduce an aerodynamic shaping of the deck elements, beyond the simple presence of distributed voids, to achieve better stability and drag performance.

The first attempt in such a direction, optimal from a number of viewpoints, saw a twin deck scheme with two stiffened plate boxes and a single large gap in between. The rail platform was split in two, with one track per box, initially located along the outside edges close to the cable planes, and subsequently moved to the inside edges on either side of the central gap. (Figure 6.15)

This scheme was optimal for weight, local structural behaviour and maintenance, but revealed a decisive weakness in the fact that non-symmetrical rail loading would have caused absolutely unacceptable transversal cross-falls due to the high distance of the heavy loads from the bridge centre line. This was true for both the external and internal track location so the designers were obliged to abandon this configuration.

At the same time it was understood that to be effective the intermediate gap had to be very large, of the order of one third of the overall width, resulting in a very large platform. Note that, as already mentioned and within a different framework, twin deck configurations with intermediate gaps had been studied, almost at the same time, for the Akashi Bridge, and the small size of the gaps envisaged in that case was indeed one of the causes for its inadequate response, leading to the scheme being abandoned.

All this work revealed two fundamental points:

a. To limit transverse cross-falls, the rail tracks must be placed as close to the bridge centre line as possible.
b. To achieve optimal aerodynamic properties, the design must include air gaps and box shapes which suit the air pressure distributions due to wind. Such a statement might seem obvious, but this idea was not immediately introduced into early deck designs. In other words, if the air gaps were placed evenly across the width or concentrated in a single spot (e.g. the centre for evident reasons of symmetry), the total air gap width had to be a very large percentage of the whole, while selecting the "best" locations for the gaps could reduce their total width.

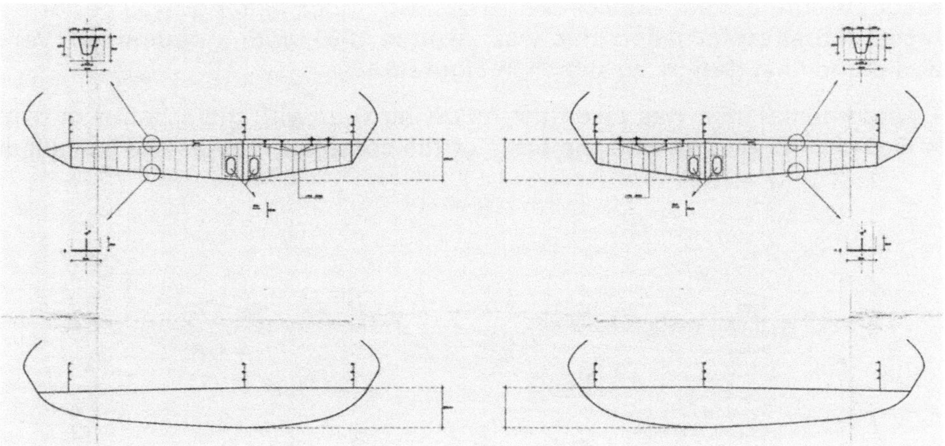

◀ Figure 6.15
Twin deck scheme, depth ~3.50 m, average self-weight ~0.35 t/m², (1987).

Such considerations led to the conclusion that the optimal number of boxes for a road and rail bridge was three, one for rail at the deck centre and two for each road direction. The first such scheme, as initially proposed, is shown in Figure 6.16.

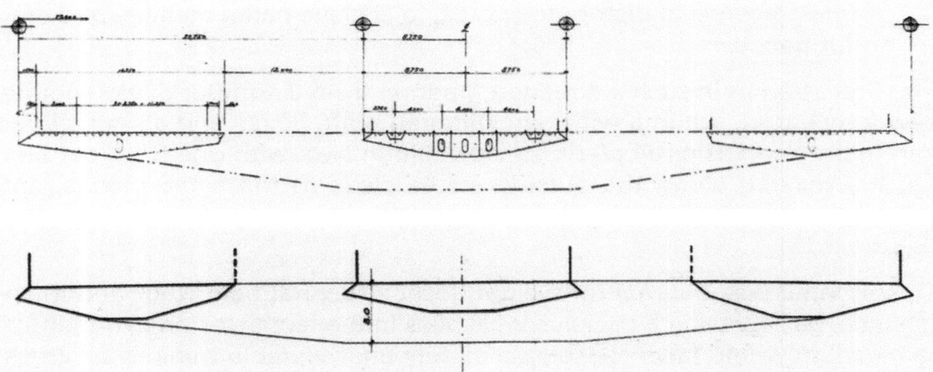

Figure 6.16 ► Initial triple box development scheme (1988) depth ~5.00 m, self-weight ~0.40 t/m².

Such an arrangement obviously solved the cross-fall problem, but also showed a number of other improvements:

a. The windward gap was located in a high pressure area, with improved effectiveness.
b. The box edges could be shaped to favour the air flow through the higher pressure windward gap, resulting in further improvements.
c. The centre rail box was located in an area of comparatively low vertical pressures and well in the wake of the windward road box, hence becoming almost "neutral" with regard to wind forces. To enhance this effect, its depth was reduced with respect to the road boxes. In simple terms, the effectiveness of the arrangement was close to that of a twin deck scheme with a single central gap, as if the rail box was absent altogether.

This produced an excellent basis for further development, towards an optimisation of the overall design, including the aerodynamic performance. In fact, the scheme needed significant structural improvement, as it was rather naïve in terms of plate arrangements (e.g. too many plates and angles), with room for improvement in structural effectiveness and constructability.

By the time the feasibility stage was completed and work had progressed towards an actual complete design of the bridge to act as the basis for construction, it had become clear that there was absolutely no need for the mixed truss scheme and that the basic configuration of Figure 6.16 showed adequate stiffness and service properties with much better overall performances. A first optimisation step was taken in 1991, with a preliminary version of the final design, as shown in Figure 6.17.

A fundamental step was taken the following year with the scheme shown in Figure 6.18. This formed the basis of the complete design (the so-called

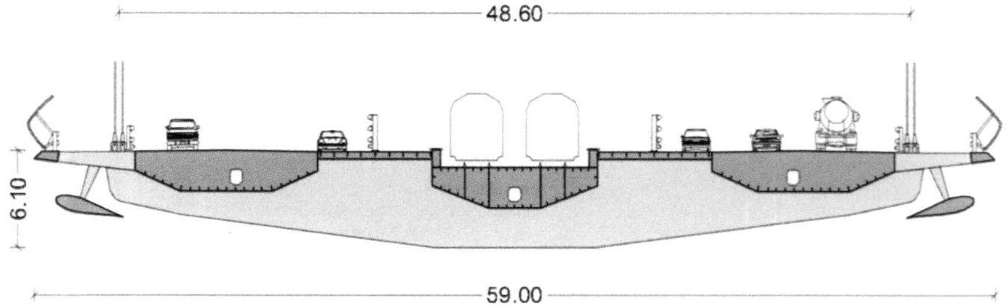

Figure 6.17 ► SdM development three deck design (1991) Depth ~6.10 m, Average self-weight ~0.45 t/m².

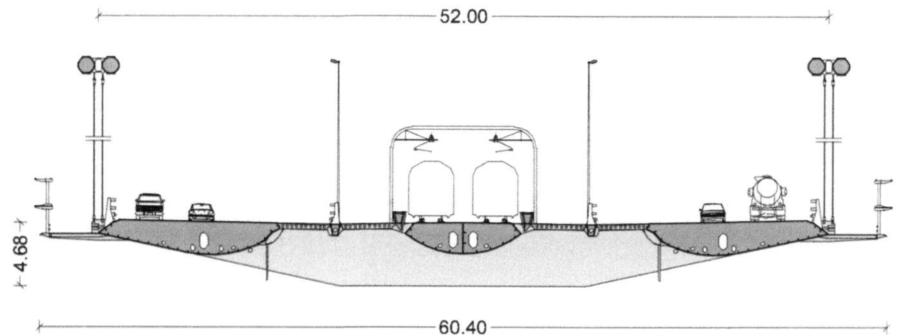

◀ Figure 6.18
SdM final deck design for approval ("Progetto di Massima") (1992)
Depth ~4.70 m
Average self-weight ~0.350 t/m².

"Progetto di Massima") which was submitted to the authorities responsible for examining the proposals and granting approval prior to any decision by the Italian Government to proceed to construction.

The Progetto di Massima comprises two railway tracks and six running road lanes on solid surfacing. Emergency strips, rail inspection areas and external inspection paths are located on open grids, hence making use of gap space also. Weight and stability are excellent, with a critical wind speed over 80 m/s. Furthermore, the deck is equipped with external windscreens to protect the traffic from direct wind action. These windscreens incorporate small horizontal aerofoils that are the evolution of the larger ones shown in Figure 6.17.

After a ten year period of approval procedures and further considerations by various succeeding governments, the project was re-considered. The basic deck concept was found still to provide excellent performance and was therefore left entirely unchanged. However, it was decided to improve certain aspects concerned with runnability and user comfort and with structural complexity and ease of maintenance. This resulted in the complete elimination of the grids, except for the external inspection strips, thus avoiding a number of small local struts supporting the grids themselves and the safety barriers. The entire road platform was located on a solid surface, and it was considered possible to eliminate the continuous rail inspection lanes. Four emergency stop areas were also added along the length of the bridge.

The resulting platform, shown in Figure 6.19, is equipped with four road lanes only plus emergency lanes. The traffic lanes are wider than the 1992 ones, being now up to full modern day standards, and are fully capable of handling predicted traffic volumes. The design has thus lost somewhat in total transport capacity but is largely optimised for maintenance and comfort.

The design based on the above deck formed the basis of the invitation to tender on a Design and Build basis in 2004. The tender conditions did not allow changes in the deck concept, but local optimisation was permitted. It is interesting and important to note that the submitted tender designs completely confirmed the effectiveness of the concept, introducing only minor design variations. These were concerned with details of the shape of the boxes and the optimisation of stiffened and orthotropic plate fabrication and fatigue performance. See Figures 6.20 and 6.21.

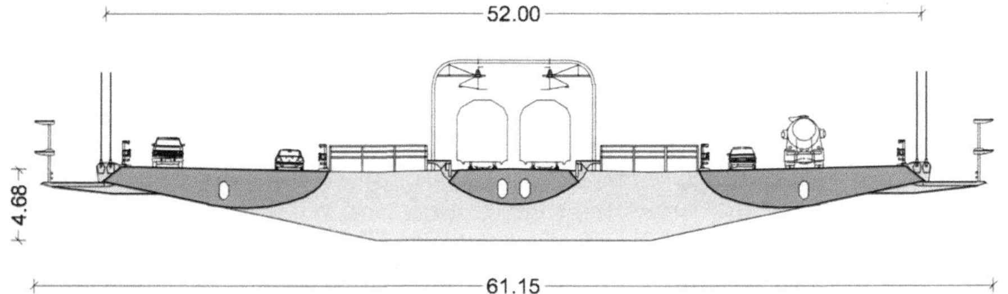

◀ Figure 6.19
SdM Preliminary Design (2002) used as the basis for the invitation to tender.
Average self-weight ~0.35 t/m².

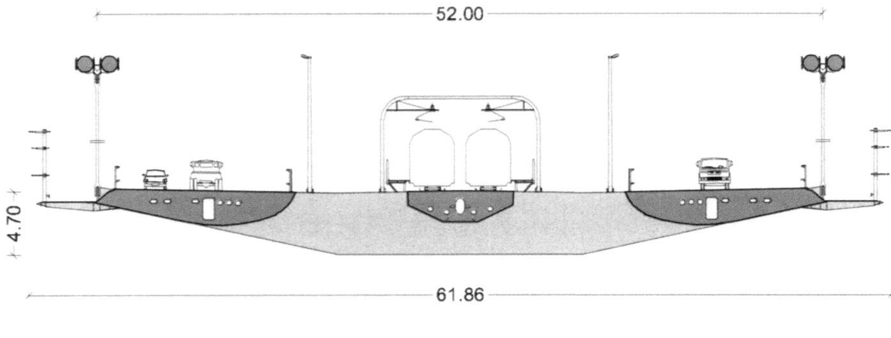

Figure 6.20 ► Impregilo JV tender design, winner (2005).

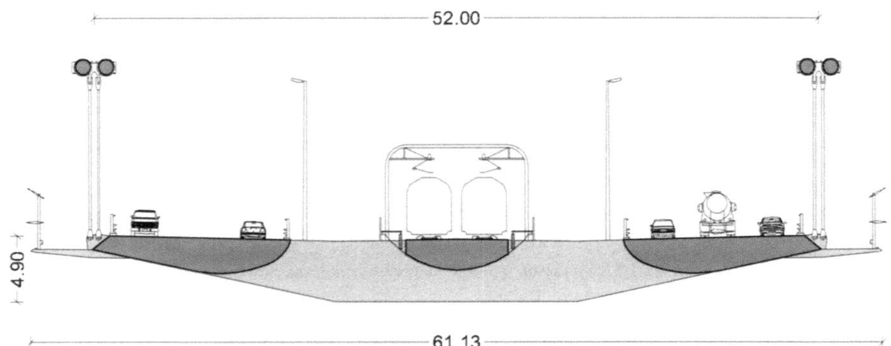

Figure 6.21 ► Astaldi JV tender design (2005).

Thus the multiple deck concept has been shown to be not only the optimal solution for the present state of the art within the Messina crossing project, but also an accomplishment of general effectiveness and interest toward even larger bridge spans, that can find useful application throughout the world. Examples supporting such an ambitious statement will be discussed further in Chapter 8.

6.2 Wind and aerodynamics

As we have already seen, the behaviour in wind is the major issue conditioning the design of a super long suspension bridge.

Wind action on the bridge produces aerodynamic forces which must be well identified in order to understand all the problems related to wind interaction. Consequently, and for the benefit of the non-expert reader, the first part of this section presents some basic information regarding the aerodynamic forces acting on a bridge due to wind action.

This section is arranged in the following sub sections.

6.2.1 Wind aerodynamic forces
6.2.2 Different types of problems related to wind effects
6.2.3 Static loads
6.2.4 Dynamic instability (Aeroelastic problems)
6.2.5 Vortex shedding
6.2.6 Buffeting

6.2.1 Wind aerodynamic forces

When a body or a general cross section as in Figure 6.22 is subject to a wind speed of velocity V the pressure distribution and friction action due to the fluid produces on the body a resultant force that can be resolved into a drag force F_D parallel to the wind and a lift force F_L normal to the wind direction.

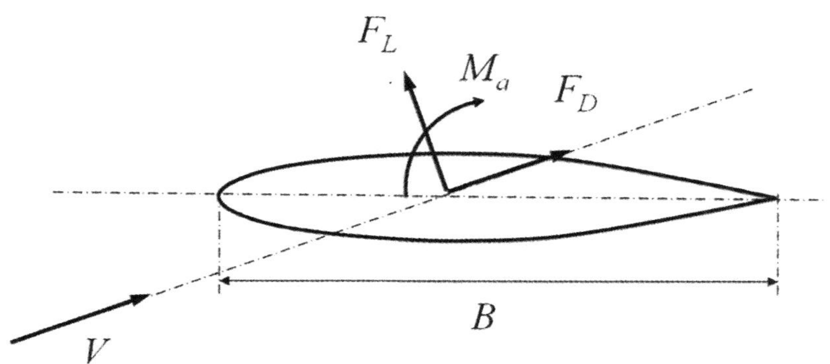

◀ Figure 6.22
Wind actions on a section and forces representation conventions.

An aerodynamic moment M_A is also applied to the body, acting around a reference point C where the F_D and F_L forces are applied.

The drag, lift and moment forces have the following expressions:

$$F_D(\alpha) = \frac{1}{2}\rho V^2 S\ C_D(\alpha)$$
$$F_L(\alpha) = \frac{1}{2}\rho V^2 S\ C_L(\alpha)$$
$$M_A(\alpha) = \frac{1}{2}\rho V^2 SB\ C_M(\alpha)$$

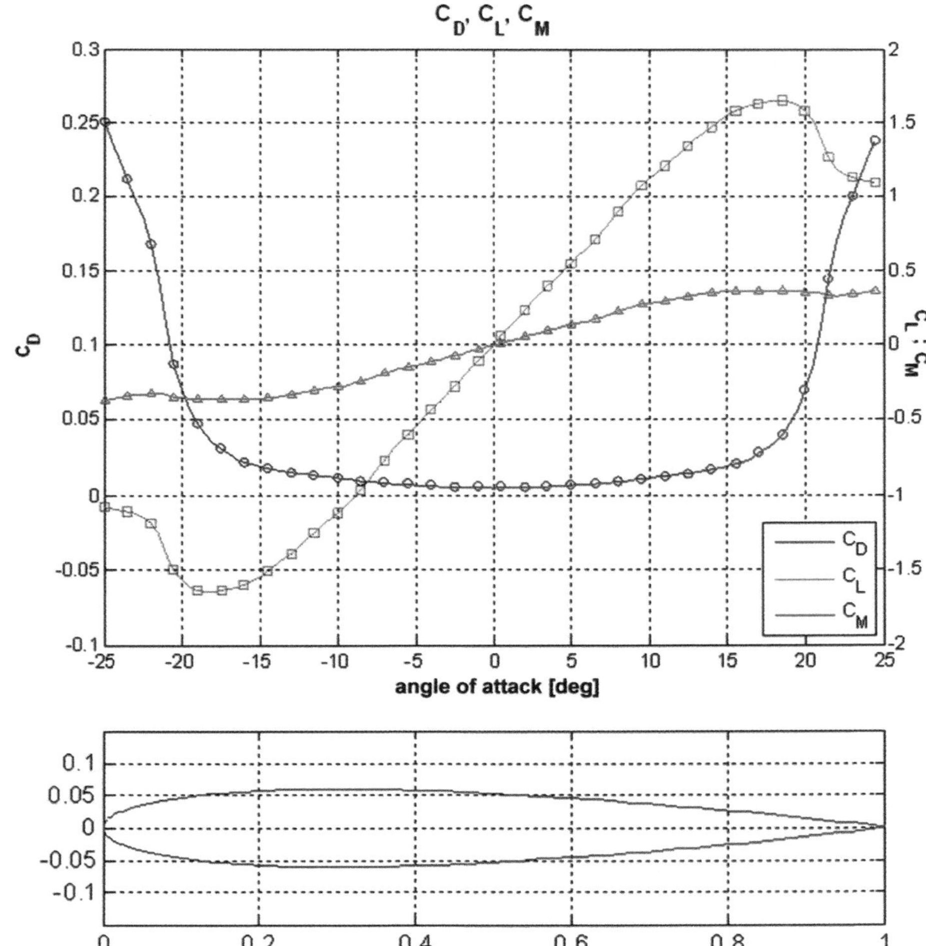

◀ Figure 6.23
Symmetric airfoil aerodynamic coefficients. The reference area is S = B*L, where L is the airfoil length.

In the above formula ρ is the air density, V is the air velocity, S is a reference body surface, B is a reference body dimension, and $C_D(\alpha)$, $C_L(\alpha)$, $C_M(\alpha)$ are the drag, lift and moment coefficients which are functions of the angle of attack α of the wind direction in respect of a reference axis.

Figure 6.23 shows an example of the variation of the $C_D(\alpha)$, $C_L(\alpha)$, $C_M(\alpha)$ coefficients for a symmetric airfoil similar to that illustrated in Figure 6.22.

The value of $C_L(\alpha)$ is increasing almost linearly with the angle of attack until a maximum value of around 1.6 is reached after which flow separation occurs on the surface of the airfoil as shown in Figure 6.24. Vortices or eddies are detached and the drag force increases as a consequence. The angle at which the maximum value of C_L is reached is defined as stall angle.

In a bridge, the deck is the most sensitive and important part for defining the behaviour of the structure in the wind, and different values of the $C_D(\alpha)$, $C_L(\alpha)$, $C_M(\alpha)$ coefficients are obtained depending on the deck shape.

Wind tunnel tests have to be performed to identify the aerodynamic properties of the deck. A sectional model of the deck, such as shown in Figure 6.25, is exposed to a constant wind flow, and the lift, drag and moment

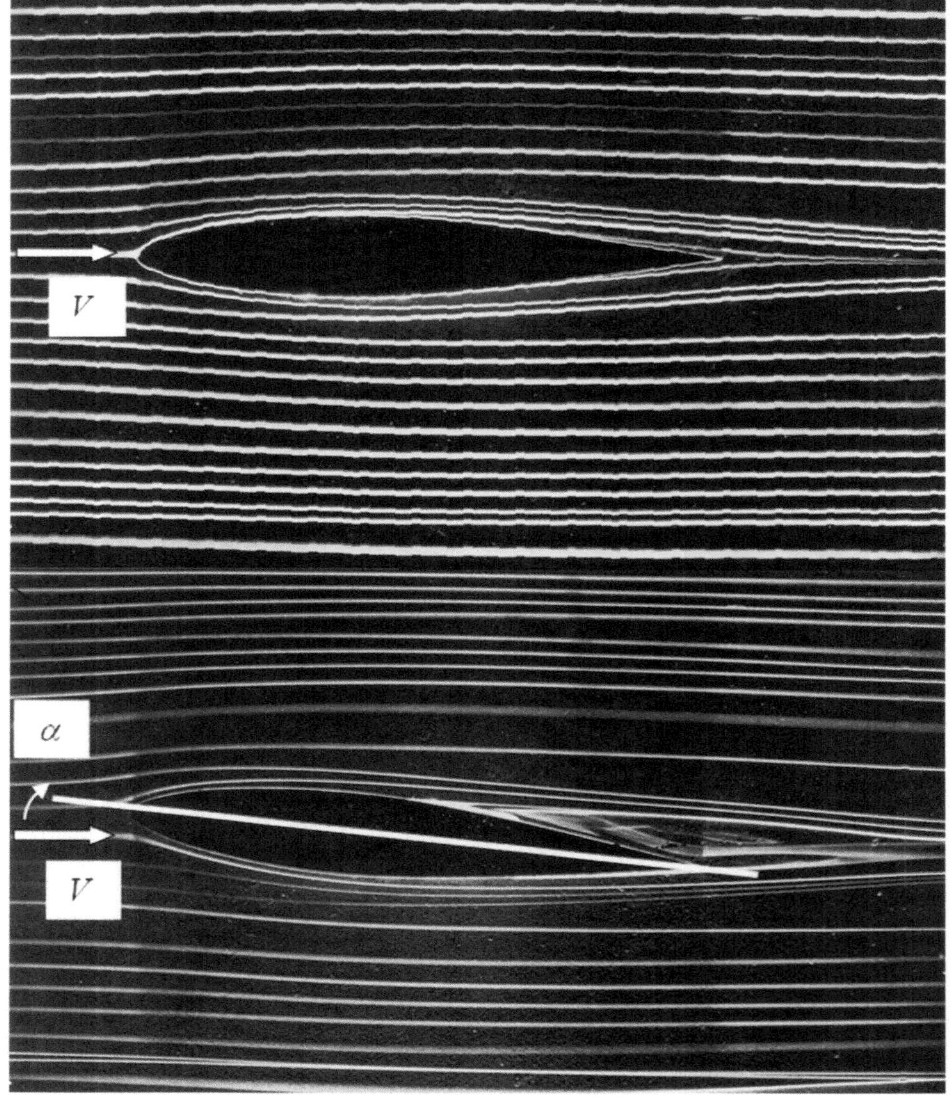

Figure 6.24 ► No separation on a wing profile at small angle of attack α. Separation and Vortex shedding at stall angle (large α).

◄ Figure 6.25 Dynamometric section model for wind tunnel tests.

coefficients are measured as a function of the angle of attack of the wind, i.e. the deck section rotation α.

Figure 6.26 shows the C_D, C_L and C_M coefficients for different types of deck section (Flat Plate, Messina, Akashi, Humber, Tacoma) taken as a reference, according to the conventions of Figure 6.22.

It can be seen that, if these decks have wing-like profiles they have similar lift and moment coefficients, whose values increase almost linearly with the angle of attack, as in the case of an airfoil or a flat plate. The $C_L(\alpha)$ and $C_M(\alpha)$ coefficient curves show a positive slope. The drag $C_D(\alpha)$ is low.

Decks with a large frontal section (like Tacoma), which are not similar to an airfoil, show a different trend for the $C_D(\alpha)$, $C_L(\alpha)$ and $C_M(\alpha)$ curves, characterised by a negative slope, indicating that they typically suffer for one degree of freedom instability. In the case of the Tacoma Narrows bridge, it will be seen later that a torsional instability led to collapse of the bridge.

Now, if for any reason the body is moving with a given velocity across the wind flow, the forces applied to the body are functions of the relative velocity V_{rel} of the wind in respect to the body and the expressions given can be applied introducing V_{rel} instead of V.

If for instance a deck is moving vertically with a given velocity $\dot{z}$, as in Figure 6.27, and is subject to a wind of velocity V, then the relative velocity, as experienced by the deck, is V_{rel} with an angle of attack α which is a function of $\dot{z}$. If the angle of attack is a function of $\dot{z}$ then also the drag, lift and moment forces will be functions of $\dot{z}$ or of the motion of the deck.

Similarly, if the deck is for any reason subject to a rotation as shown in Figure 6.28, the rotation ϑ of the deck will change the angle of attack α, being $\alpha = \vartheta$. So we can say more generally that any motion of the deck will produce a variation in the angle of attack, and with reference to the diagrams representing $C_D(\alpha)$, $C_L(\alpha)$ and $C_M(\alpha)$ it can be seen that this will

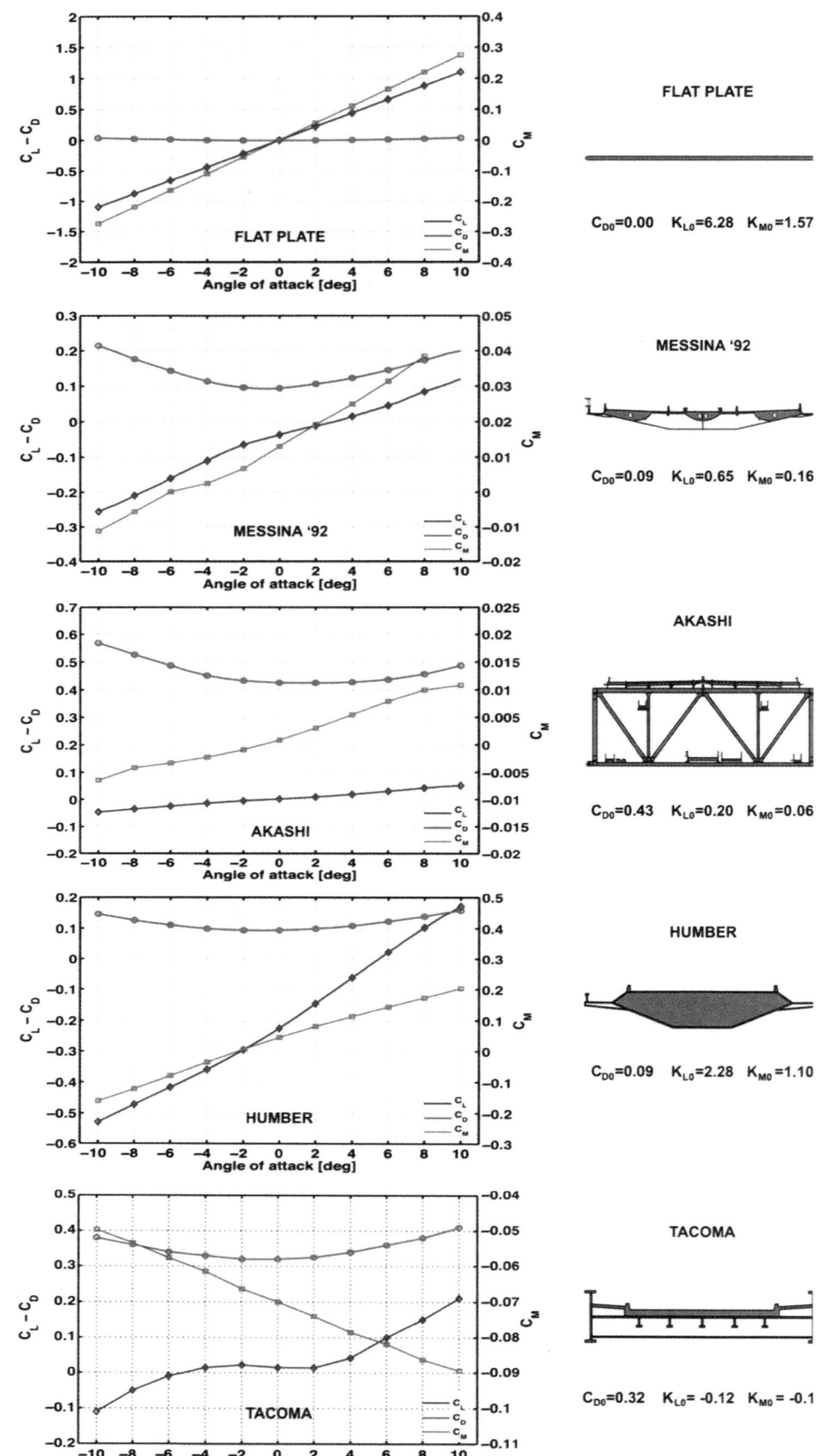

Figure 6.26 ► Aerodynamic coefficients for different deck types (Akashi, Humber, Tacoma) compared with Messina and with the typical flat plate. Reference area: S = B*L.

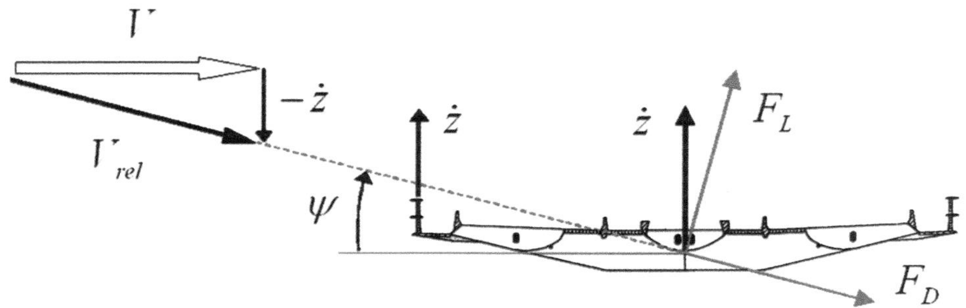

◄ Figure 6.27
Deck section moving vertically.

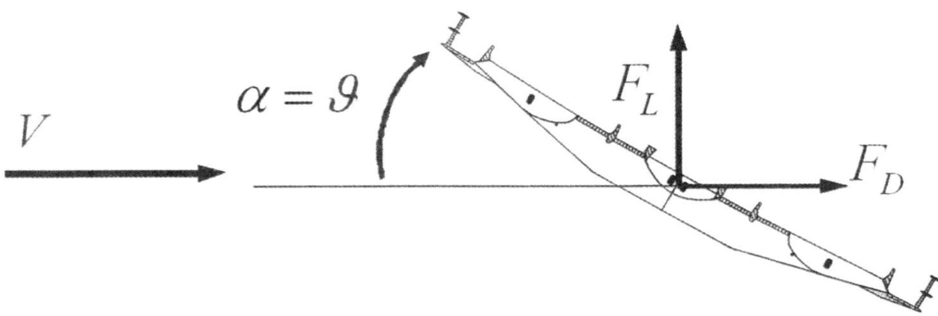

◄ Figure 6.28
Deck section moving torsionally.

produce a variation of the aerodynamic forces which become as a consequence functions of the motion. From this we can summarize that the aerodynamic forces are functions not only of the incoming wind, but also of the motion of the deck. In other words the aerodynamic forces are motion dependent, as well as being a function of the shape of the deck. If these aerodynamic forces produced by the motion act in favour of the motion (or introduce additional energy) then the motion is amplified and the bridge becomes unstable.

Different types of instability can occur as will be seen below. Appendix 1 presents in more detail an analytical definition of the motion dependent aerodynamic forces and the different approaches for identifying the wind forces using wind tunnel tests.

6.2.2 Different types of problems related to wind effects

The wind blowing on a bridge can produce different types of problems, related to the structure of the incoming wind and the particular type of bridge, both from a structural point of view and from an aerodynamic point of view.

The structure of the wind is characterized by how the nature of the bridge location affects the level of the turbulence.

The wind blowing on a structure like a bridge is characterised by an average component called the mean speed V_{Av} and by a fluctuating component which is a function of space and time, as illustrated in Figures 6.29 and 6.30.

The fluctuating horizontal $v(t)$ and vertical $w(t)$ components represent the level of turbulence in relation to V_{Av}. The V_{Av} direction for a bridge is generally in the horizontal plane, so that the fluctuating vertical component $w(t)$ of the wind turbulence produces a fluctuation of the angle of attack $\alpha = \psi(t)$ of

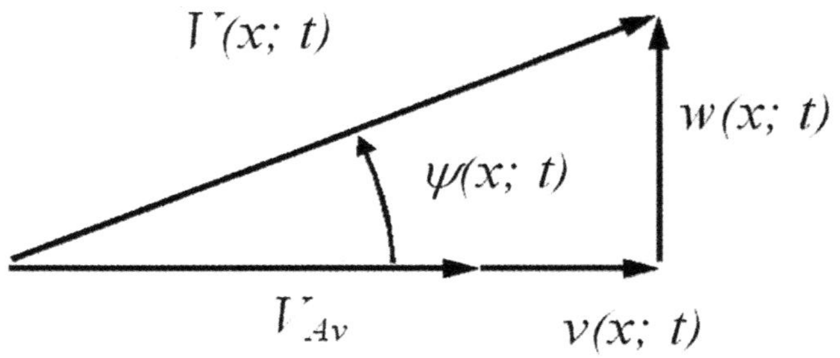

Figure 6.29 ▶ Average and turbulent components of the wind.

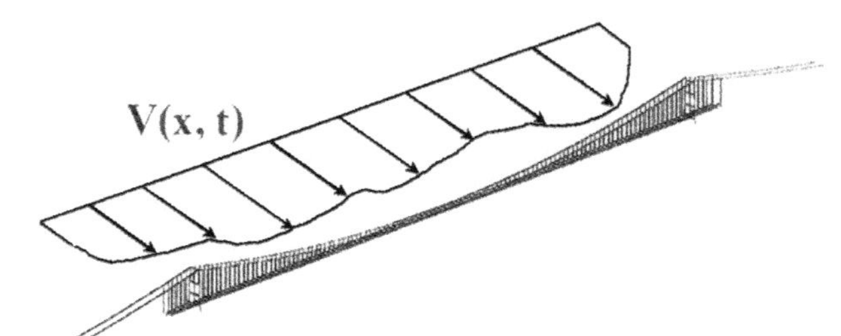

Figure 6.30 ▶ Wind turbulence: space variation of the wind speed along the span in the same instant of time.

the type shown in Figure 6.31. These fluctuations of angle of attack produce variations of drag, lift and moment forces due to turbulence.

Figure 6.32 shows the time history of the wind speed at a certain point of the bridge main span.

In the expression for the aerodynamic forces, the wind velocity $V(t)$ due to turbulence is function of time, so the aerodynamic forces applied to a bridge are also functions of time because the incoming turbulence produces not only a fluctuation of the intensity of the wind $V(t)$ but also a variation of the angle of attack $\alpha(t)$ due to $\psi(t)$.

In summary, we can say that the mean value of the incoming wind speed produces static loads on the bridge that strongly influence the static design of the bridge, while the fluctuations of speed and direction with time at different points of the structure produce dynamic loads, which, also as a function of the space correlation, generate that kind of wind induced motion generally referred to as buffeting.

The motion produced by turbulence causes a variation of the aerodynamic forces due to the motion or motion-induced aerodynamic forces, as already explained in Section 6.2.1. If the motion-induced aerodynamic forces

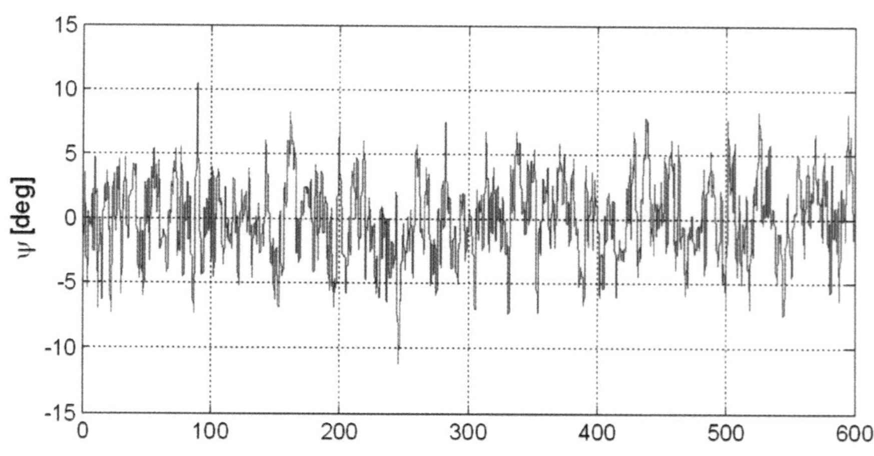

Figure 6.31 ▶ Wind turbulence: angle of attack of the wind as a function of time.

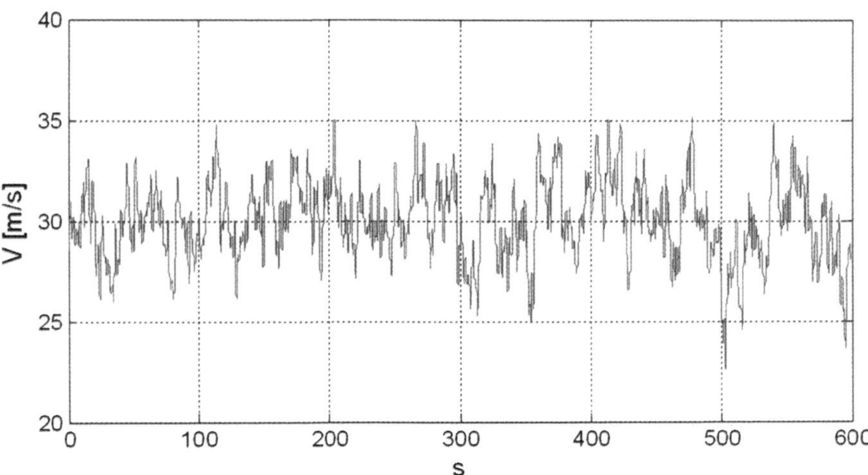

◄ Figure 6.32 Wind turbulence: instantaneous wind speed as a function of time at a given position on the main span.

introduce additional energy into the system, the motion is amplified and the system becomes unstable. Conversely, if the motion-induced aerodynamic forces dissipate energy, the system is stable.

Different forms of instability can take place, as it will be discussed later on:

- single degree of freedom vertical instability.
- single degree of freedom torsional instability.
- coupled vertical and torsional two degree of freedom instability, generally called flutter instability.

These forms of instability are related to the structural characteristics (natural frequencies and modes of vibration and the associated damping) and to the aerodynamic properties of the bridge.

For a suspension bridge, the instability, or motion-induced problem, comes mainly from the aerodynamic behaviour of the deck, principally due to the following reasons:

1. the main cables generally have an aerodynamically stable section.
2. the towers have a small influence and can generally be excited only by vortex shedding.

As will be explained in Section 6.2.4 and Appendix 1, the type of instability or motion-induced problem pertaining to a certain deck section can be deduced, in the first instance, from an analysis of the coefficients $C_D(\alpha)$, $C_L(\alpha)$ and $C_M(\alpha)$.

First of all, to avoid a single degree of freedom instability, with the sign conventions of Figure 6.22, the $C_L(\alpha)$ and $C_M(\alpha)$ coefficients derivatives must be positive.

The second observation is that wing-like profiles, as already said, have similar lift and moment coefficients, whose values increase almost linearly with the angle of attack, as in the case of an airfoil or a flat plate. The $C_L(\alpha)$ and $C_M(\alpha)$ coefficient curves show a positive slope. The drag $C_D(\alpha)$ is low. Due to these features, wing-like profiles do not suffer from single degree of freedom instability, while they do suffer from coupled vertical and torsional two degree of freedom instability (flutter). (See Section 6.2.4 and Appendix 1.)

Bridge decks like the Tacoma Narrows Bridge, (see Section 6.2.4), or decks with a large frontal section, show a different trend for the $C_D(\alpha)$, $C_L(\alpha)$ and $C_M(\alpha)$ curves, characterised by a negative slope, indicating that they typically suffer from single degree of freedom instability. In the case of the Tacoma Narrows a torsional instability led the bridge to collapse.

In summary then, the main problems related to wind effects are:

- Static loads, due to the mean wind speed.
- Single degree of freedom vertical or torsional instability.
- Two degree of freedom coupled flutter.
- Vortex shedding.
- Buffeting (motion induced by turbulence).

These different problems will be described in more detail and discussed in the sections below, illustrating the choices made for the Messina bridge to overcome these problems and obtain an optimised solution. (Note that the work described relates mainly to the SdM 2002 Preliminary Design and not the tender designs submitted by the bidding contractors in 2005.)

6.2.3 Static loads

Static loads due to wind aerodynamic forces are very important for a long suspension bridge and they condition all the bridge design. These loads are applied to the towers, the deck and the main cables (Figure 6.33).

In order to define the loads on the cables, towers and deck, aerodynamic tests in a wind tunnel must be carried out to identify the aerodynamic coefficients for these elements.

As an example, in Figure 6.34 the aerodynamic coefficients of the Messina bridge twin main cable are reported as a function of the angle of attack of

Figure 6.33 ► Wind incident on cables, towers and deck.

The background photo of Akashi Kaikyo bridge is a courtesy of Honshu-Shikoku Bridge Expressway Company Limited – Japan.

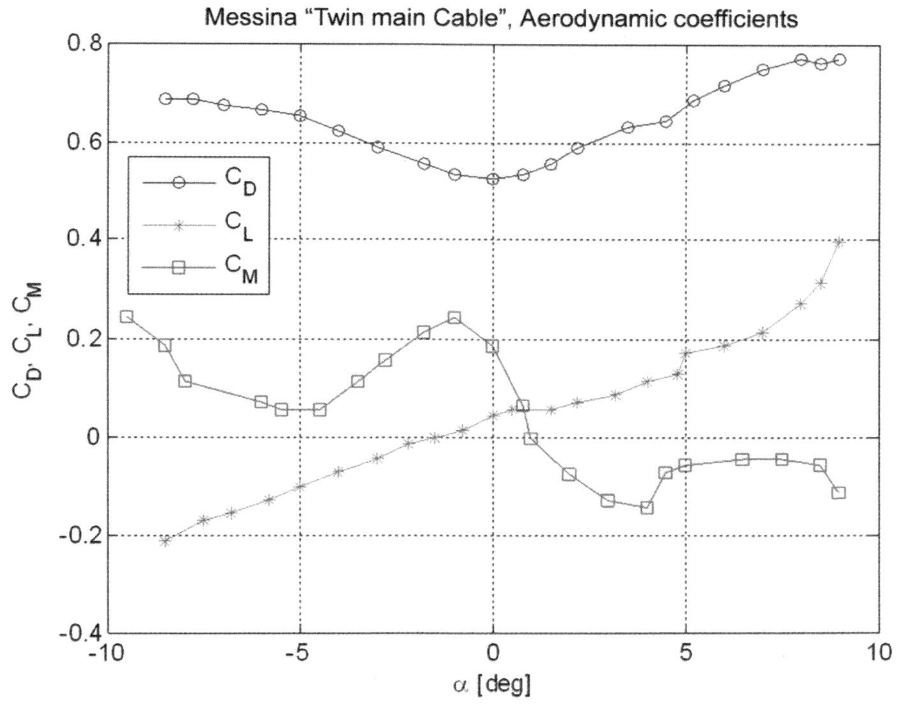

◄ Figure 6.34 Aerodynamic coefficients of the Messina bridge Twin Main Cable. The frontal area (Diameter times Length) has been assumed as reference area for the twin cable global aerodynamic force. Reference Length for aerodynamic Moment assumed equal to cable diameter.

the wind. Figure 6.35 shows the lift, drag and moment coefficients of a sectional model of the tower, with and without the screens which are proposed to improve aerodynamic performance.

The aerodynamic coefficients for different types of bridge deck have been already described in Figure 6.26. Due to the fact that the maximum static wind load on the structure has to be considered, the worst situation is generally with the wind normal to the bridge axis. Thus with reference to the previous Figures the most important parameter is the drag coefficient with angle of attack equal or close to zero. The most significant component is indicated by the deck drag coefficient, since the drag contribution of the main cables and hangers is less important due to their smaller frontal section.

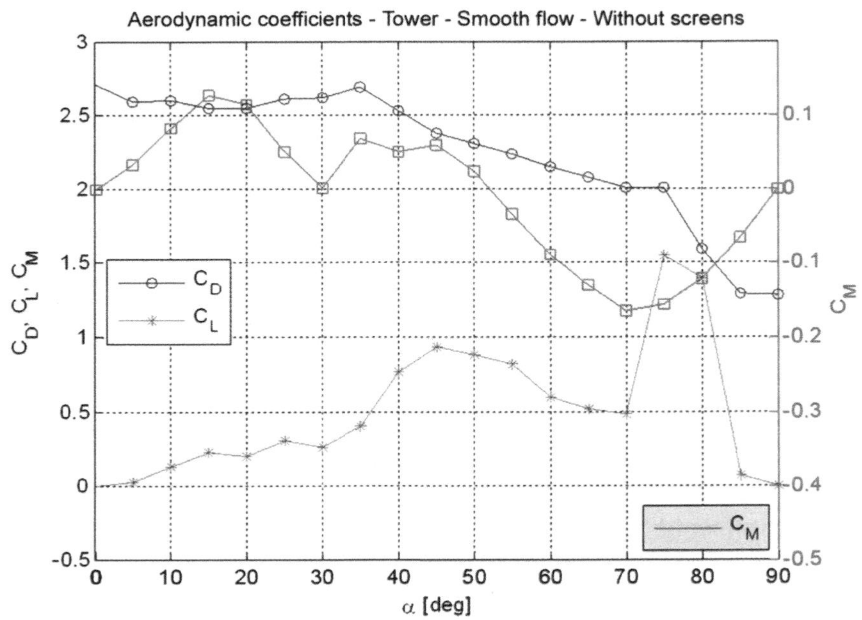

◄ Figure 6.35a Aerodynamic coefficients of the Messina bridge tower as a function of the exposure angle $\tilde{\alpha}$. The reference surface S for Drag, Lift and Moment is the frontal area of one tower leg. The reference length for the aerodynamic Moment is B = 64.5 m. Smooth flow without screens.

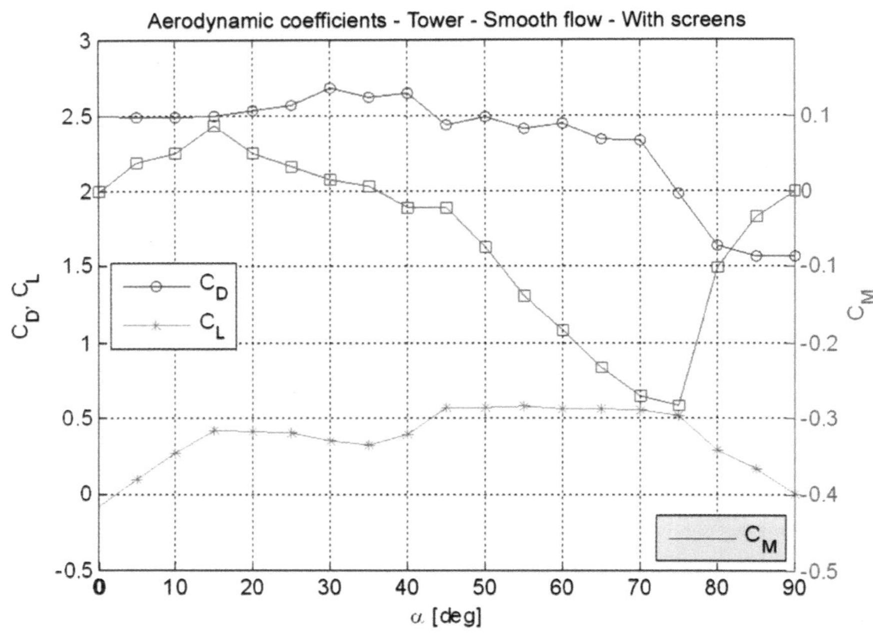

Figure 6.35b ► Aerodynamic coefficients of the Messina bridge tower as a function of the exposure angle $\tilde{\alpha}$. The reference Surface S for Drag, Lift and Moment is the frontal area of one tower leg. The reference length for the aerodynamic Moment is B = 64.5m. Smooth flow with screens.

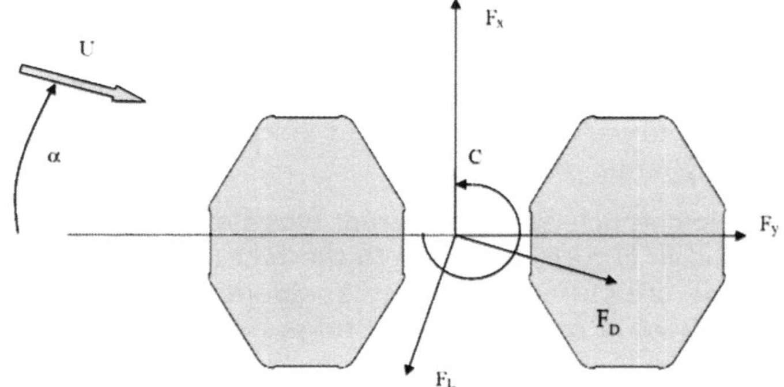

Figure 6.35c ► Scheme of the wind test: cross section of the tower legs and exposure angle $\tilde{\alpha}$.

As an example, for a mean wind speed of 10 m/s, the load on each "twin" main cable is about 45 N/m, while the load on the deck is almost 10 times greater in the order of 350 N/m.

6.2.3.1 The Messina case

As already explained, for very long suspension bridges the static wind load on the deck is transferred through the hangers to the main cables, since the deck lateral stiffness is negligible in comparison with that of the main cables.

So the overall static load (deck + cables) is applied to the top of the towers, producing a very large flexural moment on the tower. This is the most important load affecting the tower design.

From this point of view, it is clear that the deck drag must be kept as small as possible, in order to reduce the load at the top of the towers. This tower load is proportional to the cable tension and the angle β (in the horizontal plane) between the vertical plane containing the main cable at the top of the tower and the plane connecting the two towers. A measure of this angle, for a certain span length, is related to the horizontal deflections of the bridge deck and cables at mid-span, as in Figure 6.36(a).

To better appreciate the problem, we can compare the maximum static horizontal deflection of the Akashi bridge with that of the Messina bridge.

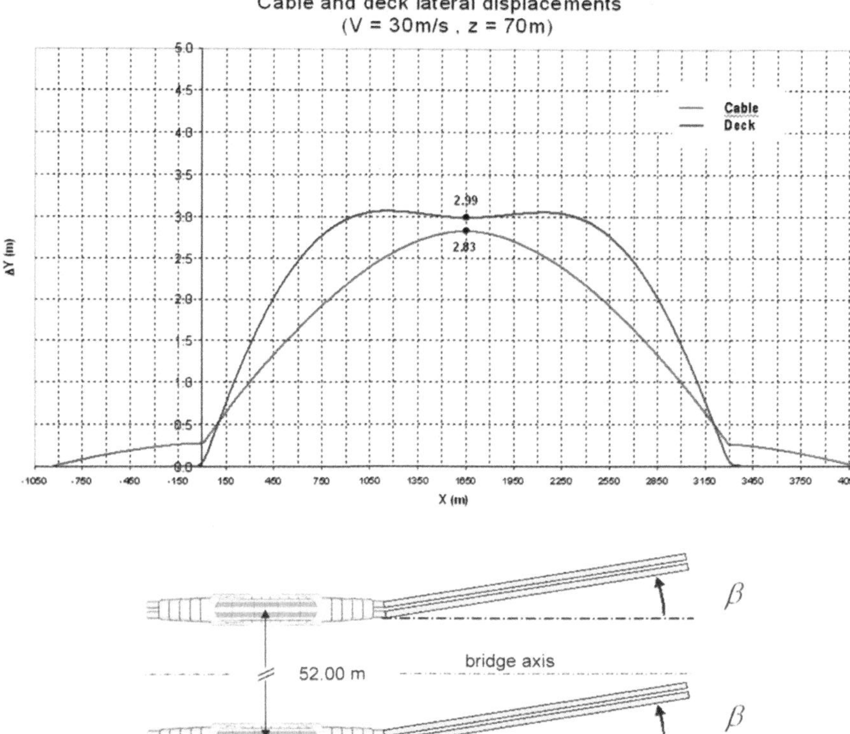

◄ Figure 6.36a
Lateral deflection of the main cables and deck. Plan view showing main cable deflection in the horizontal plane. Note lateral deflection of tower tops.

◄ Figure 6.36b
Lateral deflection of the cables. Tower top plan view showing main cable deflection in the horizontal plane and definition of the β angle.

Figure 6.37 illustrates the deflections at mid-span for a mean wind speed of 60 m/s for the Akashi and the Messina bridges. These data have been computed and also measured in the wind tunnel on aeroelastic models of the two bridges. The maximum deflection for Messina is around 10 m, compared with approximately 30 m for Akashi. These results are mainly due to the larger deck drag of Akashi with respect to that of Messina.

◄ Figure 6.37
Comparison of Akashi (left) and Messina (right) horizontal deflection due to wind in the wind tunnel. Equivalent wind speed = 60m/s.

The Akashi photo is courtesy of Honshu-Shikoku Bridge Expressway Company Limited – Japan.

We can clearly summarise this part of the analysis by stating that the deck drag coefficient is one of the parameters which must be minimized for very long suspension bridges.

6.2.4 Dynamic instability

The two forms of instability that can be excited in a bridge will be discussed in this section:

- Single degree of freedom instability.
- Two degree of freedom coupled flutter.

As already noted, these forms of instability are due to the motion-induced aerodynamic forces acting on the deck. In particular, these forces are functions of the displacement and velocity of the deck. This fact can be easily understood by analysing the simple example in Figure 6.38, in which the deck is moving in vertical direction (z) with associated vertical velocity ($\dot{z}$).

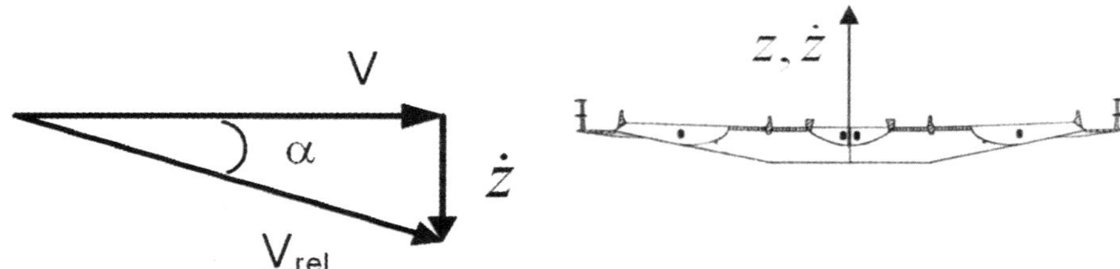

Figure 6.38 ▶ Vertical motion of the deck section and relative wind velocity.

In this case, even with a constant wind V, the relative velocity of the wind with respect to the deck becomes V_{rel} and has an horizontal component equal to V and a vertical component equal to $-\dot{z}$: the resulting wind angle of attack (α) is therefore a function of $\dot{z}$.

If a harmonic vertical motion (z) of the deck is considered, and the frequency of this motion corresponds to a natural frequency of the bridge excited by the incoming turbulence, then $\dot{z}$ will also harmonically change and the same will hold for the angle of attack α.

Analysing the $C_D(\alpha)$, $C_L(\alpha)$ and $C_M(\alpha)$ curves shown in Figure 6.26, it is observed that the coefficients change as a function of α and, as a consequence, the aerodynamic forces applied to the deck will be functions of the vertical velocity $\dot{z}$ of the deck.

As another example, the case of a time dependent deck rotation ϑ excited by the incoming turbulence can be considered. The wind angle of attack α will change with ϑ and the resulting aerodynamic forces on the deck will also be a function of the rotation ϑ.

The aerodynamic forces which are proportional to the generalized displacement are similar to the structural elastic forces, which are also functions of the displacement, and, if linearized, they give rise to an equivalent elastic matrix. See Appendix 1 for a fuller explanation.

The aerodynamic forces which depend on the velocity, are similar to the structural damping forces and give rise to an equivalent damping matrix.

For this reason, when a bridge is subjected to a generic wind speed V, its natural frequencies are also functions of the aerodynamic characteristics of the deck section, through the equivalent aerodynamic stiffness, while the bridge overall damping is a function of the aerodynamic shape of the deck, through the equivalent aerodynamic damping.

What has been defined above is an aeroelastic problem: the behaviour of the bridge depends both on its structural characteristics and its aerodynamic parameters.

Of course, if the overall damping of the bridge increases with the wind velocity then the system is stable, while if the overall damping at a given wind velocity becomes negative, due to the negative contribution of the motion-induced aerodynamic forces, the system becomes unstable.

6.2.4.1 Single degree of freedom instability

This form of instability, due to the motion-induced aerodynamic forces, arises generally in "bluff body" type deck sections, like those in Figure 6.39, which present a large frontal area.

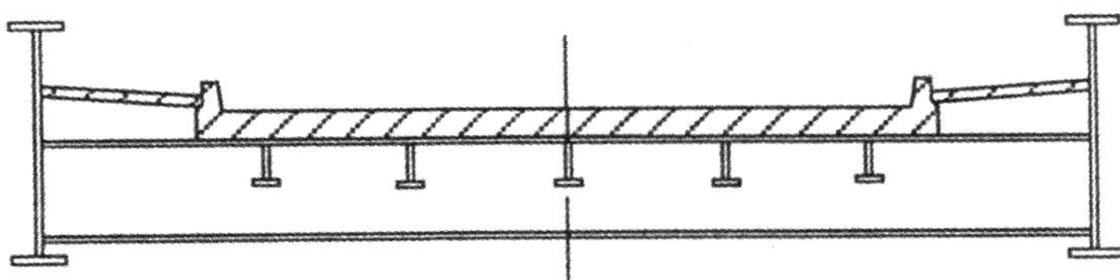

◄ Figure 6.39 Bluff body type deck section (Tacoma) showing large frontal area.

The problem can be handled using a simplified approach like the Quasi Steady Theory (QST) (see Appendix 1) which considers the angle of attack variations due to the motion and evaluates the corresponding aerodynamic forces using the static drag, lift and moment coefficients, discussed in the previous section.

It can be stated that a body is stable in general with respect to the single degree of freedom motion (vertical motion or rotational motion) if the derivatives of lift and moment coefficients are positive, with the conventions assumed.

Bodies like those represented in Figure 6.39 generally possess a negative $C_L(\alpha)$ or $C_M(\alpha)$ slope and, for this reason, they are defined as aerodynamically unstable.

Decks like the Tacoma Bridge, due to the large frontal section, tend to show negative derivatives of the aerodynamic coefficients and then can be unstable.

In fact the onset of instability is also a function of the structural damping.

More precisely, the $C_L(\alpha)$ or $C_M(\alpha)$ negative slopes produce an equivalent negative damping that, as an absolute value, increases with the wind velocity. The wind velocity for which this value becomes larger than the structural damping marks the onset of instability.

In reality this form of instability is generally associated with vortex shedding, specifically vortex shedding generated because of a large frontal section: in fact this is the case of the Tacoma Narrows failure (Figures 6.40(a) and (b)). In that case, vibrations started mainly due to vortex shedding excitation and were successively amplified by the onset of a single degree of freedom rotational instability.

Figure 6.40a ▶ Tacoma Narrows failure: single degree of freedom instability is typical of large frontal area deck sections.

Courtesy of Allan Larsen, COWI A/S – Denmark.

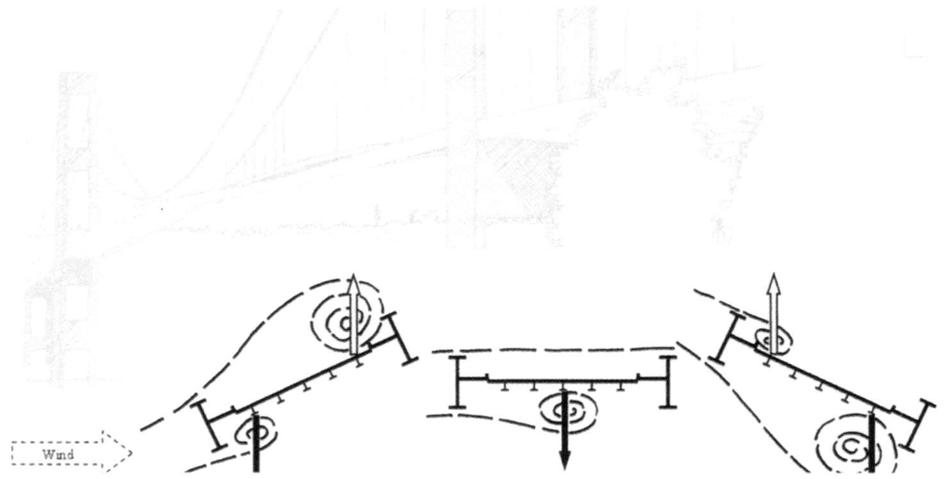

A counter clockwise vortex is formed under the deck as the bridge twists with the windward edge downwards, thus exerting a downward force (black arrow) acting on the upwind deck edge. An instant later the vortex has drifted to a middle position and the downward pull now acts close to the deck centre line. Still later, the upwind part of the deck twists upwards and the counter clockwise vortex has drifted to the leeward now excreting a down ward force on this portion of the deck. A new clockwise vortex and associated force (outline arrow) is created upwind and the process is repeated. The vortex drift process will cause instability at wind speeds at which the vortices travel faster across the deck than the time period of the twisting motion.

Figure 6.40b ▶ Tacoma Narrows failure: images of the collapse.

Stillman Fires Collection – Tacoma Fire Department – 1940 (www.archive.org).

6.2.4.2 Two degree of freedom instability

Wing like profiles suffer from two degree of freedom instability, which is the coupling between the first vertical and the first rotational (torsional) modes of vibration due to motion-induced aerodynamic forces.

The mechanism of this type of instability will be simply explained here, while more detailed explanation is given in Appendix 1, together with a description of the experimental procedures used to identify all the aerodynamic parameters necessary to perform a complete stability analysis and evaluate the response of the bridge in turbulent wind.

Referring to Figure 6.26, which shows the drag, lift and moment aerodynamic coefficients as a function of the wind angle of attack for the Humber Bridge deck, it can be seen that this wing like type of deck has an aerodynamic moment coefficient which is almost a linear function of the angle of attack (or, as equivalent, of the deck angle of rotation).

This means that an increment of deck rotation produces an increment of aerodynamic moment applied to the deck. The aerodynamic moment increases linearly with the angle of attack and has the same sign as the deck rotation angle.

On the other hand, when a rotation is imposed on the deck, an elastic restoring moment arises, mainly due to the elastic contribution of the main cables and to a lesser extent to the deck torsional stiffness. An elastic positive torsional stiffness (K_{t-s}) is the elastic restoring moment divided by the deck angle of rotation, but with opposite sign.

In the same way, a negative equivalent torsional stiffness (K_{t-a}) can be associated with the aerodynamic moment, proportional to deck angle of rotation and with the same sign.

This equivalent stiffness, proportional to the slope of the moment aerodynamic coefficient, can be expressed through the following relation:

$$k_{t-a} = -\frac{1}{2}\rho S V^2 \frac{\partial C_M(\alpha)}{\partial \alpha}$$

where $\frac{\partial C_M(\alpha)}{\partial \alpha}$ is the derivative of $C_M(\alpha)$.

When no wind is applied to the bridge, the first rotational frequency of the bridge is related to the structural parameters only: i.e. the inertia of the bridge and the rotational stiffness K_{t-s}.

The first rotational frequency of the bridge is defined as the first natural frequency for which the bridge exhibits a rotational (torsional) motion. In case of a suspension bridge, the first rotational frequency generally has a node at mid-span and two antinodes at 1/4 and 3/4 span.

When the wind is blowing onto the bridge deck, the overall rotational stiffness K_{t-t} is made of structural and aerodynamic contributions: $K_{t-t} = K_{t-s} + K_{t-a}$, with K_{t-a} being negative and proportional to V^2.

For this reason, K_{t-t} decreases with the wind velocity, together with the first rotational frequency which is related to the ratio between rotational stiffness and bridge moment of inertia (deck + cables).

Figure 6.41 shows how the first rotational frequency of a bridge decreases with the wind speed V.

On the other hand, the first vertical frequency is not practically changing with wind speed, so as wind speed increases the first rotational frequency can become equal to first vertical frequency. (Figure 6.41)

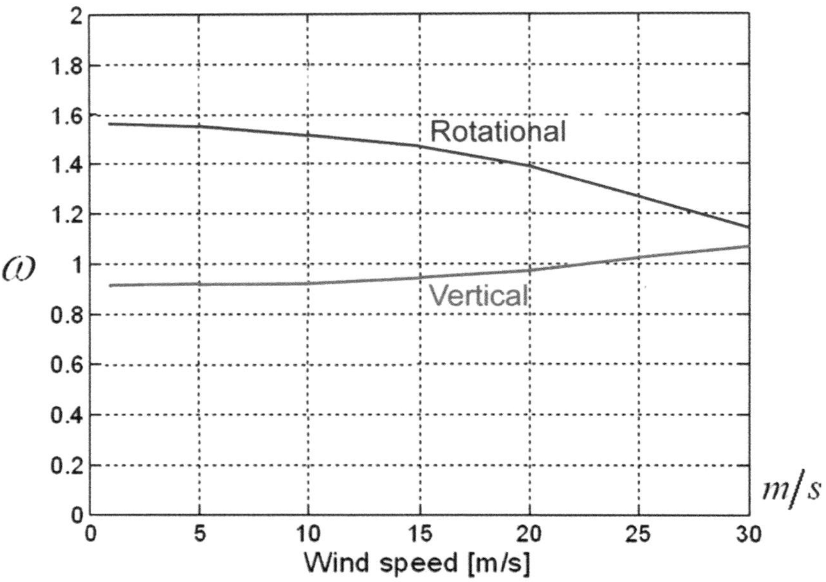

Figure 6.41 ► The first rotational frequency of a bridge decreases with the wind speed. Coupling of the rotational and flexural modes gives rise to the flutter type instability.

When this happens, a synchronized vertical and rotational motion may take place. In this type of motion the aerodynamic force (lift force) can introduce energy into the deck and can produce increasing amplitudes, giving rise to a two degree of freedom instability, or coupled flutter (Figure 6.42). The wind velocity producing this synchronization mechanism is called the critical flutter velocity.

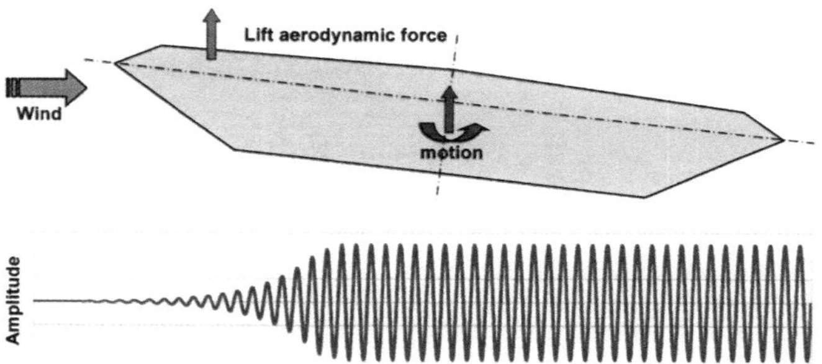

Figure 6.42 ► Two degree of freedom coupled flutter mechanism.

Such flutter instability is the major problem conditioning the design of very long suspension bridges and becomes worse and worse with increasing span length. This is due to the fact that the ratio between the first rotational frequency and the corresponding vertical one is decreasing with increasing span length, as shown in Figure 6.43.

The reason for this has already been explained in Section 6.1 and is mainly due to the increasing contribution of the main cables in the overall bridge dynamics, as the span length increases.

In any case, when the vertical and rotational frequencies become closer, the equivalent rotational stiffness related to the aerodynamic contribution easily makes the frequencies equal and gives rise to flutter instability at lower wind speed.

This can be further understood through an example.

The Storebaelt bridge has a rotational to vertical frequency ratio equal to 2.79 with a flutter velocity of about 60 to 65 m/s.

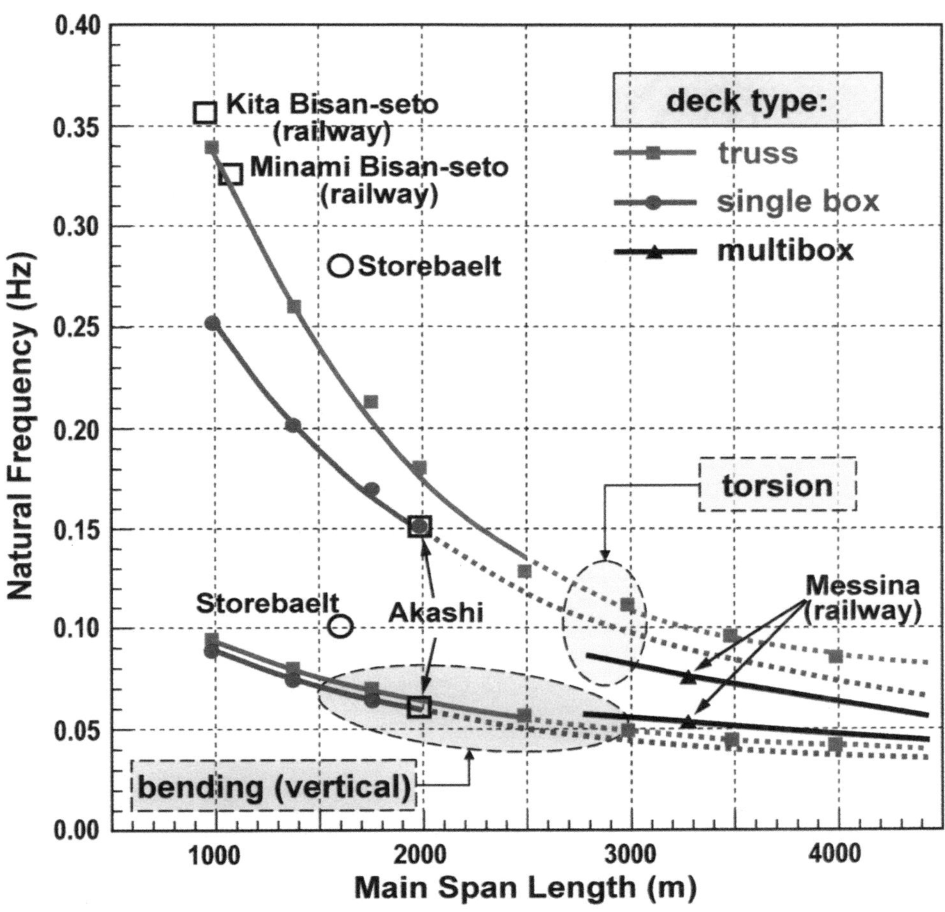

◄ Figure 6.43 Reduction of natural frequency as a function of the span length for different types of deck profiles.

The Messina bridge has a rotational to vertical frequency ratio equal to 1.36. If the same aerodynamic coefficients of the Storebealt deck were considered, the flutter velocity would result around 40 m/s.

This is a big problem for super long suspension bridges, and there are two available ways to control it:

1. The first is to work on the structural design of the bridge, in order to increase the ratio between first rotational and vertical frequencies.
2. The second way is to work on the aerodynamic properties of the deck.

1. *Structural modifications*
   Many attempts have been made to increase the ratio between rotational and vertical frequencies by structural means, and some of them are here reported:

   a. To increase the horizontal distance between the main cables and try to concentrate the deck weight along a central strip. This action reduces the contribution of the deck to the overall rotational moment of inertia.
   b. Use of transversely inclined cross hangers together with or instead of vertical ones. (Figure 6.44) This is another proposed solution, but it presents practical difficulties, including restricted headroom clearance near the centre of the span.
   c. Some projects have proposed a mono cable solution. (Figure 6.45) In this case the pendulum like motion of the deck becomes the one that, combined with the vertical motion, can give rise to the flutter instability.

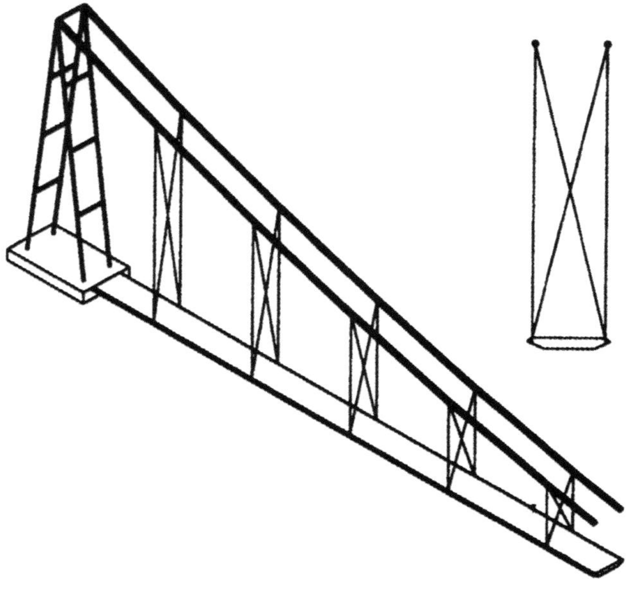

Figure 6.44 ▶ "Cross Hangers" structural scheme.

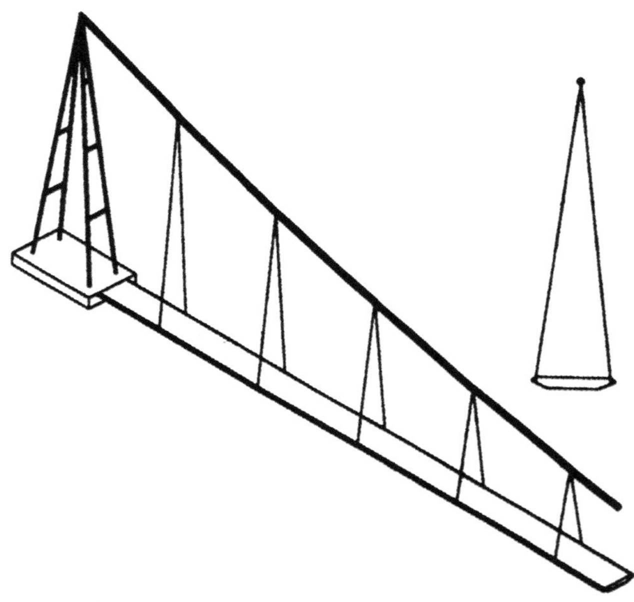

Figure 6.45 ▶ "Mono Cable" structural scheme.

2. *Aerodynamic modifications*
   This work starts from the consideration of the following points:
   a. the flutter velocity is defined as the value of the wind speed which makes the rotational and vertical frequencies equal;
   b. the decrement of the rotational frequency shown in Figure 6.41 is due to the negative value of the equivalent aerodynamic rotational stiffness K_{t-a};
   c. the value of K_{t-a}, using the QST, is defined by: $k_{t-a} = -\frac{1}{2}\rho S V^2 \frac{\partial C_M}{\partial \alpha}$.

One way to increase the flutter velocity is to reduce the value of the $C_M(\alpha)$ derivative $\frac{\partial C_M}{\partial \alpha}$. As a matter of fact, this problem is more complex and a fuller explanation is given in Appendix 1.

If the value of this derivative was zero, no variation of the rotational frequency with the wind speed would occur, and no flutter instability would occur.

However care must be taken in reducing this derivative because it must in any case remain positive for single degree of freedom stability, and if it is reduced, the effects related to Reynolds number have to be taken into account and a major sophistication in the approach would be required, as reported in Appendix 1.

Multi box sections, such as the one developed for the Messina deck, are characterised by a low slope of the moment coefficient $C_M(\alpha)$, compared to a flat plate or a single box section like the Humber.

Figure 6.46 compares the Messina aerodynamic moment coefficient (C_M) derivative $K_{Mo} = \left.\frac{\partial C_M}{\partial \alpha}\right|_{\alpha=0}$ (slope of the C_M coefficient) with those of a flat plate and the Humber bridge.

Figure 6.46 accounts for the fact that the Messina flutter velocity is around 80 to 90 m/s, while it would be around 40 m/s if the same aerodynamic coefficients of a single box section, like the Humber, were to be considered.

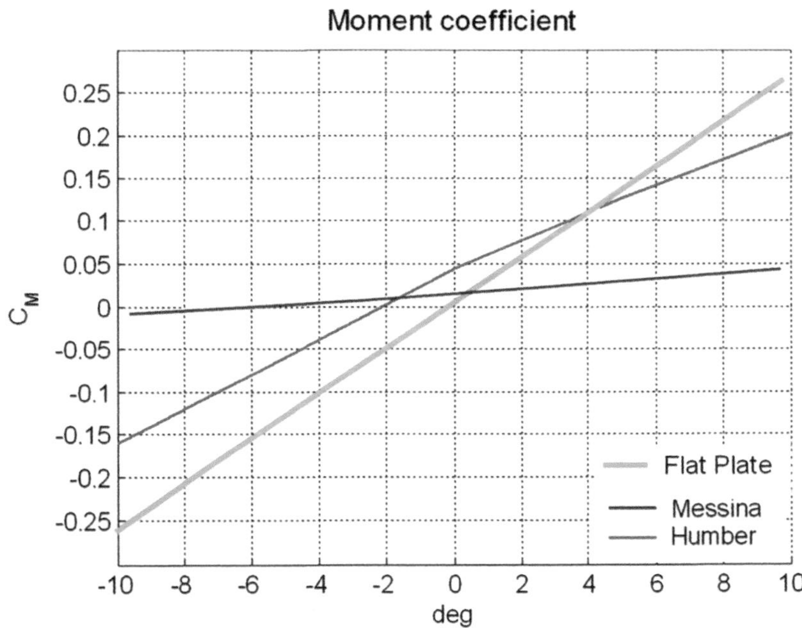

◄ Figure 6.46 Messina (Km_o = 0.16) aerodynamic moment coefficient derivatives compared with those of the Flat Plate (Km_o = 1.57) and Humber Bridge (Km_o = 0.9).

6.2.5 Vortex shedding

As already observed, vortex shedding is an important problem both for the deck and the tower design. Vortex shedding excitation will be separately considered for the towers and the deck.

6.2.5.1 Tower

The cross sectional shape of the tower legs is generally that of a bluff body, as shown in Figure 6.47. The shape of the corner edge produces flow detachment and then vortex shedding.

When the vortex shedding frequency becomes equal to one of the natural frequencies of the tower, either in the stand alone configuration during construction or in service with all the restraints applied by the cables and deck,

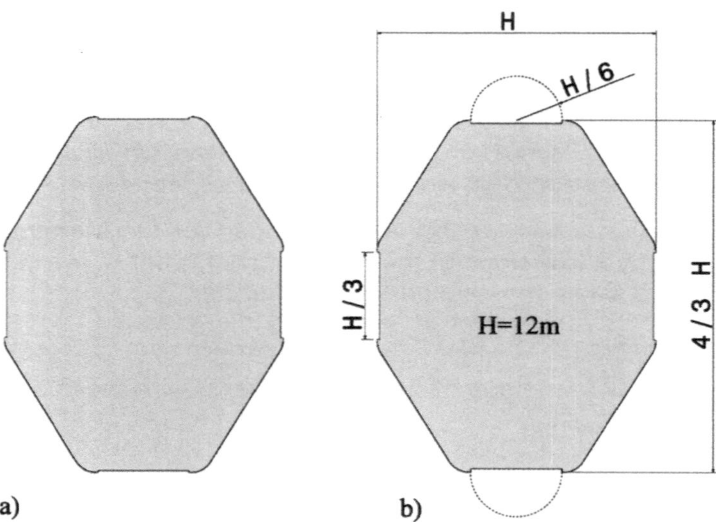

Figure 6.47 ► Messina Tower section. a) Original Configuration "Without Screens". b) Configuration "With Screens".

the vibrations can be excited with vibration amplitudes that depend on the structural damping of the tower. The amplitudes are high if the damping is low, or, using a more representative figure for vortex shedding severity, if the Scruton number is low.

The Scruton number is defined as,

$$Sc = 2\pi \frac{m\xi}{\rho D^2}$$

where m is the tower linear mass, ξ the non-dimensional damping, ρ is the air density and D a reference sectional dimension of the tower leg.

In the stand alone configuration during the construction stage it is relatively easy to prevent high tower vibrations by installing dampers or attaching temporary stays connected to dampers.

When the construction is complete and the tower reaches its final condition, the natural frequencies of the tower are generally higher than in the stand alone configurations due to the restraint from the main cables. The wind speed at which severe vortex shedding vibrations may be excited increases, so the probability of occurrence decreases. Nevertheless this problem must be fully investigated.

For instance, for the Messina bridge, the wind velocity at which the tower first vibration mode is excited is shown in Figure 6.48 to be around 12 m/s in the free standing configuration, whereas it is much higher at around 57 m/s for the condition when the main cables are attached. This is due to the main cable restraint which raises the first mode frequency by a factor of four.

These tests have been performed on an aeroelastic model of the tower without the porous screens, reproducing in the model scale the natural frequencies and modes of vibration of the real tower.

It can be seen that the wind velocity at which vortex shedding may be excited can occur both in the stand alone and in the final configuration, although with different probabilities.

Many attempts have been made to control this problem, common to the towers of all long span bridges.

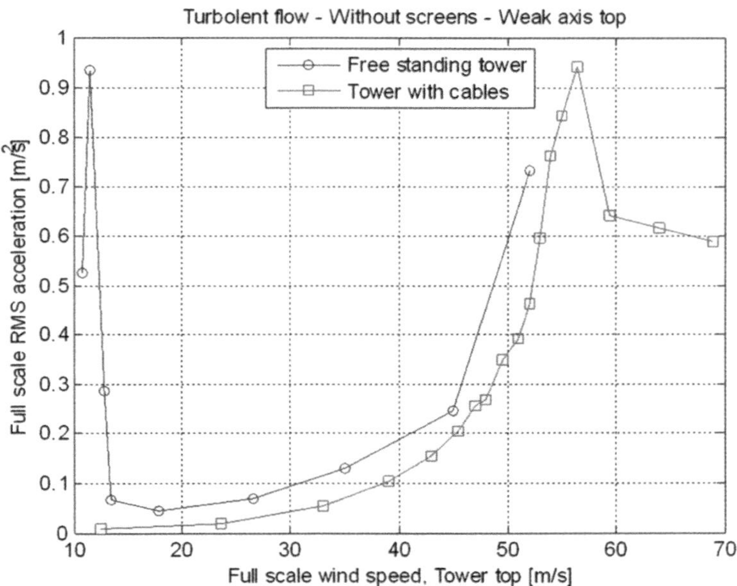

◄ Figure 6.48 Wind speeds at which are excited the first tower modes in stand alone & final construction stage configurations. (Without porous screens.)

In the 1992 design of the Messina towers, porous screens were placed at the edges of each leg section (Figure 6.47b) in order to mitigate vortex shedding excitation. Figure 6.49 shows that, due to the very effective role of the porous screens, no vortex shedding excitation is observed in either the free standing or final construction stage configuration.

This solution however has some disadvantages, such as the maintenance required by the screens and also the uncertainties introduced which are related to the difficulty of accurately representing Reynolds number in the model scale.

For the final tower design it was decided to control vortex shedding excitation through the addition of large Tuned Mass Dampers (TMDs) in such a way as to increase the Scruton number and avoid dangerous vibrations of the tower. The same strategy was adopted on the Akashi Bridge towers,

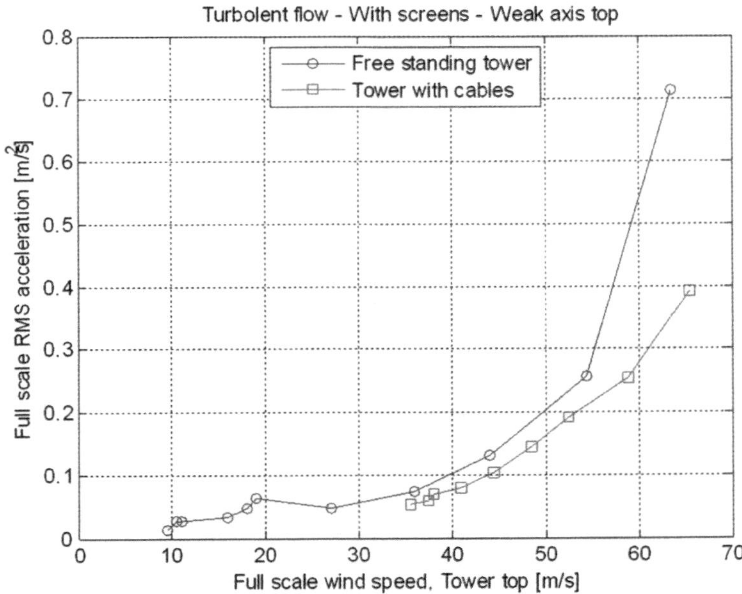

◄ Figure 6.49 Messina Tower section – with porous screens installed to mitigate vortex shedding excitation.

placing several TMDs in different locations within the tower legs. Figure 6.50 shows one of the Akashi tower TMDs being tuned before installation.

All the tests required to be performed in the wind tunnel, together with the maximum allowed accelerations at the top of the towers, are described in the technical specifications issued to the bidders for the international tender won eventually by the Impregilo Group. These define a limiting maximum value of $a_{max(x;y)RMS} < 0.3 \div 0.5$ m/s².

Figure 6.50 ► One of the Tuned Mass Dampers for the tower legs of the Akashi Bridge.

Courtesy of Honshu-Shikoku Bridge Expressway Company Limited – Japan.

6.2.5.2 Deck

All bridge deck sections suffer more or less from vortex shedding. The decks with large frontal sections or sharp edges are more prone to this problem than streamlined wing type box sections.

Figure 6.51 shows the vortex shedding visualization obtained for the Messina bridge deck model in the wind tunnel.

Vortex shedding excitation is associated with the separation of flow from the upstream edge of the deck to produce a vortex that moves downstream. Generally at the downstream edge, or in another point depending on the deck shape, another vortex can be detached. Any corners in the deck shape can produce flow separation and vortex shedding.

Depending on the relative position of the two vortices and at a given wind speed (or in non dimensional form at a given reduced velocity $V^* = \frac{V}{fB}$) a motion amplification may occur.

Critical values of reduced velocity tend to be around 1 or 0.5.

Figure 6.51 ► Messina deck section: Vortex shedding visualization.

As an example, the Storebealt deck section (Figure 6.52) experiences vortex shedding from the bottom deck corners.

If wind barriers are placed along the edges of the deck they can trigger vortex shedding, as can any other similar obstacle or detail close to the deck edges.

For the Messina bridge, boxes with flat plane surfaces and sharp corners as shown in Figure 6.53 produced poor vortex shedding behaviour and were therefore replaced by boxes with curved surfaces.

The solidity of the wind barriers, while needing to be high enough to reduce the wind flows across the bridge deck to reduce the wind forces on the vehicles, must be relatively permeable in order to limit vortex shedding and reduce drag. Airfoils are also incorporated in order to increase the aerodynamic damping of the deck.

In the 1992 design, a trip wire was placed under the two road boxes to control the derivatives of the moment aerodynamic coefficient. However wind tunnel testing showed that a large vortex shedding excitation was caused by these trip wires, so they were eliminated from the deck design.

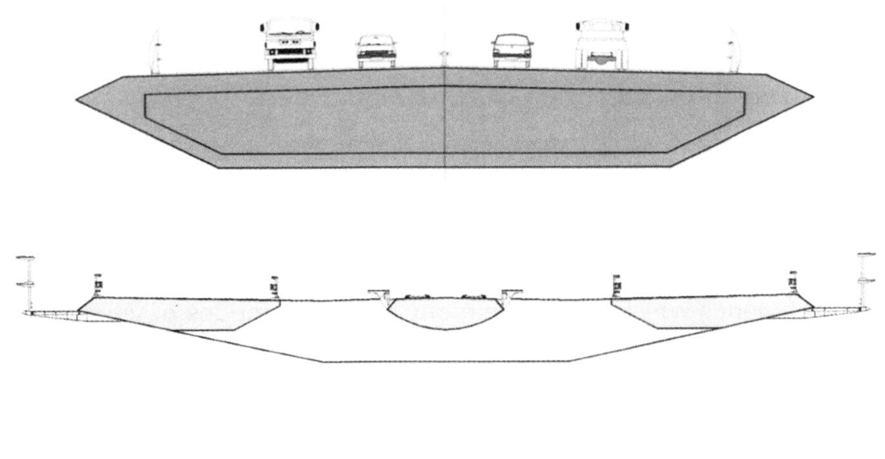

◄ Figure 6.52 Storebaelt deck section.

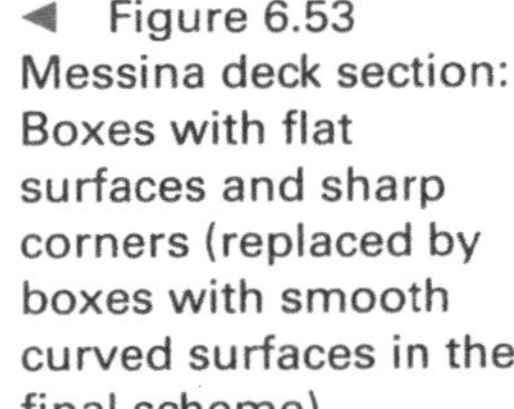

◄ Figure 6.53 Messina deck section: Boxes with flat surfaces and sharp corners (replaced by boxes with smooth curved surfaces in the final scheme).

An acoustic barrier for the train noise and a transparent barrier to prevent people from jumping from the bridge have also been introduced to control the $C_M(\alpha)$ and $C_L(\alpha)$ derivatives.

The final version of the deck design was thus optimised, controlling vortex shedding excitation in such a way that vortex induced vibrations produce accelerations not greater than $a_{maxRMS} < 0.1$ m/s^2 measured at the leading and trailing edges of the deck. These limits are defined in the technical specifications as the governing criteria for this type of problem.

6.2.6 Buffeting

As already explained, the fluctuations of wind due to turbulence produce a motion of the bridge which is called buffeting, or bridge motion induced by turbulence. This is always present in all bridges, and generally in all structures and cannot be avoided. The vibration amplitudes associated with buffeting can be controlled by increasing the aerodynamic damping or equivalently, by increasing the stability of the bridge, as will be explained in the following.

It has been shown above that when the wind is blowing on a bridge, the motion of the structure produces an equivalent stiffness and damping related to the motion-induced aerodynamic forces. It has also been shown that in order to avoid the single degree of freedom instability, the lift and moment coefficients must have positive derivatives, in which case only the two degree of freedom type instability can occur.

Figure 6.54 shows the overall damping of the Messina bridge for its first vertical vibration mode, which is the mode that couples with the rotational one to give rise to flutter. The overall damping comprises both the contributions from the structural and the aerodynamic parts.

It can be seen that at zero wind speed the overall damping is due only to the structural part, which in this case has been assumed equal to 5‰.

As the wind speed increases, the contribution of the aerodynamic forces becomes very important and at 60 m/s, which is the maximum design wind speed for the bridge, the overall damping becomes as high as 5%, i.e. ten times larger than the purely structural damping level.

A further increase in wind speed causes a reduction in overall damping. The wind speed at which the overall damping becomes negative is defined as the flutter velocity, and in the case of the Messina bridge its value is around 90 m/s.

Buffeting is more or less a forced motion with a wide band input spectrum due to turbulence which can excite many structural modes of vibration and mostly the lowest ones, as can be seen in Figure 6.55, where the first three lateral modes are excited.

If the overall damping is increased, for a given excitation in resonance conditions, (i.e. when the input spectrum due to turbulence and the bridge natural frequencies coincide) the vibration amplitudes will be proportionally reduced.

From this point of view it is important not only to have a high critical flutter velocity to be conservative on stability conditions, but also to increase the aerodynamic damping so as to reduce the turbulence-induced motions which are those conditioning the life of the structure and its performance under wind action, for example in terms of train runnability and user comfort.

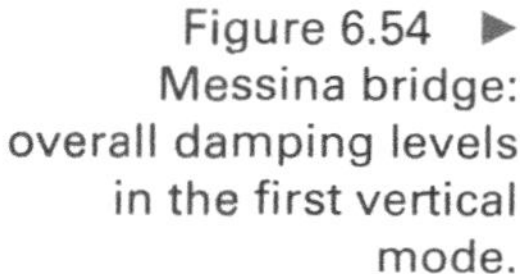

Figure 6.54 ▶ Messina bridge: overall damping levels in the first vertical mode.

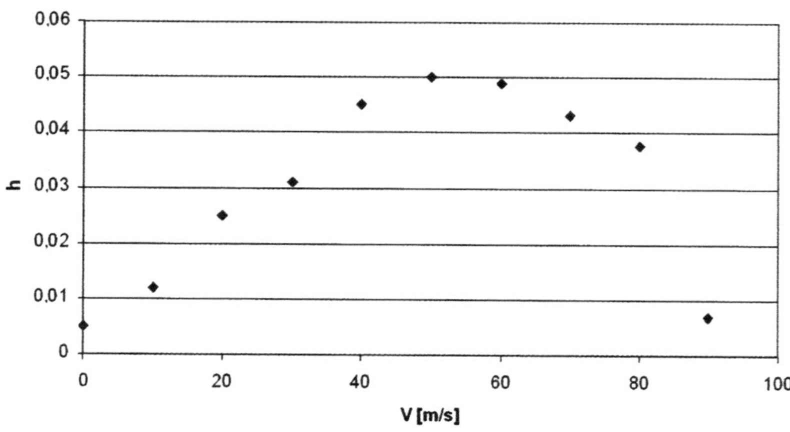

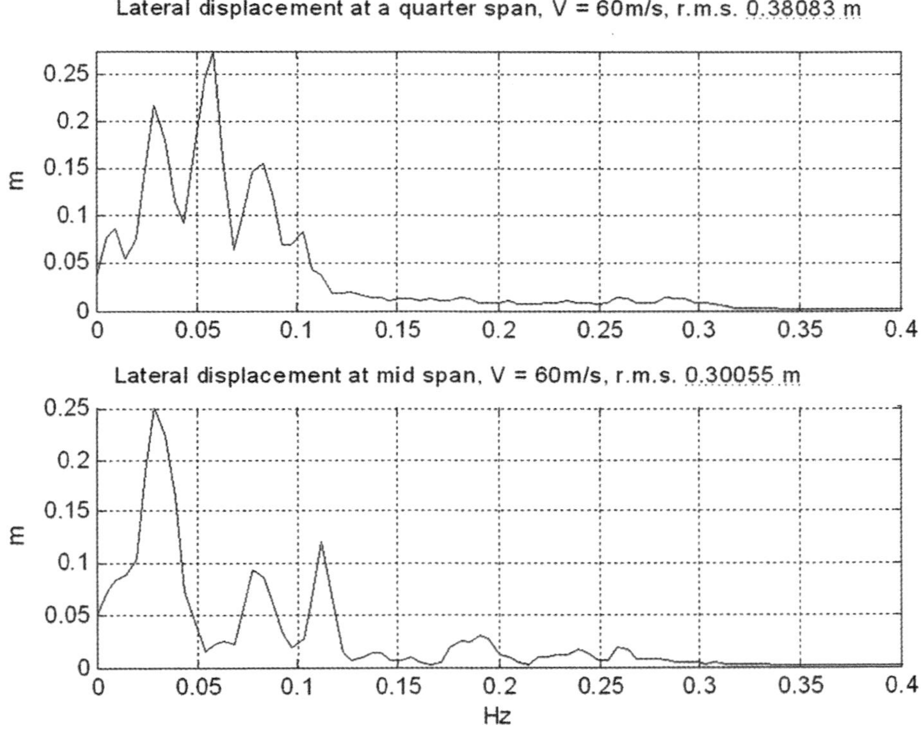

◄ Figure 6.55 Messina bridge: spectrum of horizontal displacement at a quarter span and at midspan for wind speed of 60 m/s. (see Figure 6.5 for the modes natural frequencies).

6.3 Role of earthquake events

The Messina Strait is a highly seismic zone, as discussed in Chapter 5. Careful attention to the potential earthquake response is therefore essential for any construction in the area. When examining this matter for a very large suspension bridge, distinct features are found for the steel superstructure and for the foundations and anchor block.

6.3.1 Seismic response for the superstructure

The principal requirement for the superstructure design is that the main bridge modes, with high participation factors, are not coincident in frequency with the seismic ground motion. As already seen, first modes involving the cables and the deck have vibration periods (the inverse of the frequency in Hertz) from about 10 to over 30 seconds. This is a range in which the seismic input energy is virtually nil. Local deck or cable modes with periods of only a few seconds (close to natural seismic input periods) are so high that the participation factor is practically zero, so no significant response occurs. In other words, the bridge superstructure acts as a mechanical filter for ground motion: while the soil shakes, the bridge towers flex and the body of the superstructure stands still.

The only elements of a long suspension bridge which are sensitive to seismic excitation, with local modes with periods of a few seconds and significant local participation factors, are the towers. (Figures 6.56 and 6.57)

The longitudinal tower design is definitely dominated by the simultaneous presence of the high axial force and of the flexure due to earthquake loads, while the transversal one is dominated by axial force plus wind.

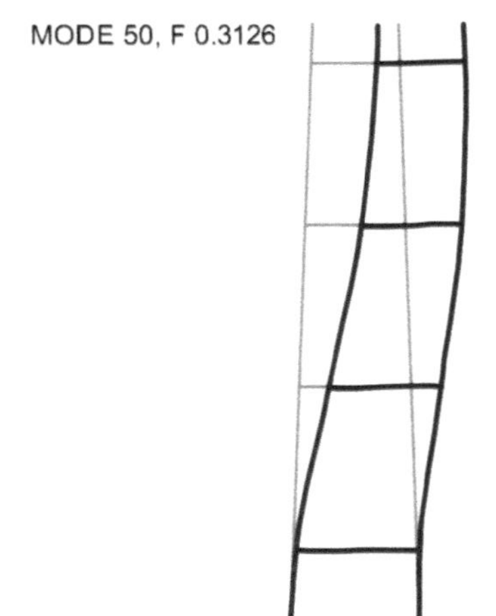

Figure 6.56 ► Transversal tower mode, period 3.20 sec.

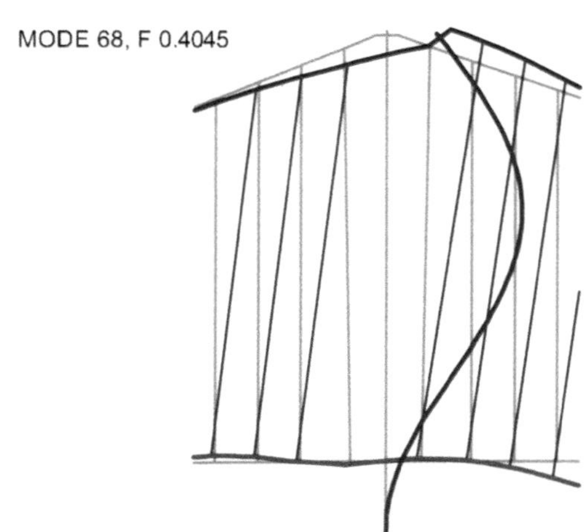

Figure 6.57 ► Longitudinal tower mode, period 2.47 sec.

6.3.2 Seismic response of foundations and anchor blocks

The seismic actions are important for the design of foundations and anchor blocks, but the problems they generate do not differ from those of other classes of massive large concrete structures on ground or underground, and are not particularly specific or dominant for suspension bridges. The work done for the Messina design in this respect is presented in Chapter 7.

6.4 Road and rail actions

6.4.1 Static behaviour & slopes

The deformability of suspension bridges and its implications for railway runnability is a classical engineering problem related to long spans which has been debated for a long time in the technical-scientific sphere.

That suspension bridges are flexible structures, undergoing significant displacements under heavy loads, is something of which designers have always been aware since the beginning of the modern age of long span steel bridges. This is specifically true for railway loading, which is characterised by large total loads applied over comparatively short lengths.

In the beginning, some doubts arose about the runnability of railway trains on long suspension bridges, mainly in relation to traction problems. It was thought that an excessively steep slope of the track could develop and cause unacceptable wheel slipping or even difficulties in re-starting after an unintended stop.

In some historical cases this awareness contributed to the choice of different structural solutions, leading to peculiar and famous works such as the rail crossing of the Firth of Forth close to Edinburgh, for which a number of suspended solutions all stiffened by additional stay systems were proposed before choosing the final familiar and impressive lattice girder solution.

When the feasibility studies for the Messina crossing were being carried out, only a few and not very significant suspension bridges carrying a railway already existed (the Manhattan Bridge in NewYork is an example), but three suspension bridges designed to have four rail tracks on the Kojima-Sakaide route of the Honshu-Shikoku project in Japan were under construction and were due to be opened in 1988. These three works, which have spans up to 1100 m (the Minami Bisan-Seto bridge), have been shown to perform in service as predicted by their design, evidencing no problems and demonstrating satisfactory performance.

Moreover, a clear and rational evaluation of the problem had been undertaken in the feasibility analyses carried out by the Gruppo Ponte di Messina, and subsequently continued by SdM. This showed how the deformability of suspension bridge decks in terms of longitudinal slopes strongly depends on the span length, and also how with increasing span length the geometrical stiffness due to the main cable tension gradually becomes more important in the deck-cable interaction, progressively reducing the deck longitudinal slopes.

In particular, it was shown how significant and maybe excessive longitudinal gradients could occur in structures up to about 1000 m, thus justifying the historical concerns with existing rail suspension bridges at that time, while for longer spans at around 1500 m the slopes decrease significantly, and the problem could be considered largely inapplicable over 2000 m. The early preliminary design solutions presented for the Messina crossing in the eighties (with one or two spans) fully satisfied the railway traffic specifications.

Figure 6.58 shows the variation of maximum railway slope with bridge span for different train lengths and sag-span ratios, as well as different seasonal conditions as thermal displacements also play a significant role in the process. The deck is here assumed to have a straight horizontal profile.

While a detailed treatment of the topic can be found in [2], it is just worth noting here that the results come from an interplay of several factors, with the most significant being the rail convoy length in proportion to the bridge span. Train loads are in fact limited both in weight per unit length (due to the maximum capacity of rail fittings) and in total weight (due to engine traction capacity limits). For long bridges this produces either loaded lengths much shorter than bridge lengths or comparatively low unit loads, and in either case the resulting gradient reduces by comparison with shorter spans. Finally, optimised bridge deck profiles can further contribute to optimum performance.

6.4.2 Train bridge interaction

A train running on a bridge produces dynamic movements in the deck, which in turn influence the train dynamics. Thus a dynamic interaction is generated between train and deck and this interaction strongly affects the railway runnability.

The analysis of railway runnability in terms of safety and comfort requires that this complex bridge-train interaction mechanism is investigated. This problem is well understood for viaducts and short span bridges, and in such cases well established procedures allow resonance problems to be avoided, mainly in relation to bridge natural frequencies and train excitation frequencies, which are in turn related to the spacing of bogies and train velocity.

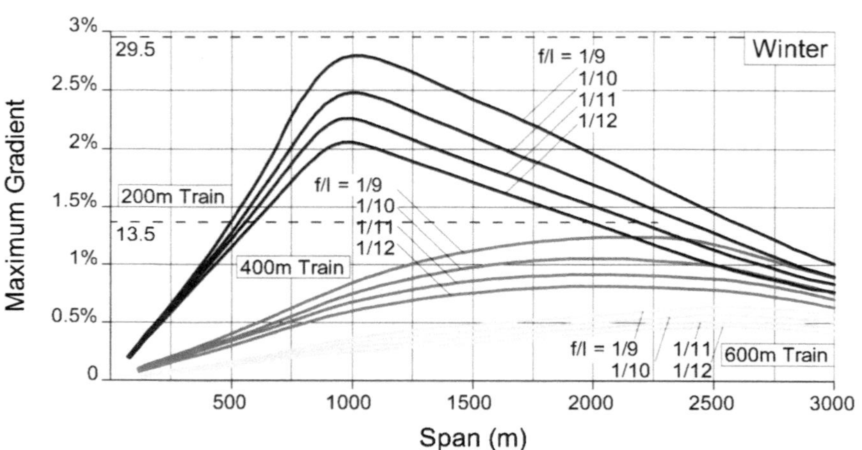

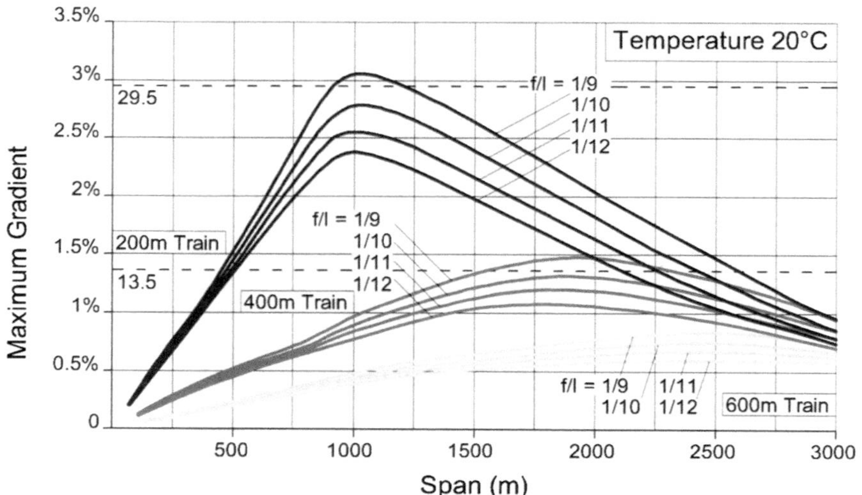

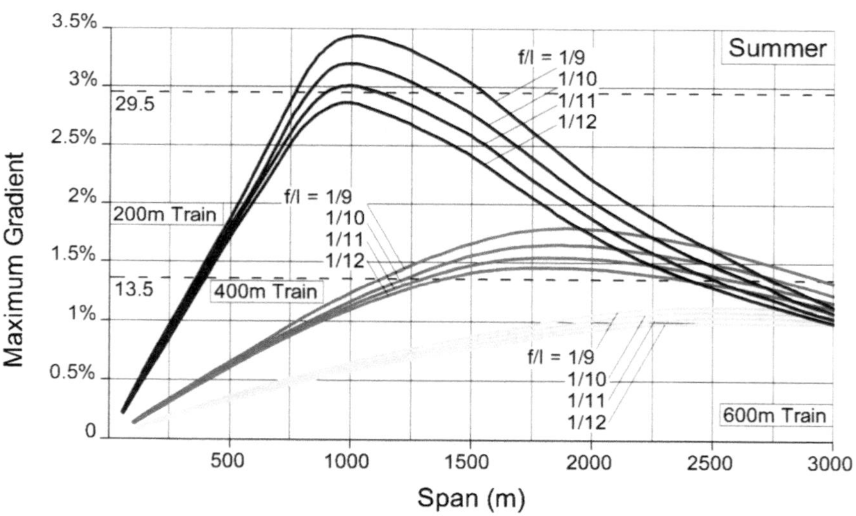

Figure 6.58 ▶ Variation of maximum longitudinal gradient with span for different train lengths and sag to span ratio: a – Winter. b – 20 degrees Celsius. c – Summer.

For long span bridges, and in particular for very long suspension bridges such as Messina, a detailed analysis reproducing the bridge-train interaction is necessary. This must be done in such a way as to evaluate all the problems connected to the railway runnability, leading to the definition of the safety and comfort indices [2, 4, 8, 12].

Two safety indices can be defined: the first is related to derailment and the second to rollover.

The derailment index is defined as the ratio between the lateral force Y on a single wheel and the corresponding vertical wheel load Q. The derailment index Y/Q must be lower than 1.2.

The overturning index C_{ovt} is defined by the expression

$$C_{ovt} = \frac{Q_L - Q_{unL}}{Q_L + Q_{unL}}$$

Where Q_L is the vertical force on the loading wheel and Q_{unL} is the vertical force on the unloading wheel of the same wheelset. The C_{ovt} must be lower than 0.9.

The ride comfort is defined as the maximum value of the vertical and horizontal accelerations measured on the rail car-body under different running conditions. This value must be lower than 1 m/s^2.

A more sophisticated comfort index based on the RMS acceleration values measured on the rail car-body can be defined using filtering functions.

Two types of problems must be handled in the analysis:

a. problems related to global deformation of the bridge.
b. problems related to local bridge deformability.

6.4.2.1 Global deformations

This problem relates to the deformation of the bridge caused by the weight of the train acting in a quasi steady situation, which produces deck deformations which move along the bridge with the train. (Figure 6.59)

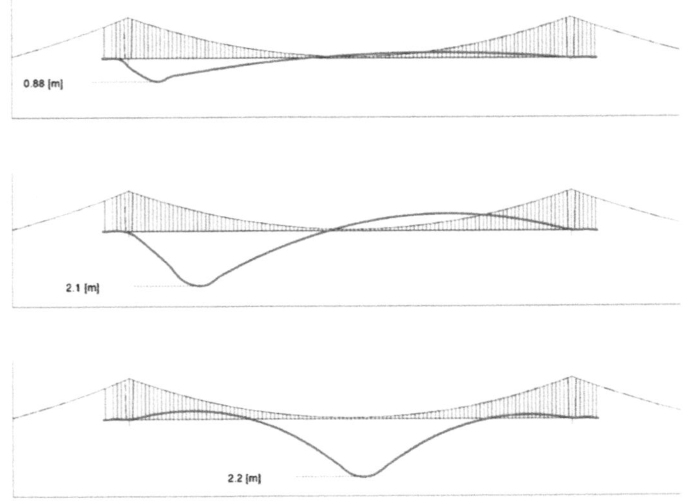

◄ Figure 6.59 Global deformation under the weight of a travelling train.

The slopes in the vertical and horizontal planes affect railway runnability and have been already discussed in the preceding sections.

Global bridge deformation is caused not only by rail and road traffic but also, in the horizontal plane, by wind action, as shown in Figure 6.60.

These global deformations cause rapid changes in slope, or cusps, at the ends of the deck which may compromise railway runnability. These cusps arise both in the vertical plane due to rail and road traffic and in the horizontal plane due to wind action. These singularities occur in the zone where the expansion joint is located and demand difficult detailed design in these areas to ensure smooth passage of trains.

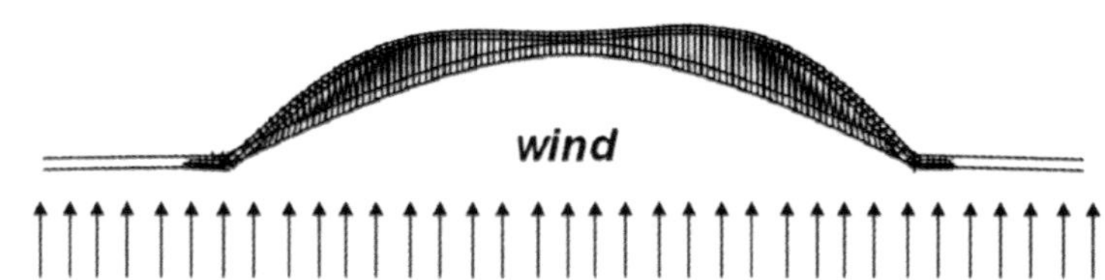

Figure 6.60 ► Wind contribution to the bridge global deformation (view in the horizontal plane).

6.4.2.2 Local interactions

These problems are associated with the local interaction between the deck structure and the train.

While global effects are quasi steady phenomena exciting the train dynamics mainly in the cusp zones at the expansion joints, local interaction effects between the train and deck can occur anywhere, influencing the train dynamics over the entire length of the superstructure. The relevant frequencies tend to be those related to the distance between hangers or cross beams and the running speed of the train. This last aspect of the problem affects the local deck design, and in particular its fatigue life.

The dynamic local train-bridge interaction affects train safety and passenger comfort, and influences, with the associated local modes of vibration, the problem of noise produced by the passing train.

These problems have been analyzed through experimental observations on existing bridges and through mathematical models reproducing the Messina bridge using finite element models. Simple large scale schematisations are used to reproduce the global effects (Figure 6.61), while more sophisticated detailed schematizations are used to reproduce the local effects, such as the deformability of the upper plate of the railway box girder, whose finite element model is shown in Figure 6.62.

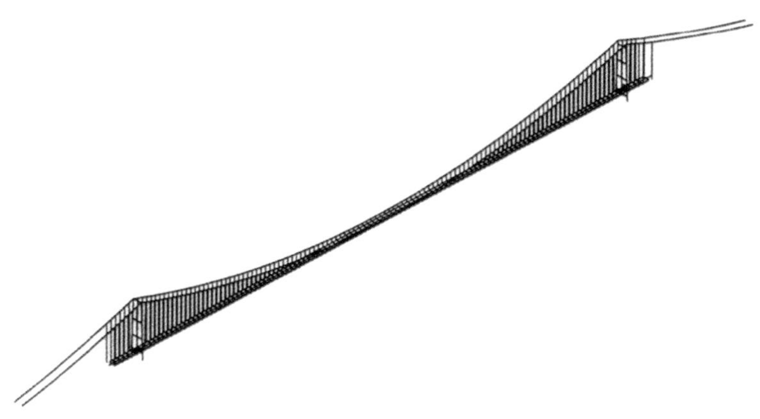

Figure 6.61 ► F.E. schematisation used to reproduce global effects.

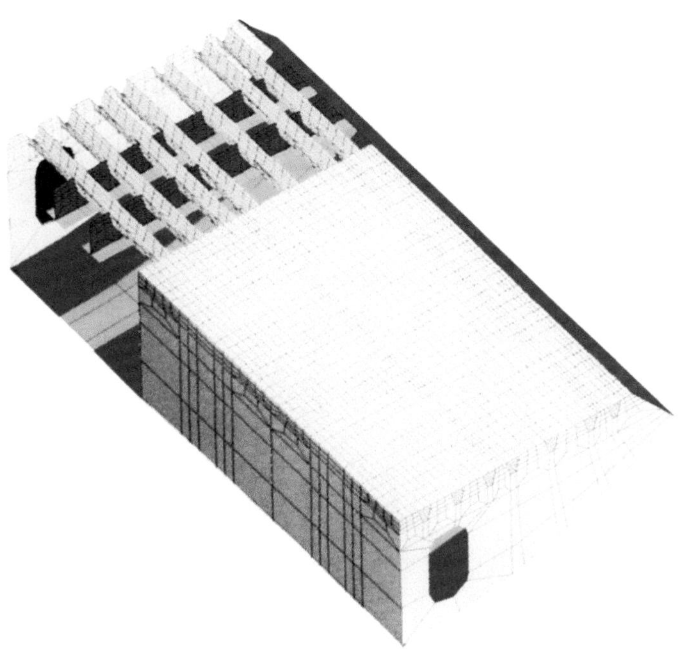

◄ Figure 6.62
F.E. schematisation used to reproduce local effects: finite element model of a portion of half railway box girder.

The train itself is modelled using a multi-body approach, taking into account the vertical and horizontal motion of the train and the complex mechanism of interaction between the wheels and the rail.

Thus the analyses necessary to define the railway runnability for Messina have been performed, and some relevant results are reported below.

6.4.2.3 Actions related to global effects

As an example, the space history of the wheel-track forces for the first wheel set on a part of the central span is represented in Figure 6.63 for the case of a lateral wind with a 50 year return period and a relatively frequently occurring traffic condition corresponding to a 1 year return period.

Figure 6.64 shows the time history of the lateral acceleration of the car-body centre of gravity (c.o.g.).

Derailment and rollover indices are derived from the wheel-track forces, such as those given in Figure 6.63, while the comfort index is derived from the car-body acceleration values, such as those shown in Figure 6.64.

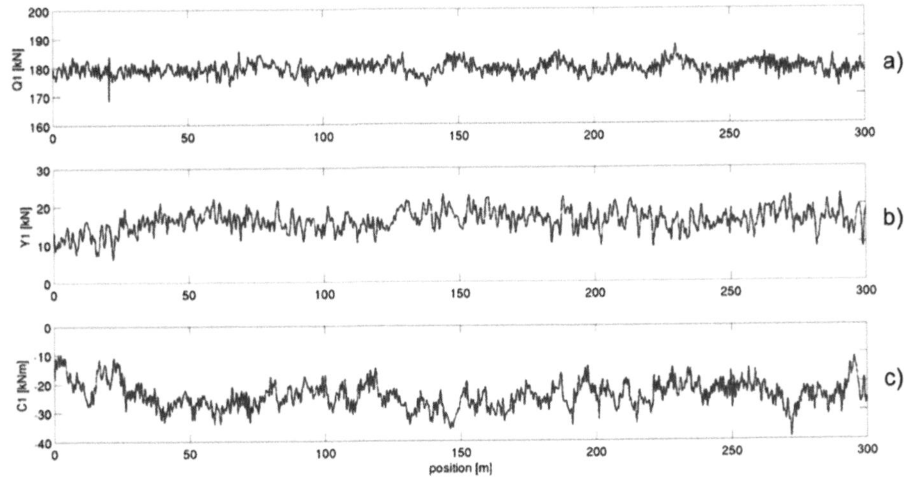

◄ Figure 6.63
First wheel set wheel-track forces. Train velocity: 130 km/h.
a) vertical wheel-track forces;
b) horizontal wheel-track forces;
c) couple wheel-track forces.

By repeating the simulations under different running conditions, assuming a wind speed of 130 km/h and a traffic condition corresponding to a one year return period, the safety and comfort indices can be derived. These are reported in Table 6.2.

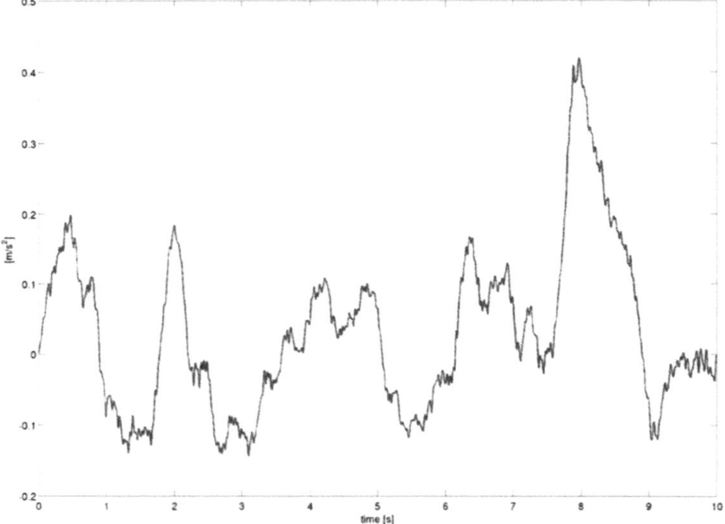

Figure 6.64 ► Time history of car body c.o.g. lateral acceleration.

Table 6.2 ► Safety and comfort indices.

Position		In span	At the joints				
Speed	*(kph)*	130	130	90	72	50	20
Carbody max acceleration							
vertical	*(m/s²)*	0.19	0.85	0.66	0.56	0.27	0.07
horizontal	*(m/s²)*	0.45	1.21	0.76	0.53	0.43	0.45
C_{ovt} max value	*(%)*	25	38	34	31	29	25
C_d max value	*(%)*	13	81	53	53	40	26
Max vertical shock	*(m/s²)*	–	2.20	0.88	0.25	–	–
Max horizontal shock	*(m/s²)*	0.50	2.40	1.00	0.65	0.60	0.66

6.4.2.4 Actions relating to local effects

The wheel-track forces and the car-body accelerations have been analysed, based on the Preliminary Design rail box girder shown in Figure 6.65 and two types of rail track detail as shown in Figures 6.66a and 6.66b. The results are shown in Figures 6.67a and 6.67b respectively.

The rail track support details are very important for the determination of local stresses and noise emissions.

Figure 6.67 shows the forces transmitted from the track to the upper plate of the railway box girder for the two types of rail track support, while Figure 6.68 shows that a significant improvement in the vibration levels and noise emissions is obtained using the Type b track support compared to Type a.

Figure 6.69 shows the acceleration levels at different frequencies for a direct comparison between the two types of rail track support.

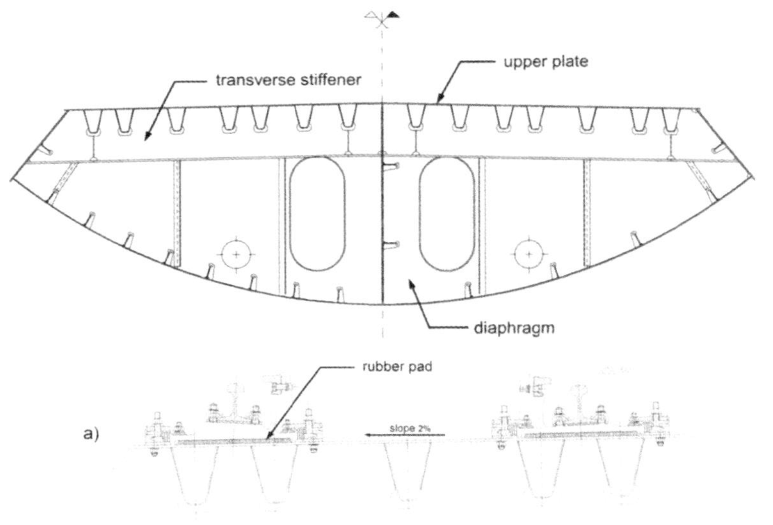

◄ Figure 6.65
Preliminary Design railway box girder section.

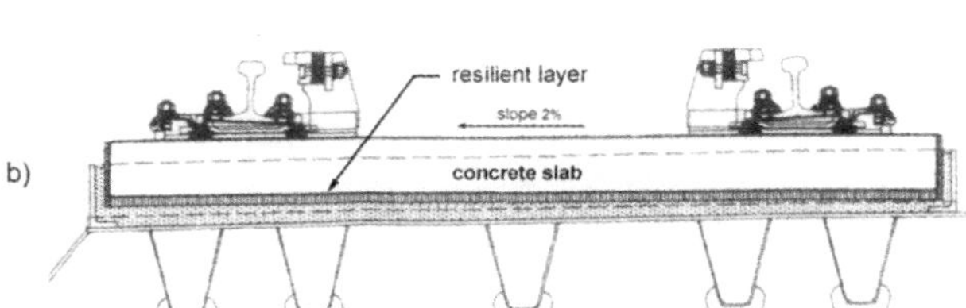

◄ Figure 6.66
Preliminary Design rail track support details:
a. direct fastening.
b. slab track solution.

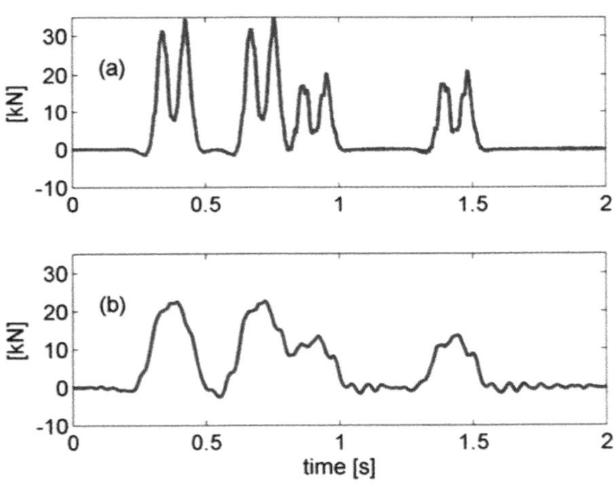

◄ Figure 6.67
Time history of the forces transmitted to the box girder:
a) direct fastening.
b) slab track solution.

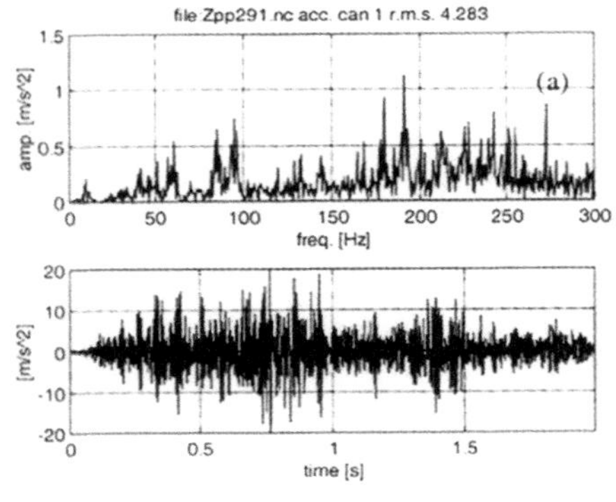

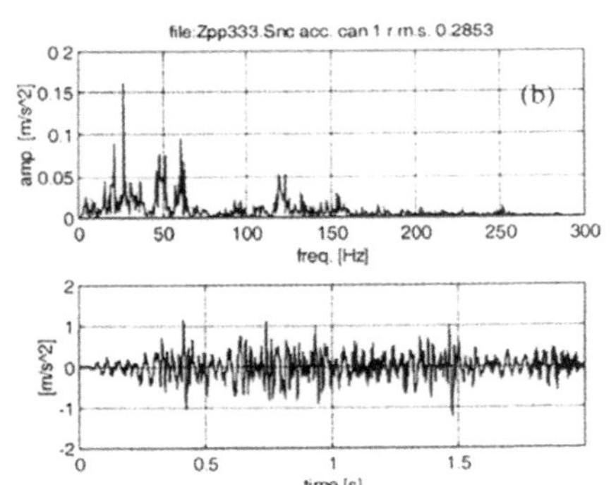

◄ Figure 6.68
Acceleration of the upper plate of the railway box girder:
a) direct fastening.
b) slab track solution.

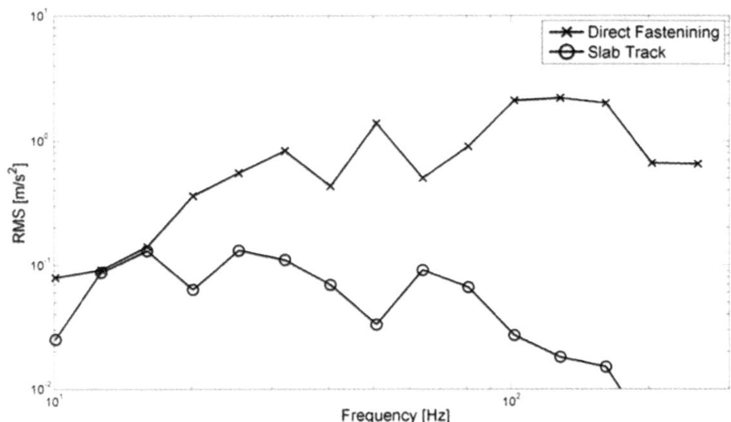

Figure 6.69 ▶ Acceleration level spectrum the upper plate of the railway box girder.

This type of analysis allowed for the optimization of the rail box girder details at the Preliminary Design stage and also for the choice of the most suitable type of rail track support. After many studies and analyses, the track support system which was found to represent a good compromise between reducing noise and vibration and minimizing the weight was the embedded rail.

The winning tender design received from Impregilo proposed a modification to the railway box with a continuous longitudinal web under each rail track. This reduces local deformation of the top deck plate and will modify the results presented above. Although not illustrated here, this modification is expected to provide significant improvements in fatigue performance and will be developed in the final design.

6.4.2.5 Railway runnability in presence of wind

Another important aspect affecting railway runnability is the train behaviour when subjected to wind action. Besides causing a global bridge deformation (as already discussed), the wind affects, with its turbulence, the train and bridge dynamics.

Figure 6.70 shows the derailment and rollover safety coefficients evaluated for the case of an ETR 500 train crossing the bridge at a constant speed of 130 kph, in the presence of a transverse wind of 36 m/s (50 years return period).

It should be noted that, due to the presence of the wind barriers and noise barriers, the wind velocity experienced by the train is much lower than the free stream velocity, and this fact explains the low values of the C_{ovt} and C_D

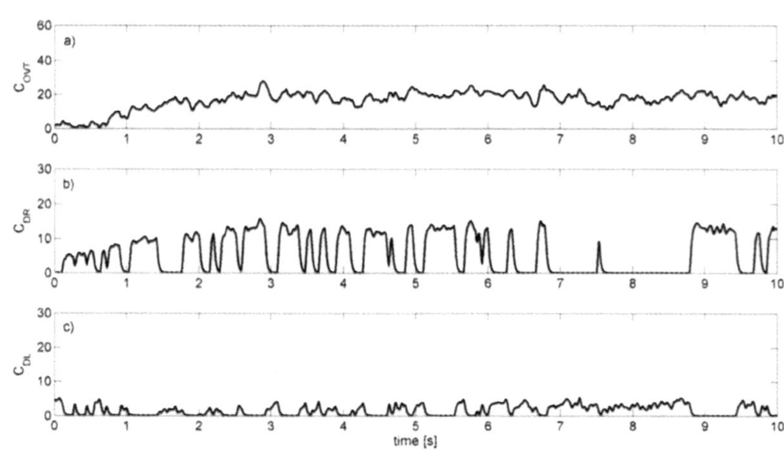

Figure 6.70 ▶ Safety coefficients time history: overturning (a), left (b) and right (c) derailment coefficients. (first column of Table 6.2).

coefficients, which are lower for the train running on the bridge than for the train running on a viaduct without wind-barriers, for the same wind speed.

6.4.2.6 Railway runnability in presence of seismic action

For combined railway and seismic loading, an earthquake with an intensity corresponding to a return period of 50 years has been considered, and its influence on a train running on the bridge and on an embankment has been compared. Figure 6.71 shows the lateral car-body acceleration for the two cases, and Figure 6.72 shows the corresponding wheel-track contact forces and safety coefficients.

It is clear that the behaviour is generally better on the bridge than on the ground. This is due to the fact that the bridge itself filters the earthquake maximum accelerations, thanks to its low natural frequencies.

6.4.2.7 Global and local impact factors

From the results of the analyses and simulations described above the impact factors have also been evaluated.

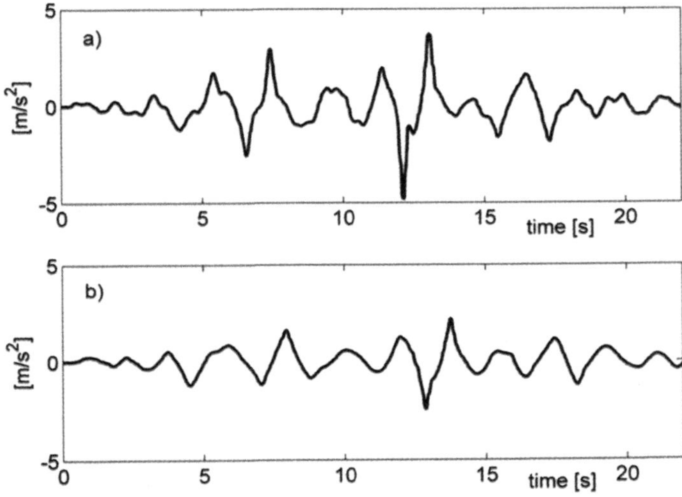

◄ Figure 6.71 Lateral car-body acceleration: (a) running on the ground (b) running on the bridge.

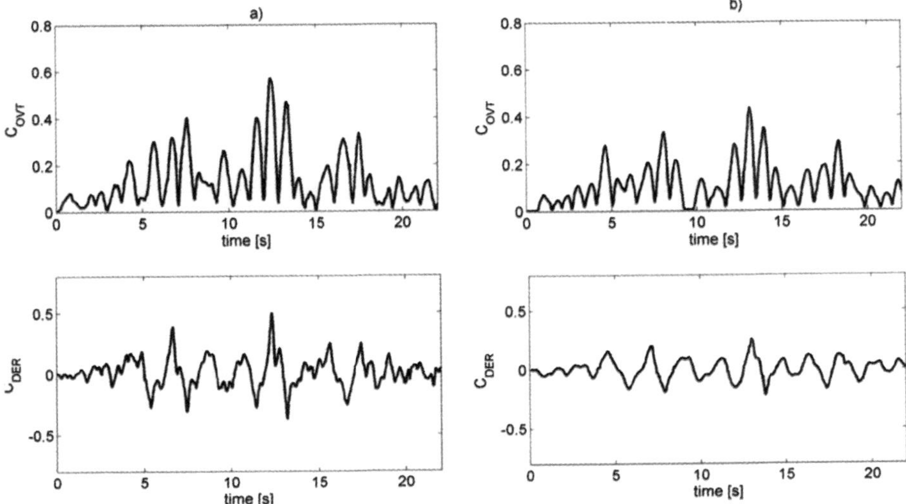

◄ Figure 6.72 Safety coefficients when running on the ground (a) and on the bridge (b).

The impact factors are defined as the ratio between the stresses in the structure taking into account the dynamic train-bridge interaction and the corresponding stresses arising simply as a result of the train acting as a static force. Global impact factors have been calculated for deformations in the vertical plane and are equal to 1.2 for the flexural moment My, and 1.1 for the shear force Tz. Table 6.3 gives the calculated local impact factors for different types of track support.

Table 6.3 ► Local impact factors.

Solution	Direct fastening		Track slab	
Reinforced	no	yes	no	yes
Vertical force on deck	1.25	1.24	1.14	1.14
Longitudinal stress in deck	1.14	1.12	1.03	1.03

The conclusion is that global impact factors are small, demonstrating that a long span bridge does not suffer significantly from this aspect, compared to a short span bridge. On the contrary, the local impact factors depend on the local deck and track support details, and are independent of the bridge span length.

References and Further Readings

[1] N. Gimsing: *Cable Supported Bridges, Concept and Design,* WILEY-VCH, 2 edition, 1998.

[2] F. Brancaleoni: *Verformungen von Hängebrücken unter Eisenbahnlasten,* Der Stahlbau, Darmstadt, Febbraio 1979.

[3] F. Brancaleoni, A. Cheli, G. Diana: *The Railway Runnability of Long Span Suspension Bridges,* Proc. In. Congress of the In. Ass. for Bridges and Structural Engineering, IABSE, Session 6 'Long Span Structures', Helsinki, 1988.

[4] G. Diana, F. Cheli: *Dynamic interaction of railway system with large bridges,* Vehicle System Dynamics, Volume 18, Number 1-3, 1989, pp. 71-106.

[5] F. Brancaleoni: *The Construction Phase and its Aerodynamic Issues,* in Aerodynamics of Large Bridges, ed. A. Larsen, Balkema, 1992.

[6] F. Brancaleoni, G. Diana: *The Aerodynamic Design of the Messina Straits Bridge,* International Journal of Wind Ingeneering and Industrial Aerodynamics, n. 48, 395-409, 1993.

[7] J. Brownjohn, M. Bocciolone, A. Curami, M. Falco, A. Zasso: *Humber Bridge Full Scale Measure Campaigns 1990-1991,* Journal of Wind Engineering and Industrial Aerodynamics 52 (1994) 185-218, Elsevier Science Publisher The Nederlands 1994.

[8] G. Diana, F. Cheli, S. Bruni, A. Collina: *Super long suspension bridges runnability,* 15th IBASE Congress, Copenhagen, Denmark, 1996.

[9] G. Diana, F. Cheli, A. Zasso, S. Bruni, A. Collina: *Aerodynamc Design of Very-Long Span Suspension Bridges,* proc. IABSE Symposium on Long-Span and High-Rise Structures, Kobe, Japan – September 2-4, 1998.

[10] G.L. Larose, A. Zasso, S. Melelli, D. Casanova: *Field Measurements of the Wind Induced Response of a 254 m High Free-Standing Bridge Pylon,* Journal of Wind Engineering and Industrial Aerodynamics, Vol. 74-76, 1998 pp. 891-902.

[11] G. Diana, M. Falco, F. Cheli, A. Cigada, A. Collina, A. Zasso: *The Bridge Over The Messina Strait [aeroelastic stability and train runnability],* proc. Brobyggardagen 2000, Göteborg, 31 January 2000.

[12] G. Diana, F. Cheli, S. Bruni: *Railway runnability and train-track interaction in long span cable supported bridges, Advances in Structural Dynamics* Vol. I, Elsevier Science, 2000.

[13] Cheli F., Cigada A., Diana G., Falco M.: *The Aeroelastic Study of the Messina Straits Bridge*, Kluwer Academic Publishers, Natural Hazards, pp. 79-105, Vol. 30 (2003), N° 1.

[14] F. Brancaleoni: *The Messina Straits Bridge – Past and Future*, Keynote Lecture, 2004 October Seminar "Bridge Structures: Assessment, Design & Construction", Bridge Engineering Association, NY, 2004.

[15] Diana G., Resta F., Belloli M., Rocchi D.: *On the vortex shedding forcing on suspension bridge deck*, Journal of Wind Engineering and Industrial Aerodynamics 94 [2006], pp. 341-363.

[16] Diana G., Belloli M., Rocchi D., Resta F., Zasso A.: *Sensitivity analysis of different aerodynamic devices on the behaviour of a bridge deck*, proc. of the 12th International Conference on Wind Engineering ICWE-12, Cairns, Australia, 1-6 July 2007.

7 – DESIGN OF THE BRIDGE

Contents of this chapter:

- **Global design concepts**
- **Design basis**
- **General bridge outline**
- **Suspended deck**
- **Deck restraints and expansion joints**
- **Suspension system**
- **Towers**
- **Foundations and anchor blocks**
- **Operation & maintenance strategies**
- **Approach infrastructures**
- **Strategies and provisions for construction**

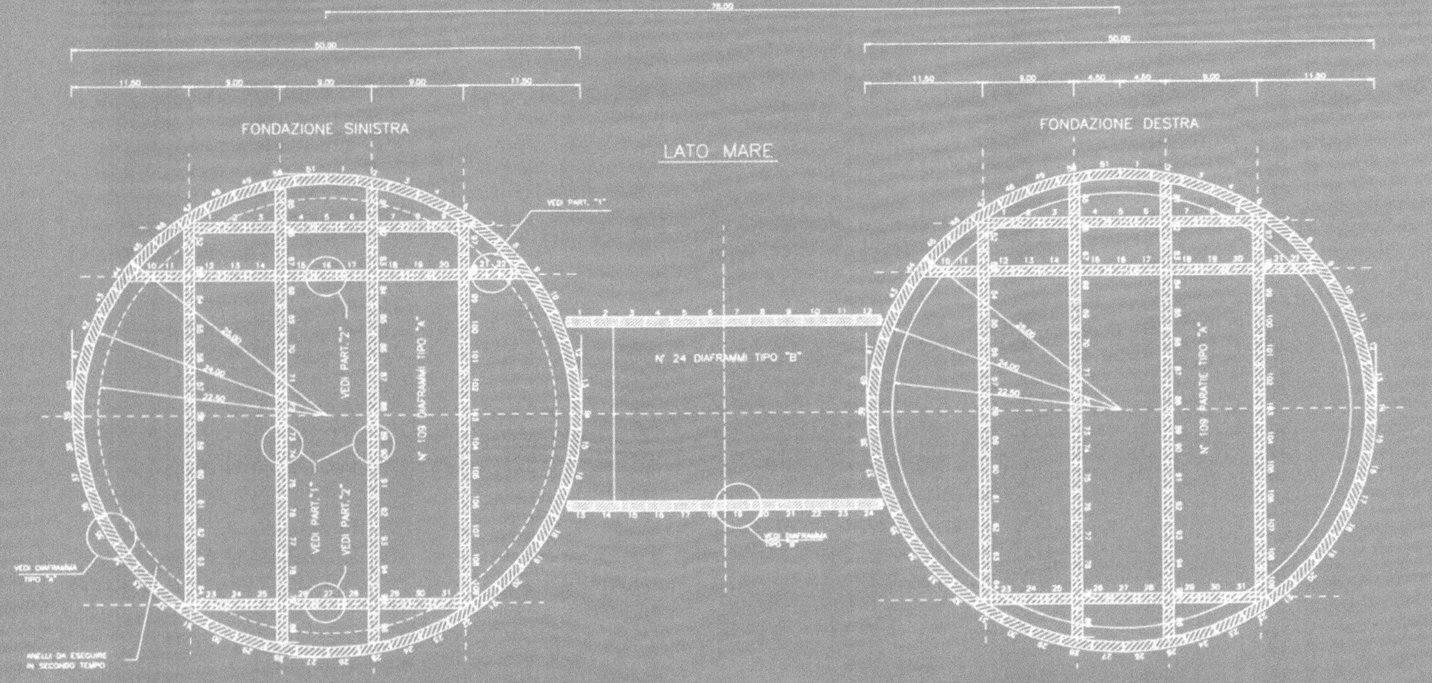

Achieving a sound design concept on which to base the aerodynamic stability and the feasibility of the 3300 metre suspended span was the beginning of the journey towards tackling a wealth of further challenging engineering issues, with a variety of structural components to be conceived and optimised, as well as behavioural aspects to be understood and mastered. This chapter explores some details of this process, both for certain general aspects and for particular structural elements, and describes the design as it stands following the selection of the General contractor.

7.1 Global design concepts

As a fundamental principle, it is essential that such a critical piece of infrastructure should set as a general design target the achievement of a robustness and performance level at least as high if not higher than that internationally achieved for other such key elements of transportation infrastructure. In other words, the bridge has to be at least as safe, functional, reliable, durable and efficient, as other major bridge crossings built recently throughout the world.

Consequently, the design has been based on the most advanced international state of the art and on current best international practice in the design and construction of such works, taking into account existing relevant knowledge and current research and development.

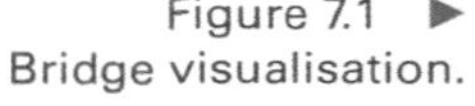
Figure 7.1 ▶ Bridge visualisation.

Within the above framework, a number of concepts were selected as a basis for the design:

a. The structural concept should be based on the scheme of a "classic" suspension bridge, with well understood behaviour, backed by the experience of the many such realizations all over the world (although at a smaller scale) over more than a century.
b. The construction of the main bridge components (deck, suspension system, towers, foundations and anchor blocks) should be based on proven materials, technologies and methods, whose performance has

already been experienced on other works with comparable importance. (See Table 7.1 for a comparison of key parameters with those of the Akashi Bridge, the current longest suspension bridge in the world.) In other words, no further scale effects should be introduced, besides those that could not be avoided as discussed in Chapter 6.

c. Any interference with the sea, navigation and the marine environment should be minimised, both during construction and in service.
d. The aesthetic value of the bridge should be defined by the intrinsic beauty of the scale and slenderness of the bridge and performance-optimised structural forms, as befitting the best example of its bridge type, without unwanted sculptural additions.
e. Environmental and territorial compatibility and sustainability should be held in strict focus throughout, both during and after construction.
f. As far as aerodynamic performance is concerned, adequate global safety should be achieved through the intrinsic, passive characteristics of the design, without the need for any additional active control device. Such devices would be accepted only to control non-destructive phenomena involving the durability of single elements, such as limited amplitude tower and hanger vibrations for example, or to manage certain limited serviceability aspects, or as temporary measures taken only during the erection stages.

	Messina bridge	**Akashi-Kaikyo**
Tower		
– height above sea level (m)	380	297
– leg cross section (m)	12×16	7×15
– steel panel thickness (mm)	up to 65	up to 70
Main cables		
– diameter of a cable (m)	1.24 (four cables)	1.12 (two cables)
– length between anchorages (m)	5,270	4,070
– total weight (t)	$4 \times 41{,}650$	$2 \times 25{,}250$
Deck		
– length (m)	3,666	3,911
– total steel weight (t)	67,000	90,000
Tower foundation		
– volume (m^3)	$72{,}400 \div 86{,}400$	350,000
Anchor block		
– volume (m^3)	237,000	$140{,}000 + 420{,}000$

◄ Table 7.1 Dimensions are huge but not out-of-scale compared to the current world record span.

7.2 Design basis

As for any such major structure, national and international standards were integrated with specially derived project-specific specifications, to form a dedicated design basis for the crossing. This was the outcome of a process that evolved over more than 30 years throughout the feasibility and design phases. At each stage, design specifications were prepared for the crossing in general, including tunnel, floating tunnel and single or multi-span bridge schemes. These evolved, focussing in the end on the single long span option, together with an increasing knowledge and understanding of the bridge physical environment, the growing traffic demand and critical aspects of the structural behaviour. All this closely followed and in some case anticipated the best relevant international practice. For example, the step from allowable stress to limit state design was taken in the early eighties, possibly for the first time for such a structure.

At the root of the design basis was the decision to adopt a long design life of 200 years, thus resulting in action return periods and overall safety levels significantly higher than normal. Long term scenarios were related to a 2,000-year return period for "extreme" actions (in particular wind and earthquake), i.e. ten times the design life. Other limit state requirements were defined in relation to a sequence of shorter return periods (long-medium-short time) of 200, 50 and 10 years respectively.

A summary of the fundamental design basis figures, as stated in the 2004 tender documents, is given in the following table.

Table 7.2 ▶ Summary of basic design specifications (2004).

	Serviceability limit states		**Safety limit states**	
Levels	**1**		**2**	**3**
Required performances				
Limit state	Serviceability 1 SLS1	Serviceability 2 SLS2	Ultimate ULS	Damage-Ultimate D-ULS
Serviceability	Road & Railway full service	Railway traffic guaranteed	temporary loss of serviceability	complete loss of serviceability
Admitted structural damage	no damage	no damage	minor damage only	main structures survive
Design actions				
Return period	50 years	200 years	2,000 years	>2,000 years
Wind average speed (at + 70 m elevation)	44 m/s	47 m/s	54 m/s	60 m/s
Flutter critical speed	–	–	>75 m/s	–
Seismic action (free-field PGA)	0.12 g	0.27 g	0.58 g	0.64 g
Total roadway load	3,300 t	16,500 t	24,750 t	–
Total railway load	13,200 t	26,400 t	39,600 t	–

7.3 General bridge outline

Within the framework described, decisions were taken to define the general bridge outline. Among the most significant were the following:

- *Bridge span and location.* The reasons leading to the selection of a shore-to-shore single span bridge have already been discussed in Chapter 1. The above decision restricted the plan location to a narrow corridor, about 800 m wide, where the distance between the two coasts is a minimum. Among the alternatives considered, both single and multi-span, shown in Figure 7.2, the final location was selected so as to minimise interference with the built and natural environment, such as examples of the local cultural heritage (e.g. the Ganzirri monumental cemetery) and important natural resources (e.g. the two small shallow lakes, Pantano Grande and Pantano Piccolo

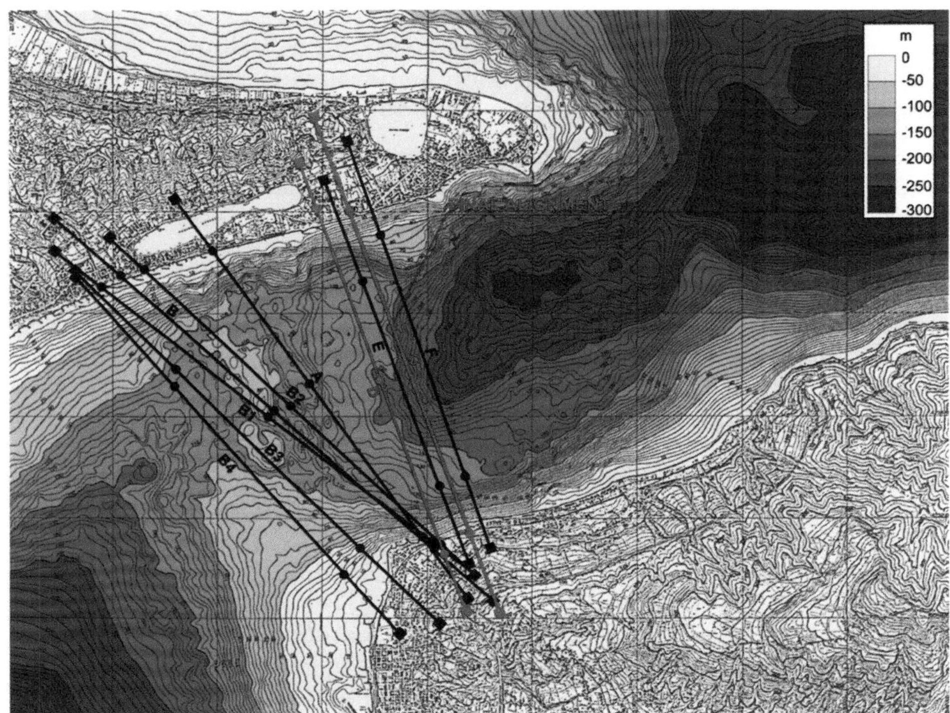

◄ Figure 7.2
Design corridor for single & two span suspension bridge alternatives.

in Sicily). The figure also shows different alignments considered for the discarded two-span scheme, south east of the final one, where the shore-to-shore distance is greater but the sea depth lower. (See Chapter 1.)

- *Sag to span ratio & tower height.* It has already been stated in Chapter 6 that to minimise cable weight the sag to span ratio should be kept as high as possible, consistent with service performance requirements. The need for a high stiffness for railway runnability purposes, coupled with the wish to restrict the tower height for landscape impact reasons, led in 1992 to the selection of a rather tight solution, with a ratio of 1/11. This value has afterwards never been changed, but it is believed that the overall design could benefit of a higher ratio. (See also the considerations on this matter discussed in Chapter 8.)
- *Profile.* The navigation requirements have been met allowing for a central channel 600 m wide with 65 m clearance plus two wider lateral channels to the shore with a reduced 50 m clearance. These clearances must be ensured under specified combined thermal and imposed live loading conditions defined within the service performance requirements in the Design Basis. The resulting deck level, giving consideration to maximum railway gradient aspects, varies from about 63 to 77 metres.

The final overall configuration is given in Figure 7.3. Note the two short side spans that are part of the solution for tackling the complex problem of terminal expansion joints and tower to deck connections, discussed in

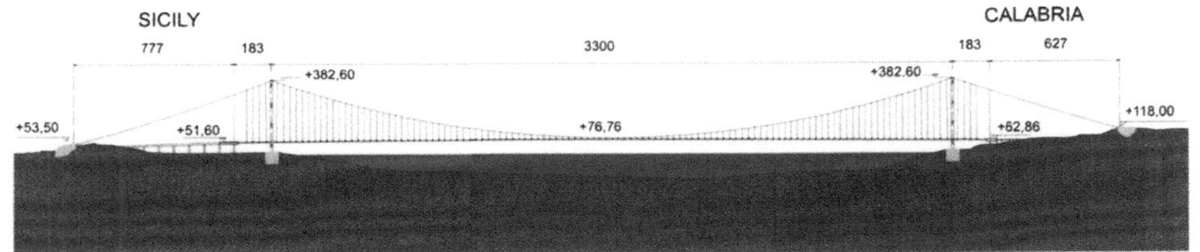

◄ Figure 7.3
General bridge outline.

the following sections. No normal side spans were possible due to the topography and the need to move the road and rail alignments off the straight bridge plan alignment as soon as possible in order to make the connections to the infrastructure systems on the two coasts.

The following sections are devoted to a more detailed discussion of the single components of the bridge superstructure, listed in the sequence corresponding to the load flow from deck to ground: deck & expansion joints – suspension system – towers.

The design has been developed over several years and has been subject to evaluation and checking by several leading international companies and experts. Among them a prominent role was played by Steinman-Parsons, appointed as advisor to the Italian Government, who concluded their independent scrutiny of the design as follows:

> *"...In accordance with our agreement with The Ministry of Public works, we have completed our in-depth technical studies of the preliminary design for The Messina Strait Bridge as prepared by Stretto di Messina S.p.A. in 1992 and as further supplemented by additional studies and reports prepared in 1997.*
> *We find that the level of development for this preliminary design is significantly more advanced than would commonly be done in worldwide practice, and that the work has been performed to a very high level of professionalism, using state-of-the-art engineering methods.*
> *There are no fundamental design issues that would preclude proceeding to final design. The strength and serviceability of the bridge have been validated by comparison with design standards and practices for long span bridges in other parts of the world.*
> *We have included specific recommendations concerning optimization of the bridge's geometry, structural details and details affecting future maintenance and operations. Suggestions to be considered for incorporation into the final design specifications are also provided..."*

Figure 7.4 ▶ Plan alignment of bridge and approaches.

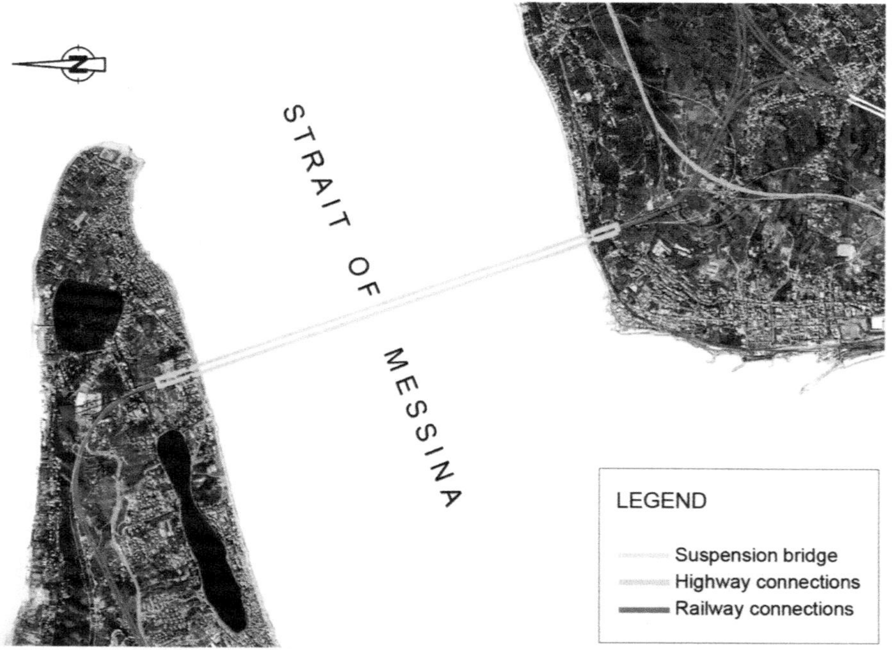

7.4 Suspended deck

As discussed in Chapter 6, the suspended deck is formed by three streamlined longitudinal boxes, the lateral ones for the two road carriageways and the centre one for the railways.

The two road platforms comprise a total of four 3.75 m wide road lanes plus emergency lanes. Inspection and maintenance lanes are located externally, on light cantilevering elements at the side of the road boxes, outside the hangers. The centre box carries two railway tracks plus inspection strips. The large open areas between the boxes bring the total deck width to about 61 m, with a 52 m distance between the cable planes. (Figure 7.5)

The three longitudinal boxes are in turn carried by transverse box girders spanning between the hanger planes, forming a grid of boxes for the whole suspended deck structure. The deck structure is equipped with four inspection and maintenance gantries, two in the centre span and one in each of the side spans.

The final configuration (Figures 7.5 and 7.6) was reached by taking into consideration a blend of needs, ranging from aerodynamics to structural behaviour and maintenance. Among the most significant:

- For best aerodynamic efficiency, the centre box had to be as shallow as possible, so as to maximise the effectiveness of the total centre void area. Hence, the depth of the railway box was restricted to ≈2.20 metres, which was considered the minimum for internal inspection and maintenance.
- Having defined the rail box depth, the spacing of the transverse girders was maximised with regard of the span of the rail box itself, taking into account the strength and railway dynamic response aspects, and making reasonable assumptions regarding steel plate thicknesses and mechanical properties. This resulted in an optimum distance between transverse girders of 30 m. The intersections between longitudinal and transverse stiffened plate boxes present certain difficulties from a fabrication and behaviour viewpoint, with complex welding and a large number of small elements to be interconnected, hence such intersections need to be kept to a sensible minimum. This is an area where potential improvements can possibly be made in the detailed design, also in relation to different possible solutions for the transverse girders (see Chapter 8). The transverse girders form the main deck supporting structure, also bearing in mind that the heavy loads from the railway boxes are applied at the bridge centre line. They are 4 metre wide rectangular boxes spanning 52 metres between the hangers, varying in depth from about 1.3 to 4.7 metres, and formed of plates whose thickness varies from 12 to 24 mm.

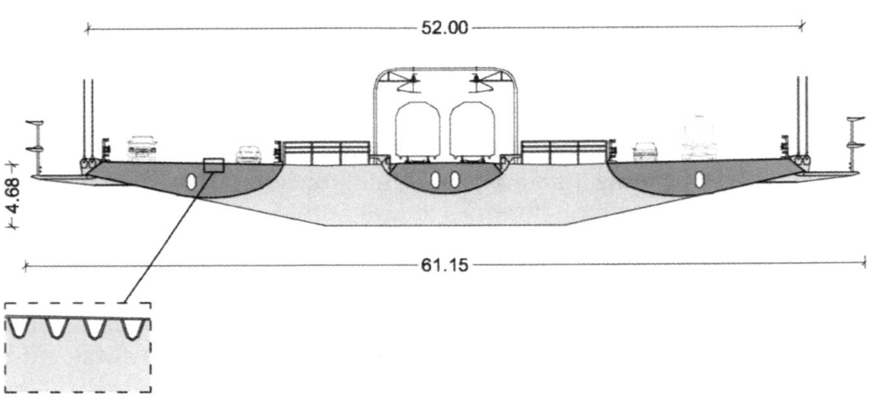

◄ Figure 7.5 Cross section of the deck as per Preliminary Design.

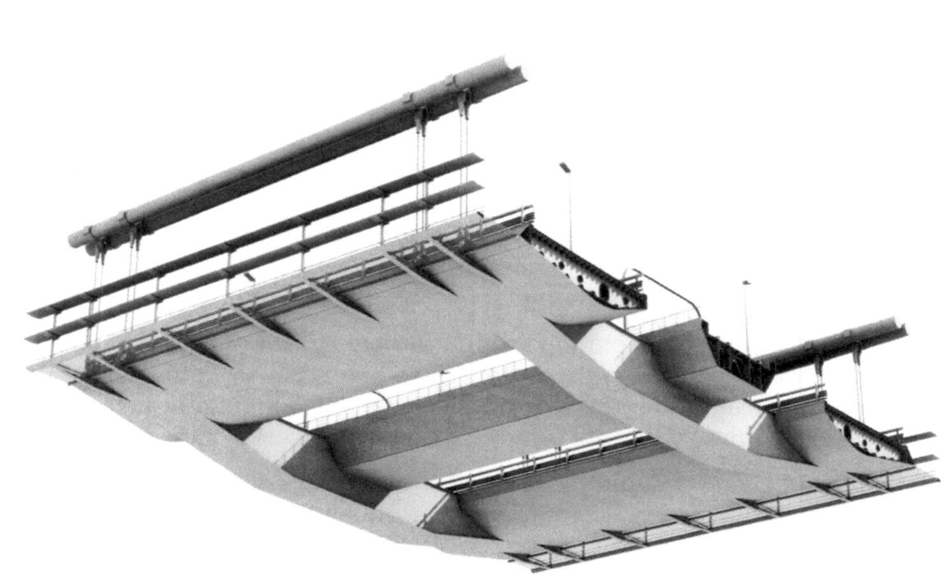

Figure 7.6 ► View of the suspended deck arrangement. This picture shows an alternative for the railway box having an optional trapezoidal shape. The section in this picture is just over 30 m long and represents about one half of the typical unit expected to be transported to site on barges.

- Next, the road boxes were made deeper than the rail box consistent with the need to achieve a minimum bottom flange plate thickness to minimise weight while maintaining durability and local stability. The minimum external plate thickness was established at 9 mm, resulting in a ≈2.80 metres depth for the road boxes. The larger box depth was also needed to increase the overall deck torsional stiffness, which is useful in the global behaviour although unable alone to provide adequate aerodynamic stability, and to obtain an optimal aerodynamic behaviour by ensuring that the centre box lies fully in the wake of the outer boxes. This provides regularity of the aerodynamic moment curve. The upper orthotropic deck plate configuration was on the contrary dominated by local strength and fatigue behaviour (see typical trough detail in Figure 7.5) resulting in a minimum deck plate thickness of 14 mm and 6 mm for the trapezoidal trough.
- Shape details were optimised for best aerodynamic efficiency, together with rail and road fittings, equipment and crash barriers. The curved shapes also play a role in improving local stability against buckling of plates in longitudinal compression.
- Lightweight design was sought for all structural and equipment elements. Roadway surfacing is based on a 38 mm thick elastomeric multilayer compound, and the rails are also embedded in an elastomeric compound located in a continuous longitudinal trough, avoiding any floating slab or transverse discontinuity details. A continuous rail is provided on either side to minimize the risk of train de-railing.

The static redundancy in the system (as also required by the Basic Design Specification) ensures that the deck would not collapse even in the extremely unlikely event that all the hangers at one end of a crossbeam were broken.

The materials for the suspended deck girder are based on standard structural steel grades: S355ML (ultimate strength >450 N/mm^2) for most of the deck, with S420ML (ultimate strength >500 N/mm^2) for special elements at the tower articulation, that will be discussed below. The total amount of structural steel in the deck is about 55,000 tonnes.

The resulting total steel girder self weight is only 18.1 t/m, with a total deck dead load of about 23 t/m. Table 7.3 shows a comparison with other notable suspension bridges.

	Messina	Akashi	Storebaelt	Tsing Ma
Main span	3,300 m	1,991 m	1,624 m	1,377 m
Total suspended length	3,666 m	3,911 m	2,694 m	1,733 m
Total service width (*road + rail*)	35.2 m	21.5 m	23.2 m	38.2 m
Weight of deck steel structure	18.1 t/m	23.0 t/m	11.1 t/m	20.8 t/m
Weight per service width	0.51 t/m^2	1.07 t/m^2	0.48 t/m^2	0.54 t/m^2

◄ Table 7.3
Deck unit weight compared to other suspension bridges, considering also the actual surface available for train and car service ("service width").

7.4.1 Vehicle traffic in wind conditions

To improve serviceability under severe wind conditions, external wind-screens form an integrated part of the overall aerodynamic design. They are equipped with comparatively small aerofoils to control the flow and act as aerodynamic dampers, while the wind protection element is a perforated metal sheet tuned so as to reduce the air flow and protect vehicles from the direct action of strong wind.

7.4.2 Durability

The interior of the boxes will be dehumidified to control corrosion, and external steel surfaces will be painted with a high build multi-layer system for maximum durability. Access manholes and internal walkways are envisaged to allow easy maintenance access to all bridge areas.

7.4.3 Erection

As far as construction techniques are concerned, the design envisages extensive use of prefabrication, as normal for such structure types. Stiffened steel panels will be prefabricated in offsite workshops and assembled into complete box sections in remote yards. Large deck sections (60 x 60 m) weighing about 1100 tonnes are then envisaged to be transported to site on barges and lifted into position using gantry cranes running on the main cables.

7.4.4 Alternatives proposed by bidders in 2005

On the basis of the Preliminary Design and documents developed by SdM, the bidders proposed various alternatives, aimed at optimising a number of aspects, all within the defined overall design constraints. The winner (Impregilo JV with main designer COWI) proposed the following main variations:

- The curved lower plating for the railway box has been replaced by a simpler traditional trapezoidal shape (Figure 7.7). The box has a continuous longitudinal web under each rail which passes through the crossbeams and all detailing is such as to minimise the potential for fatigue damage. The bottom flange plates and webs are stiffened with large closed troughs.
- Transverse diaphragms of longitudinal boxes are typically at 3.75 m centres, and the crossbeam width is accordingly reduced from 4 m to this module.
- A reduced road surfacing thickness was proposed, based on a spray-applied acrylic surfacing system with a thickness of only 12 mm, so as to

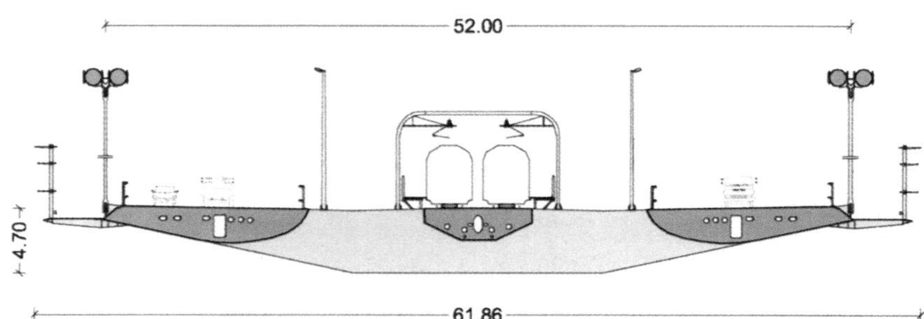

Figure 7.7 ▶ Cross section of the deck: Impregilo Tender Design variant.

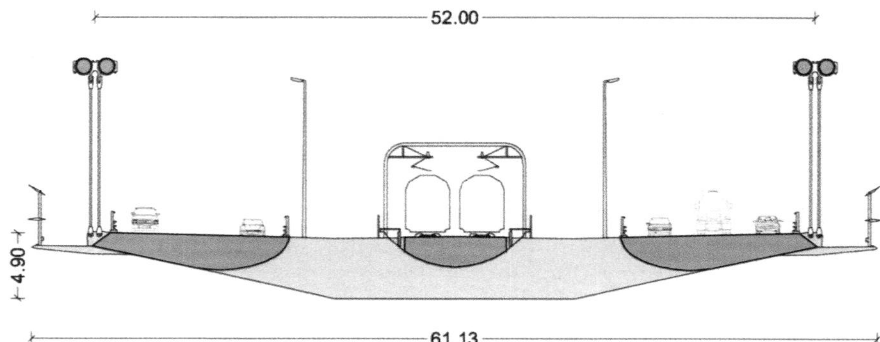

Figure 7.8 ▶ Cross section of the deck: Astaldi tender variant.

further minimise total deck dead loads. As a result of this thin surfacing, the steel deck plate thickness was increased to 16 mm for the main traffic lanes, with the troughs deepened and increased to 8 mm thickness, to improve the fatigue behaviour of the orthotropic deck.

- Steel plate grades vary to suit their location and application to achieve minimum weight and best economy. Steel grade S355 is used for about 30% of the deck structures, S420 for about 15% and S460 (ultimate strength >530 N/mm^2) for the remaining 55%, especially in the crossbeams.
- Aerofoils and wind screens were slightly modified.
- The longitudinal and transverse restraint between towers and deck is provided by hydraulic buffers, as discussed further below.

The second bidder (Astaldi JV with main designer Chodai), maintained the curved lower plate of the railway box, (Figure 7.8), and introduced a number of minor aerodynamic improvements, with a different arrangement and inclination of the external aerofoils.

As already discussed in Chapter 6, both bidders, having carried out extensive additional aerodynamic testing, fully confirmed the general arrangement of the deck, i.e. the key point for proving the technical feasibility of the bridge. This confirmation underlined the correctness of the concept, and added further confidence in the overall design.

7.5 Deck restraints and expansion joints

This section is concerned with an aspect that proved to be extremely challenging, possibly second only to the fundamental matter of achieving adequate aerodynamic stability for the bridge. The issues are as follows:

- As already mentioned, the land configuration at the site does not allow a normal configuration with two long side spans, typically close to half the main span in length, as adopted in most classic suspension bridges. On the Sicily side the site is so close to the east island point ("Torre Faro") that there is simply not enough land and the access viaducts must immediately turn southwards. On the Calabria side the coast is comparatively steep and the hills quickly become too high for a side span. Indeed the road and rail corridors go into tunnel within a short distance. Hence, the bridge is without long side spans.

- Consequently, the deck abutments must be located close to the towers, together with the associated expansion joints. When configurations with expansion joints at the towers were analysed, it was immediately found that some displacements were extremely difficult to handle. More specifically, for the railway expansion joint it was possible to accommodate large longitudinal displacements, but no system could be found to handle the large end rotations about the transverse and vertical axes arising respectively from temperature/railway loads and lateral wind loads on the deck.
- It was therefore necessary to find an arrangement that could absorb the rotational components of the bridge displacements away from the expansion joint, leaving only the longitudinal components to be accommodated at the abutment.

When this became clear in the mid eighties it was also understood that one possible solution would be the introduction of a short side span, with the deck fully continuous at the towers. In this way the larger rotations would take place at the towers on a continuous deck, where they are fully compatible, while the ends would experience a reduced rotation, with a "continuous beam" like behaviour. This scheme required the introduction of two large terminal structures on land to receive the deck ends and accommodate the expansion joints. Several variants were analysed and were included as the basis of the 1986 feasibility studies of this aspect.

While not unfeasible, such arrangements showed a number of drawbacks and none was revealed to be wholly satisfactory. There were two main problem areas:

- Although they were reduced, the rotations were still significant and not easily accommodated, needing complex and massive mechanisms.
- Secondly, and most important, very large horizontal forces arose at the deck ends in connection with transversal static wind actions. This was due to the high horizontal flexural stiffness of the 60 m wide deck, restrained in a way that caused a "short lever arm" effect in the horizontal plane. (Figure 7.9) The transverse bending effects in the continuous deck box girders at the towers were also very large as a result of this articulation arrangement.

Such large transverse horizontal forces had to be absorbed by the terminal structures, with substantial increase in their already rather massive configuration, especially on the Sicily side, where the height is considerable.

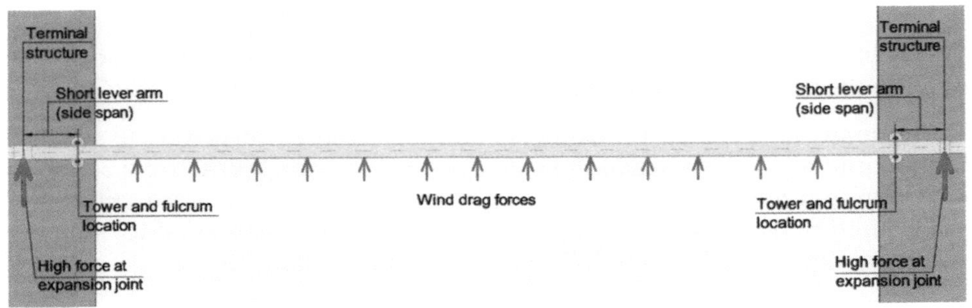

Figure 7.9
Plan showing how the lever effect of short side spans results in high transverse forces at expansion joints.

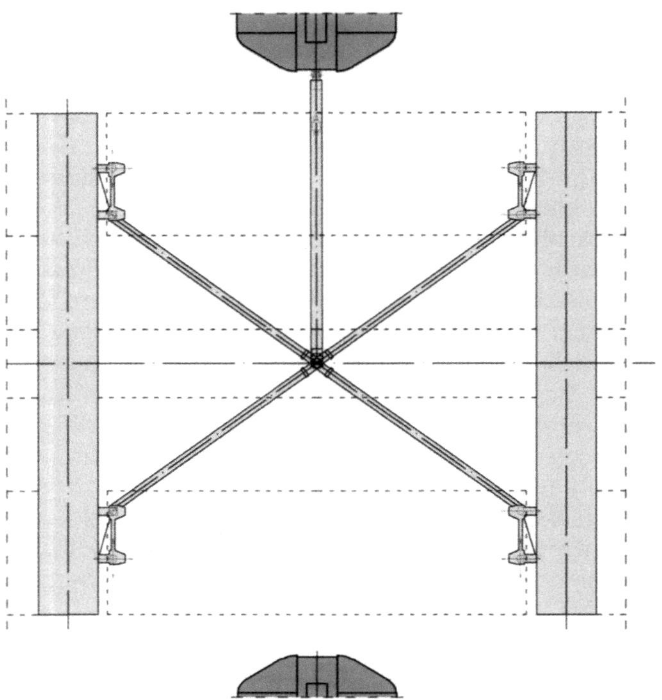

Figure 7.10 ► Articulation arrangement connecting the bridge deck to the towers as originally proposed.

This was found to be at the borderline of feasibility. An innovative idea arose in the early nineties, thanks to the contribution of Dr. W.C. Brown, who brilliantly solved the problem, based on the following measures. (See also Figure 7.10)

- The longitudinal road boxes were made discontinuous at each tower, replacing the continuous structure with two simply supported spans. The continuity is resumed immediately afterwards, in the short side span.
- The centre railway box is left continuous throughout.
- The deck structure is restrained laterally at the towers via a large lateral pendulum strut with X-bracing.

This configuration creates an extremely flexible deck segment at the towers, and as a consequence the lateral forces in the side span fall dramatically. While rotations due to transverse wind are present at the tower segment, these are not continued towards the abutment. In effect, a sort of "elastic hinge" about a vertical rotation axis is inserted into the deck.

The rotations about the vertical axis experienced in the flexible deck segment at the tower are absorbed via moderate flexure in the continuous central rail box. This does not cause forces in the simply supported outer roadway box segments which have small expansion joints. These joints do not experience the overall large span displacements but only the local relative movements associated with temperature changes over the short segment length and local kinematics from the global bridge behaviour.

This solution dealt with the problem at its root and allowed the introduction of a number of further improvements:

- The span of the special road and rail box segments at the towers to achieve the optimal balance between flexibility and size was found to be 50 m.
- The longitudinal boxes and cross girders in these areas were optimised in depth and steelwork, using high strength grade 460 steel for the rail box, aiming at an improved flexibility/strength ratio.

- Vertical displacements and rotations about horizontal axes in the short side spans were drastically reduced, by introducing a set of vertical tie ropes connecting the main cables directly to the abutment structures on land. The displacement of the main cables is in fact clearly the main source of vertical deck movements in the area.
- The hangers towards the tower area were progressively increased in diameter to further limit the same vertical displacements. Their flexibility, due to the large length, was in fact found to be significant in the local deck behaviour.

This overall arrangement, apparently complex but functionally simple, achieves the goal of substantially reducing all the displacements at the expansion joints, apart from the longitudinal movements which can be handled by large railway expansion joints. This is obtained without generating unduly large forces. The small residual rotations are easily absorbed by local movement and expansion joints mechanisms not described in detail herein.

This system was the basis for the 1992 design which demonstrated the feasibility of this aspect. It must nevertheless be recognised that the longitudinal displacements remain very large, consistent with the exceptional span, with a total movement range of about six metres. (See Table 7.4).

The type of railway expansion joint selected was of the needle type, with a flexible half-rail bent sideways by the bridge movements, characterised by a comparatively large number of parts and a rather complex mechanical arrangement. (Figure 7.11)

In the 2002 design this aspect was not further developed, considering it fully satisfactory for the general bridge behaviour. At the same time, the complexity of the abutment expansion joints was recognised, with an appreciation that the large movements led to consequently high rates of wear in the moving parts and the need for regular tuning and maintenance.

◀ Table 7.4 Calculated longitudinal displacements at expansion joints.

Load type	Min (m)	Max (m)
Road, vertical	–0.58	+0.61
Road, braking	–0.14	+0.14
Railway, vertical	–1.37	+1.41
Railway, braking	–0.90	+0.90
Thermal variations	–0.58	+0.60

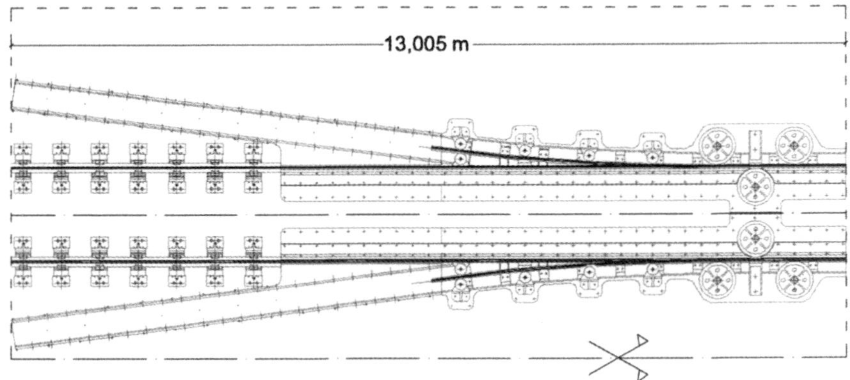

◀ Figure 7.11 Rail expansion joint schematic.

7.5.1 Alternatives proposed by bidders in 2005

During the 2005 tender, both the competing JV's devoted considerable attention to this local, but nonetheless vital aspect, as expected and indicated in the tender documents. It is very interesting to note the different approaches taken in this respect by the different designers.

The Impregilo JV, after careful consideration of a number of alternatives, concluded that the original arrangement was optimal for the global behaviour and decided to confirm it. The statement given was that *"... having studied a number of alternatives, it was concluded that the configuration providing by far the best balance among the different requirements is the one based on the disconnection of the road boxes at the towers ..."*.

Thus, having confirmed the basic configuration, the attention of the JV focussed on the matter of improving the complex expansion joint mechanisms, with consequent reductions in cost, wear and maintenance requirements, to be obtained partly through a substantial reduction of the longitudinal displacements.

The basic consideration was that the different movement sources provide contributions comparable in terms of displacements, but very different in terms of forces in restrained conditions, due to the different structural mechanisms involved. Thus, for example, thermal movements, arising as axial strain in the deck, would, if restrained, give rise to extremely high forces, while restrained movements due to vehicle braking cannot give an action larger than the braking force itself. Similar considerations apply, although with quantitative differences, to movements due to non symmetrical live loads and seismic shaking.

Such considerations were behind the concept, considered in the early nineties but not included in the Preliminary Design, of limiting longitudinal movements of the deck using passive hydraulic devices or "buffers". The Impregilo JV developed this idea very effectively in their tender design, proposing the longitudinal connection between deck and towers through two pairs of such buffers connected to short cantilever extensions of the first side span transverse cross girders. (Figure 7.12)

The devices, functioning via hydraulic control, behave elastically up to a given force limit, and then at higher longitudinal displacements maintain a constant force in a pseudo "elastic-perfectly-plastic" fashion, although no metal yield is involved. As a result:

- Only very limited restraint is given to thermal displacements, leaving virtually free movements for this aspect.
- Frequent vehicle braking forces in service are fully restrained, allowing movements only for extremely rare large braking events, such as. emergency railway braking under exceptional conditions.
- Most importantly, frequent service displacements due to non uniform rail and road loads are restrained, allowing movements only for rare traffic conditions such as road traffic jams on half the main span length or rare very heavy trains.
- Furthermore, with a secondary but very important benefit, the devices behave as a passive control system for longitudinal seismic actions, introducing a beneficial energy dissipation for the earthquake response. In other words, the deck and buffer system acts as a mass damper for the structure. Note that the term "mass damper" and not "tuned mass damper" has been used, as the system is optimised with respect to expansion joint behaviour and not towards seismic response.

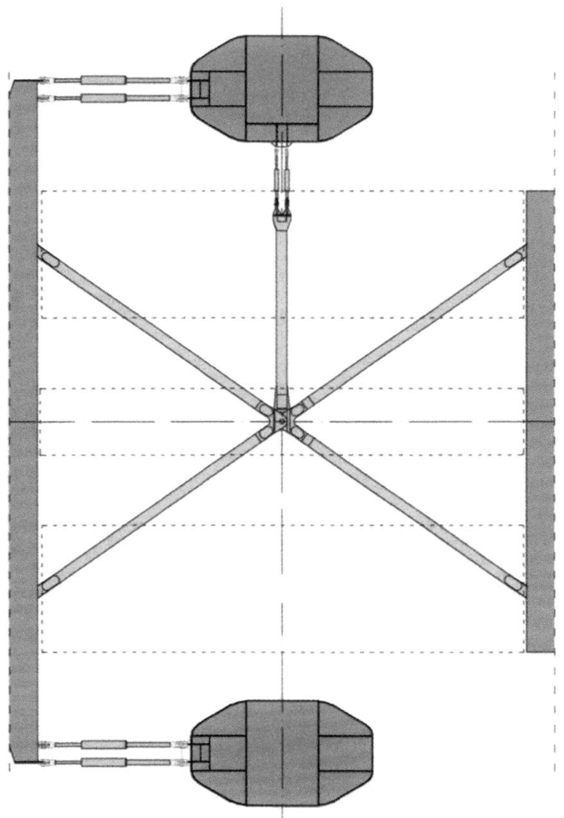

◀ Figure 7.12 Articulation arrangement at deck connection to towers, Impregilo Tender Design.

In the first place, this arrangement dramatically reduces the maximum longitudinal displacement at the expansion joint from about ±3 m for the unrestrained movement to about ±0.9 m. Secondly, most small displacements due to frequent train passages, road traffic and braking are eliminated, decisively reducing the movements experienced by the joint mechanism, with consequent reductions in joint wear and maintenance requirements. It has been evaluated by the bidder, through service life simulations, that the total movement joint excursion is reduced in the order of 40 times.

The second placed bidder, the Astaldi JV, proposed an alternative design for this aspect, concerned similarly with the restraint system at the towers, but with the expansion joint movements left fully free to take place as in the 1992 and 2003 SdM design versions.

The rationale stated for the Astaldi alternative was based on the consideration that in the original arrangement the cross bracing at the tower and the continuous railway deck absorb longitudinal forces working in parallel, hence absorbing them according to their respective stiffness. This led them to propose a Y shaped connection, located on the side spans. (Figure 7.13.) This required the original pendulum lateral restraints to be replaced with longitudinal sliders.

Such proposed arrangements are considered on the whole equivalent to the original SdM Preliminary Design, or even slightly less effective. In fact, while the relatively minor problem of the longitudinal forces in the bracing is eliminated, lateral forces are brought into the deck side span in an anti-symmetric manner, introducing somewhat higher flexure in the

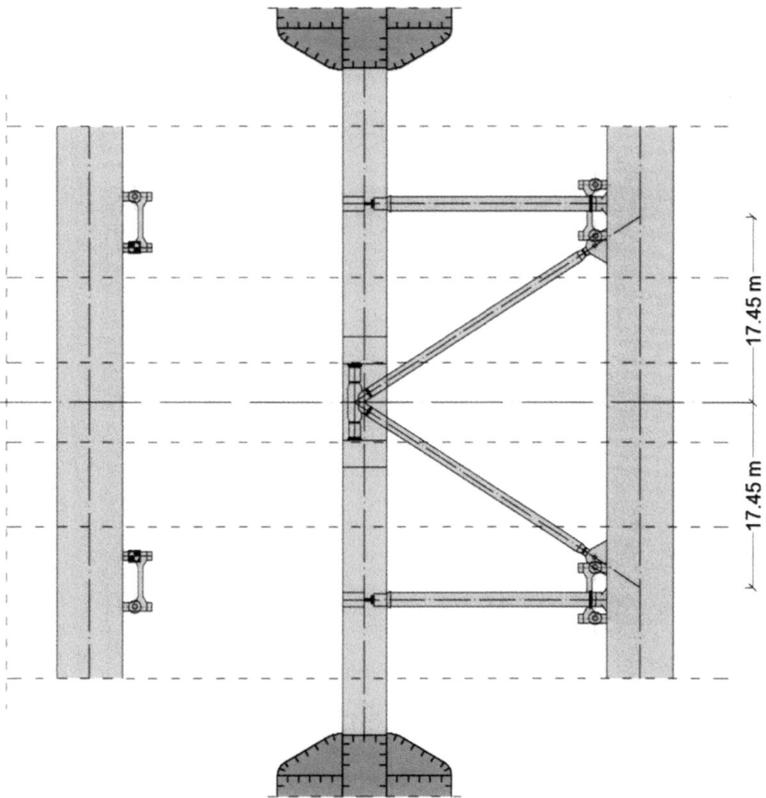

Figure 7.13 ► Articulation arrangement at deck connection to towers, Astaldi Tender Design variant.

horizontal plane. Furthermore, the sliders are more subject to wear and maintenance needs with respect to the simple pendulum.

The interest and effort devoted by both bidders to this important aspect is nevertheless emphasised, as well as the effectiveness of the original overall restraint solution.

7.6 Suspension system

7.6.1 Main cables

The suspension system comprises two pairs of main cables set 52 m apart. In the main span, each 1.24 m diameter cable comprises 44,352 wires of 5.38 mm, giving a total steel cross section area just over 1 m^2. (Figure 7.14) The total length of each cable will be 5,270 m. This brings the cable size and total length to be only slightly larger than the Akashi Bridge (1.10 m diameter and 4700 m in length). The axial force under self weight is nevertheless larger, due to the longer span, but the said configuration gives solid ground to the feasibility of the configuration selected.

The same cross sectional area of a pair could be obtained by a single cable with a 1.6 m diameter, but splitting the cables into pairs gave great advantages in terms of constructability, time saving and reduced wind drag, although with certain greater complexities in cable compaction.

The 1992 and 2002 designs both assumed the use of 1770 N/mm^2 grade wires and were developed on the assumption of construction by modified

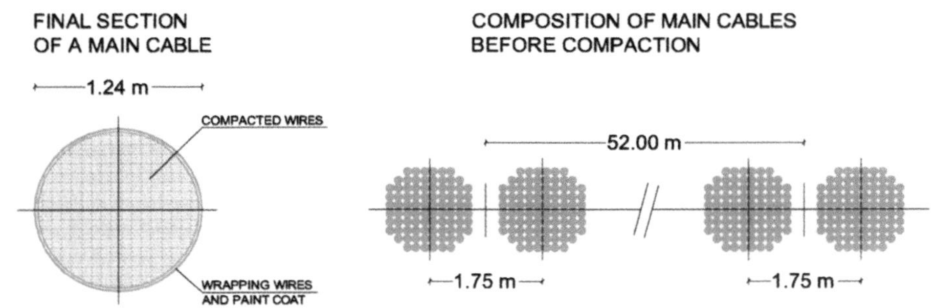

◄ Figure 7.14 Main cables as per Preliminary Design: each cable is formed by 88 strands 504 wires each.

aerial spinning (88 strands per cable, 504 wires each). As usual, individual wires are hot-dip zinc coated in the factory for protection against corrosion.

The alternative based on prefabricated parallel 127 wire strands (PPWS) was also considered and developed to a lower level of detail, leaving the contractors free to propose their preferred solution. It is interesting to note that both bidders indicated PPWS as their preferred choice, showing how the size of the project was sufficient to justify the higher investment needed for such a system.

The total amount of structural steel used in the suspension system is about 167,000 t for cable wires, 5,500 t for suspenders and 7,000 t for castings (cable clamps and saddles).

In SdM's Preliminary Design the outer surface of the cables is sealed by elastomeric paste with a high metallic zinc content. Cables are then wrapped with soft steel wire having an interlocking S-shaped cross section that creates a continuous water-tight and smooth surface. This wrapping is finally coated by an elastic paint system.

Dehumidification of the main cables is also required by the tender specifications. From the experience on older suspension bridges in the USA and elsewhere, it is known that main cable wire corrosion can create major difficulties for maintenance and rehabilitation of such bridges. (This is discussed further in Appendix 2) Similarly, saddle chambers at the tower tops and the anchorages will also be dehumidified.

7.6.2 Hangers

The hangers are in groups at 30 m centres, suspending the ends of the crossbeams from the main cables. Hanger diameters vary up to a maximum of about 160 mm, and the longest hangers next to the towers are about 300 metres long and weigh about 30 tons each. The type indicated in the Preliminary Design was based on parallel wire strands, but the contractors were left free to propose different solutions.

Large diameter spherical bearings have been envisaged at both ends of the shorter hangers to allow rotation of the pinned joints about a longitudinal axis, aiming at minimising bending effects in the hangers due relative longitudinal movement between the main cables and the deck under non uniform live loads. Other solutions based on clamping the main cable to the deck at midspan, as has been done elsewhere (e.g. for the Storebaelt Bridge), were considered but discarded due to the excessive size of the clamp needed to accommodate longitudinal seismic forces.

Stockbridge type passive tuned mass dampers, with energy dissipation provided in bending of the short spiral ropes, were introduced to handle the problem of wind and wind-rain vibrations, of considerable relevance considering the hanger lengths involved. The contractors were also in this case left free to propose alternatives, including surface roughness treatments, in connection with the type of rope/strand selected.

The "tie-down" strands, connecting the main cables to the abutment structures (mentioned in the previous section), are formed by groups of parallel wire strands, similar to those used for the hangers.

7.6.3 Cable bands

The cable bands, of rather traditional conception, have been designed to keep the distance between the main cables as close as possible while still permitting sufficient space for cable compaction and wrapping during erection. They comprise two cast steel half shells, joined by vertical threaded rods. (Figure 7.15)

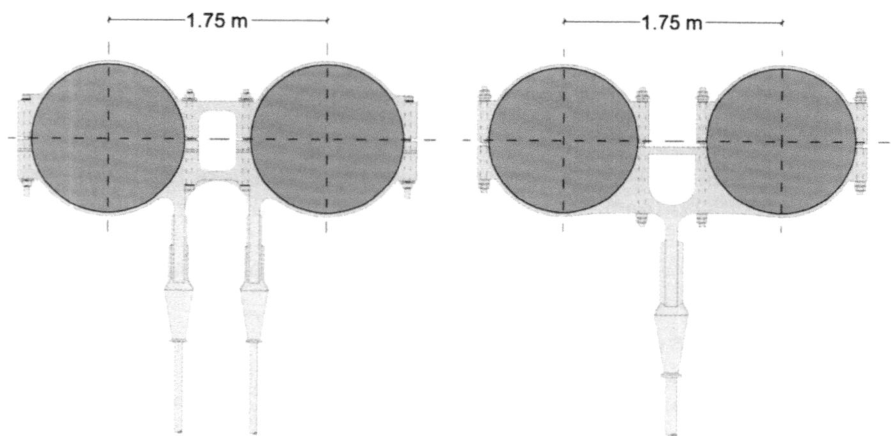

Figure 7.15 ▶ Cable clamp cross section. Left: Preliminary Design, right: Impregilo Tender Design.

7.6.4 Saddles and splay saddles

The cable tower saddles and splay saddles exhibit a rather innovative configuration. The rationale of the solution which was developed stemmed from preliminary analyses carried out on conventional schemes, showing how the radial wire pressures within the saddles was such as to require a very thickness for the side walls. This resulted in large volume castings and heavy elements, increasing the difficulty of achieving adequate material quality within the solid steel sections, as well as generating marginally higher complexity for erection.

A solution was therefore developed in which the strands are located in individual "grooves", running parallel to each other and arranged in layers. Each layer is provided through an independent casting, machine-finished, with easily controllable relatively small cast volumes resulting in greater ease of achieving the required steel properties, homogeneity and freedom from critical defects. The overall arrangement is shown in Figure 7.16.

As a further improvement in behaviour, the end sections of each layer are left free to bend in response to small changes in the geometry of the main

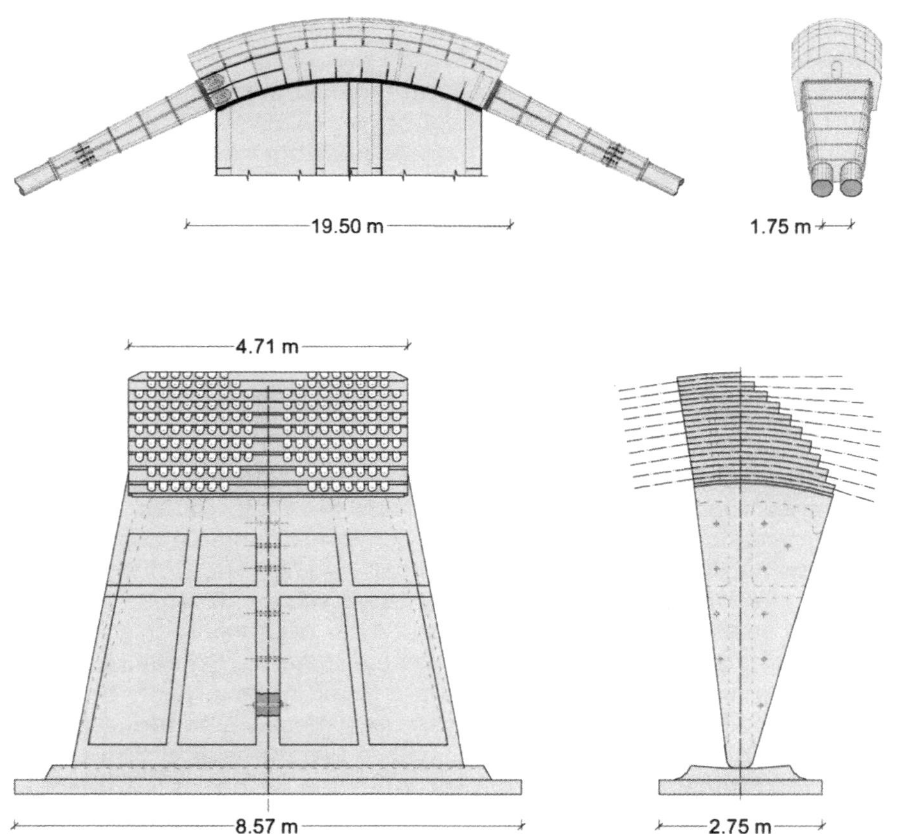

◄ Figure 7.16
Cable saddles.

cables, reducing significantly the local wire bending effects at the saddle exits, mainly connected with heavy running railway loads.

7.6.5 Alternatives proposed by bidders in 2005

Various other changes and modifications to the Preliminary Design were proposed by both bidders. These included the following:

- Both bidders (the Impregilo and Astaldi JVs) decided to use prefabricated parallel wire strands (PPWS), instead of aerial spinning for the main cables. The main reason given was connected with the shorter predicted erection time, but since both bidders also produced a price reduction, one can assume that the PPWS option was also found advantageous in terms of cost. Both bidders also optimised the cable size, according to their own detailed weight and configuration analyses. For example, the winning bidder, Impregilo, proposed a configuration with 324 strands per cable, each formed by 127 wires, about 5,270 metres long and with a weight of 130 tons including reels.
- Impregilo proposed an increased cable wire strength of 1860 MPa. The consequent reduction of cable cross sectional area results in the quantities given above and leads to overall savings in quantities in all the superstructure and sub-structures, with the exception of the suspended deck, although with a marginally reduced overall bridge stiffness.
- Dehumidification for the main cable was proposed by both bidders, with different detailing, adopting concepts similar to the Akashi Kaikyo Bridge, where dry air is pumped into the wire bundle under pressure at intervals along the cable and moist air is extracted so as to reduce the tendency for wire corrosion. To this aspect, of vital importance for achieving long structural life, further comments are devoted in Chapter 8 and in

Appendix 2. This is a fundamental aspect being developed and given considerable attention within the worldwide suspension bridge industry.

- The Impregilo JV also proposed a variation to the cable band arrangement, envisaging three parts instead of two (Figure 7.15) so as to facilitate clamp assembly and accommodate possible local minor variations of diameter between the two cables, allowing independent tensioning of the vertical connection rods.
- Impregilo also proposed to adopt larger hangers and halve their number, placing them on a single vertical plane per side instead of two, which has certain other advantages.

7.7 Towers

The towers are huge structures, rising to a height of 382 metres and sustaining the enormous vertical load applied from the main cables. Key design considerations include the challenge of economic constructability and the need to sustain substantial lateral forces deriving from horizontal wind load on the whole structure as well as very large seismic forces arising from the design earthquake conditions. It is mainly because of this last factor that high strength steel was chosen as the structural material, since alternatives using high strength concrete were found to be virtually unfeasible. A concrete tower would have been extremely massive, suffering huge lateral seismic forces in an earthquake, with a weight several times greater than the load transmitted by the cables, while the weight of the steel tower is only a fraction of that.

The steel plate material is grade 420, with thicknesses varying from 40–45 mm over most of the height to 60–65 mm in the lower three sections. The total amount of structural steel is about 55,500 t per tower.

In the Preliminary Design, the tower legs are multi-cellular structures with overall dimensions of 12 x 16 m, and connected by four crossbeams having rectangular cross section of 4 x 16.9 m (Figure 7.17 and 7.19-left).

Each leg is divided into 21 segments, about 17.5 m long, which will be preassembled in a remote yard, carried to site by barge and lifted into place with a special climbing rig capable of lifting units of more than 1,500 tonnes. Each segment is an assembly of 40 steel panel elements, each 17,5 m long and weighing less than 50 t. All joints are bolted, both between the panel elements (vertical joints) and between pre-assembled segments (horizontal joints). The latter rely on 1.6 m long M100 high strength bolt-bars that are tensioned in order to pre-stress the joint against possible opening due to tower leg bending.

The base connection of the steel legs to the concrete foundation plinth is obtained by embedding into the concrete a 12 m section of leg equipped with a large number of shear connectors.

Such tall towers will have dominant appearance on the shores of the Messina Strait, so its shaping and aesthetic design is also a very important factor. The shape of tower legs in the preliminary design was optimized for structural performance, aerodynamic behaviour and aesthetic appearance, resulting in the peculiar "lozenge" cross section (Figure 7.19-left). Such a shape exhibits better wind drag coefficients than a simple rectangular one, while at the same time eliminating high stress areas in connection with the strong multi axial bending present. This aspect, not so relevant for limit state safety is important for service conditions and connections.

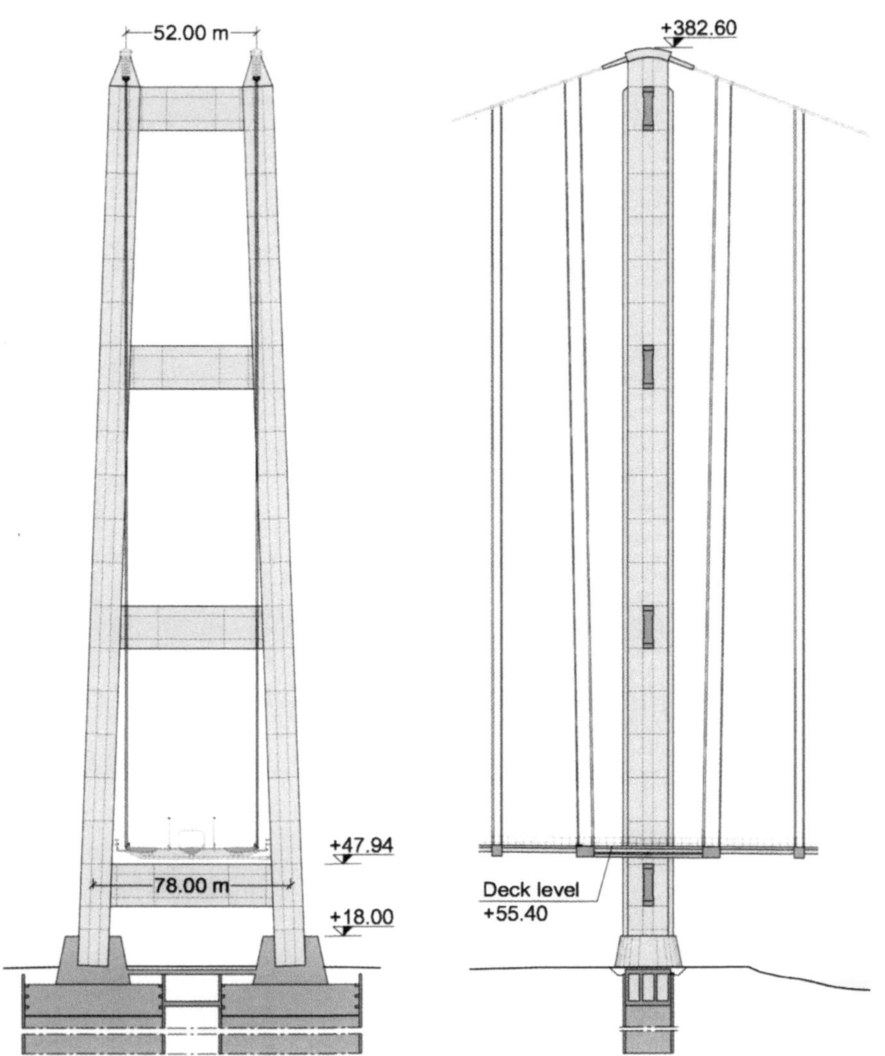

◀ Figure 7.17 Towers as per Preliminary Design.

Nevertheless, such slender and flexible structures are, as may be expected, subject to wind-induced vibrations, that is buffeting and especially vortex shedding that involves both bending and torsional motions. The phenomenon is especially significant for the temporarily "free standing" tower condition (first bending mode 0.1 Hz, critical wind speed 13 m/s), that is during construction before the installation of the main cables. But it also occurs in permanent conditions when the tower is restrained by the cables (first bending mode 0.4 Hz, critical wind speed 55 m/s). For the 1992 design, studies supported by wind tunnel tests showed that the phenomenon could be controlled by adding porous wind screens (fairings) along both narrow faces of the legs (Figure 7.19-left). The 2002 Preliminary Design left the contractors free to propose alternative solutions, and included the provision of appropriate tuned mass dampers (TMDs).

7.7.1 Alternatives proposed by bidders

Both bidders devoted considerable attention to the tower design and erection procedures, confirming them as one of the most relevant elements from the point of view of cost and construction complexity.

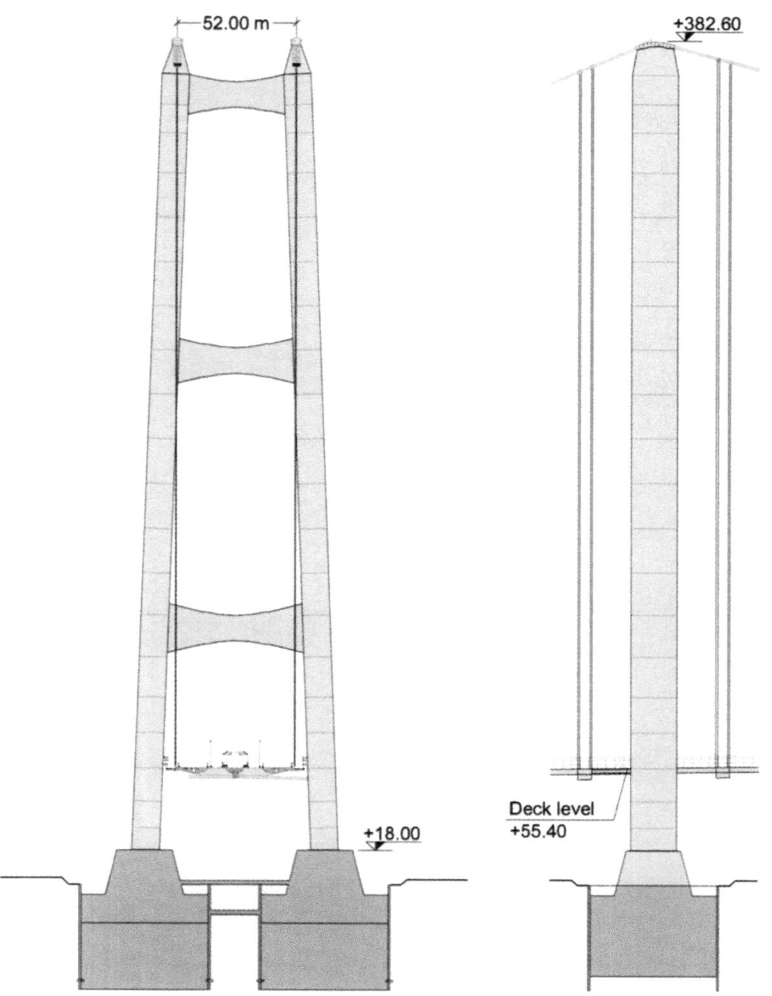

Figure 7.18 ▶ Towers according to Impregilo Tender Design.

The winner, Impregilo JV, made the following proposed changes to the Preliminary Design:

- Several changes to improve the appearance and performance of the towers. The most apparent change is the omission of the lower crossbeam and the shaping of the remaining three with curved top and bottom flanges (Figure 7.18). This, while reducing to a certain extent the structural effectiveness for lateral loads, gives undoubtedly a lighter appearance to the towers when observed from land, emphasising the continuously suspended nature of the deck.
- The steel grade has been increased from 420 ML to 460 ML (ultimate strength >530 N/mm^2).
- Each tower is equipped with large TMDs to control the amplitudes of vibrations both in the temporary free-standing condition and in the permanent condition. The installation comprises eight separate 20 tonne masses per leg mounted as pendulum dampers about 2/3 of the way up the towers and tuned to a frequency of about 0.47 Hz.
- The base connection of the steel legs to the concrete foundation is a conventional thick base plate with a group of large holding down bolts. Special measures will be required to achieve a relatively uniform bearing distribution under the base plate, possibly involving polishing the concrete top surface to achieve a close tolerance flat surface prior to tower erection.

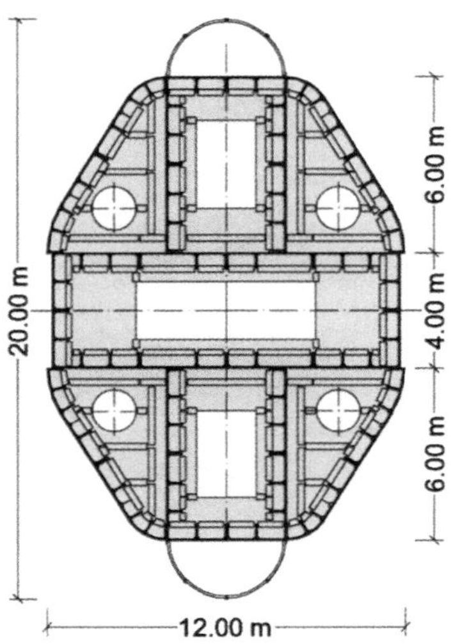

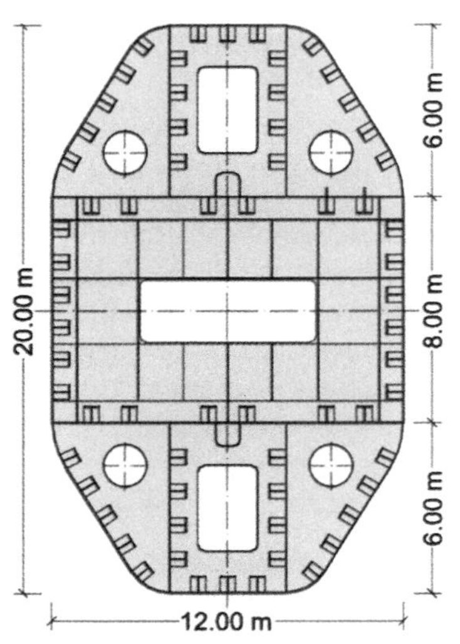

◀ Figure 7.19 Cross section of tower leg. Left: Preliminary Design, right: Impregilo Tender Design.

- The tower leg sections have been structurally increased for greater efficiency without increasing the overall dimensions or significantly changing the visual appearance, thanks to the removal of the curved fairings designed to control vortex shedding, whose effect was replaced by the afore-mentioned TMD's (Figure 7.19-right).
- As far as the construction is concerned, each tower leg is divided into 20 segments. Joints between the segments will be initially bolted and then welded. Welding the steel plates allows structural continuity between adjacent sections to be achieved without the need for very demanding machine-milled contact surfaces, which would otherwise be necessary for bolted flanges. On the other hand, the mixed bolted-welded force transmission mechanism will require further special attention during the final design development.

The second Bidder Astaldi JV:

- Conserved the original design of the towers in terms of cross sections and crossbeam and leg geometry, introducing a number of detail optimisations for connections, fabrication and erection.
- Carried out advanced soil-structure interaction studies for earthquake behaviour, that formed an essential part of their seismic tower design.
- Introduced TMDs for the control of wind vibrations, with concepts analogous to those already described for the Impregilo design.
- Introduced a corrosion protection system based on metallization of external surfaces instead of conventional painting.

7.8 Foundations and anchor blocks

The bridge main span interacts with the soil at the two towers and at the two anchor blocks locations whose foundation sizing was accomplished taking into account the load combinations summarised in Section 7.2.

7.8.1 Tower foundations

The steel towers are connected to massive concrete foundations 130 x 60 metres in plan, placed at the elevation of 10 metres below mean sea level, within an excavation supported by T-shaped diaphragm walls and waterproofed by jet grouting treatment (Figure 7.20).

The design of the tower foundations was guided by the following considerations:

- The deep excavations in gravelly soils with very high hydraulic conductivity, are located close to the shores of the Strait (Table 7.5).
- With reference to the Calabria tower, the concern that a damage Limit State earthquake can trigger a slide of the Coastal Plain (CP) deposits within the steep, circa 1:2, adjacent underwater slope which might induce an unaccepted rotation of the tower towards the Strait.
- The settlements or even the liquefaction of the CP deposits due to the excess pore pressure generated by earthquakes.

The above conditions, together with the high stresses transmitted to the foundation strata during a third level earthquake, led to the geometry of the tower foundations shown in Figures 7.21 to 7.24.

The intensive jet grouting treatment, employing 1.6 m diameter columns, foreseen beneath and around the foundations is aimed at preventing potential settlement and liquefaction due to excess pore pressure triggered by earthquake events.

The diaphragm barrettes penetrating the Pezzo Conglomerate (PC) and adopted for the Calabria foundation have been conceived to enhance the rotational stiffness of the foundation and to counteract the tower tendency

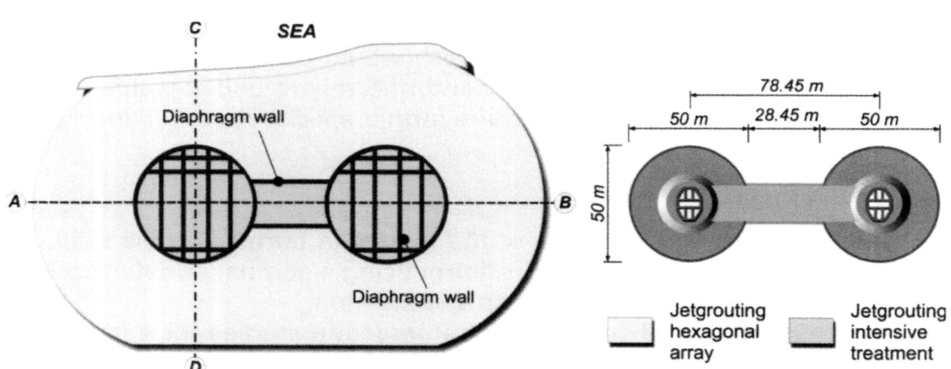

Figure 7.20 ► Calabria tower foundation – Plan view.

Table 7.5 ► Hydraulic conductivity from pumping tests.

Location	$T \cdot 10^{-3}$ (m^2/s)	$K_n \cdot 10^{-3}$ (m/s)	$S \cdot 10^{-3}$ (–)	Gravel + cobbles (%)	Fines (%)
Sicily shore	65	5.0	1.7	44 to 79	3 to 8
Calabria shore	65	2.6	1.7	12 to 80	1 to 16

T = Transmittivity; S = Storage coefficient.

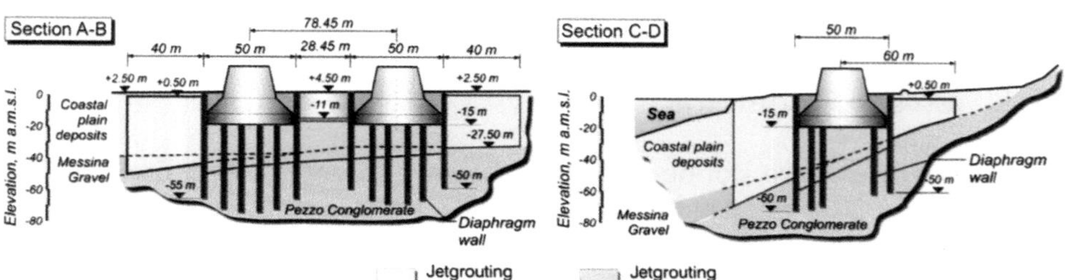

Figure 7.21 ► Calabria Tower foundation – Sections.

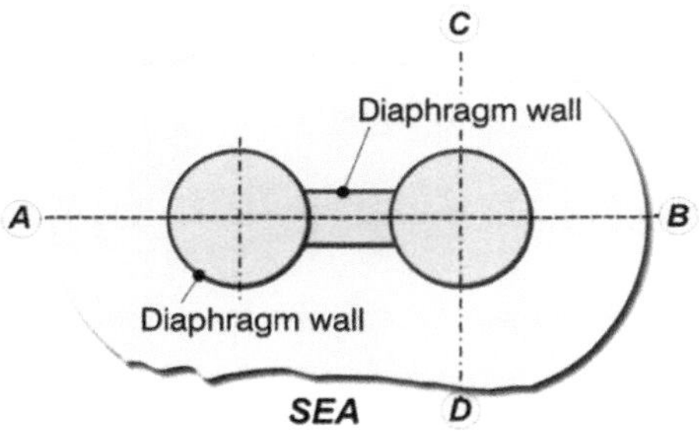

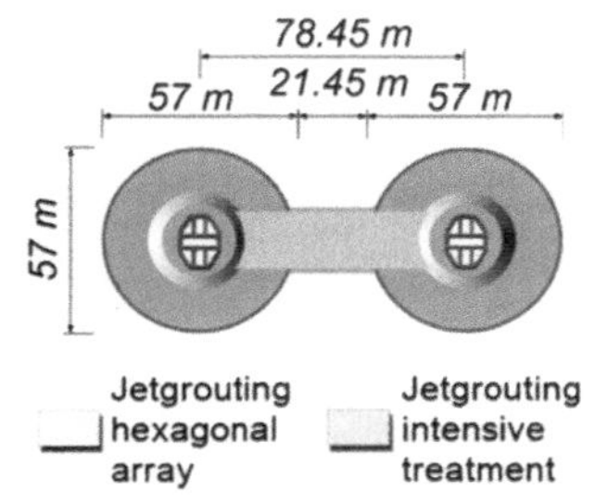

◄ Figure 7.22
Sicily tower foundation – Plan view.

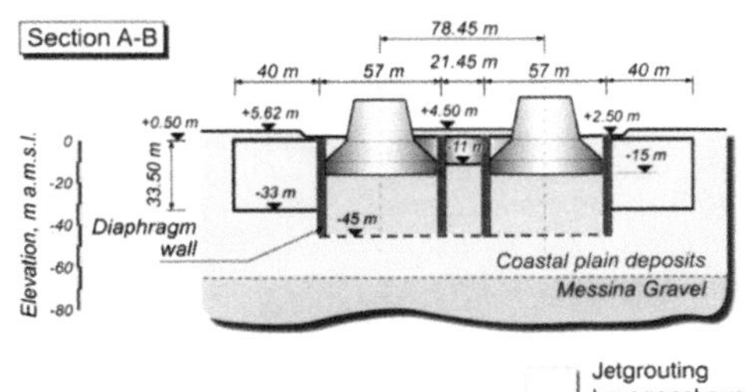

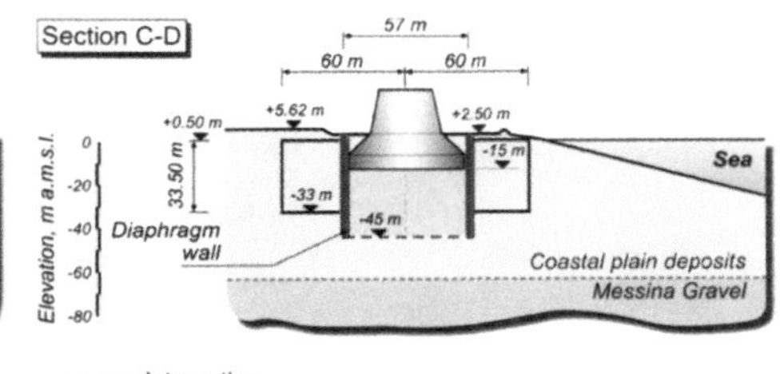

◄ Figure 7.23
Sicily tower foundation – Sections.

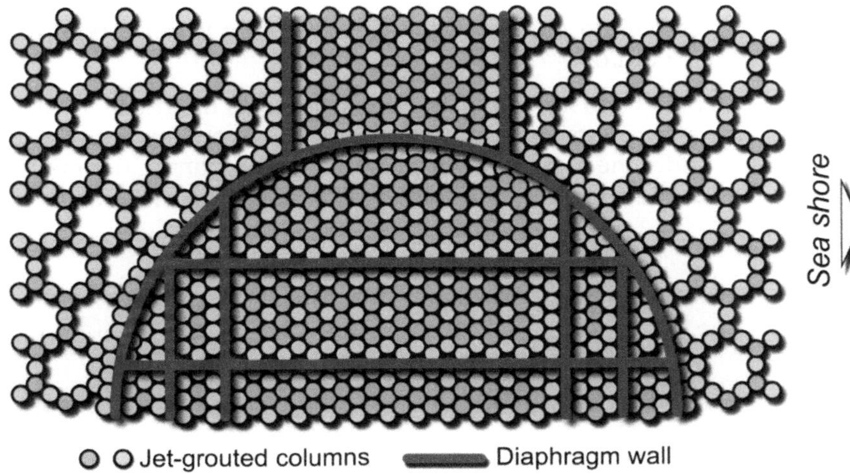

◄ Figure 7.24
Tower foundations – half plan on jet grouting treatment scheme.

to tilt towards the Strait during a third level earthquake, particularly in the case of an underwater slope slide within the CP deposits.

The performance of the tower foundations, under static and earthquake loading, was investigated by means of finite element (FEM) and finite difference (FDM) numerical models employing GEFDYN and FLAC computer codes, respectively. As to the stress-strain-time behaviour of the materials involved, the following assumptions were postulated:

- Visco-elastic, isotropic: concrete, intensive jet grouting beneath the foundations and Pezzo Conglomerate (PC).
- Visco-elastic, cross-anisotropic: external jet grouting with hexagonal array.
- Elasto-plastic, isotropic, with Mohr-Coulomb strength envelope and non-associated flow rule. Coastal Plain (CP) and Messina Gravel deposits.

Major computational efforts have been devoted to the Calabria tower foundation where the issues of seismic stability of the underwater slope and of the non-uniform soil stratigraphy have generated special concern about permanent displacements under the maximum design earthquake.

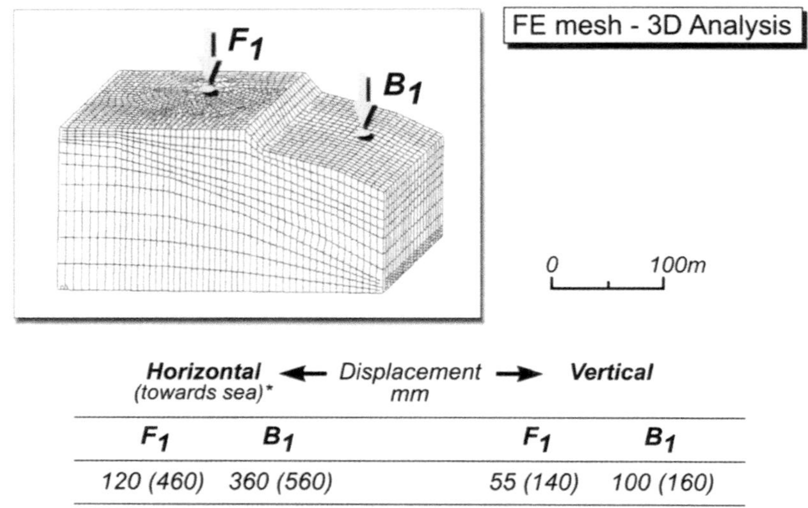

Figure 7.25 ▶ Example of dynamic FE analyses – Permanent displacements.

Figure 7.25 shows an example of the maximum and permanent calculated displacements at the centre of the foundation; (point F_1) and on the slope, after a third level earthquake.

As to the Sicily tower foundation, the deformation analyses were carried out using 2-D FDM, calibrated against both 3-D and 2-D FEM performed for the Calabria tower foundation.

Computed vertical (s_v) and horizontal (s_h) displacements are given in Table 7.6. These have also allowed the rotation (α) of the two towers towards the Strait to be evaluated under the four considered loading conditions. Under the most severe conditions, the rotation (α) is calculated to be of the order of 1/1000.

Table 7.6 ▶ Tower foundations – Computed displacements.

Loading conditions	Calabria tower				Sicilia tower	
	2-D FDA		3-D FEA		2-D FDA	
	S_v (mm)	S_h (mm)	S_v (mm)	S_h (mm)	S_v (mm)	S_h (mm)
Dead load	15.0	1.5	10.0	0.6	70.0	1.0
Dead load + live load + 3rd level wind	17.0	3.0	10.0	1.0	73.0	3.0
Dead load + 3rd level earthquake	22.0	29.0	12.0	12.0	93.0	37.5
Dead load + 3rd level earthquake lack of confinment	25.0	62.5	55.0	120.0	120.0	90.0

7.8.2 Anchor blocks

The salient features of the anchor blocks, (Figure 7.26) are summarized in Table 7.7. The different weights and geometries of the two anchor blocks are due to the different mechanical properties of the two geomaterials being considered. The Calabria block is embedded in the PC, whereas the Sicily

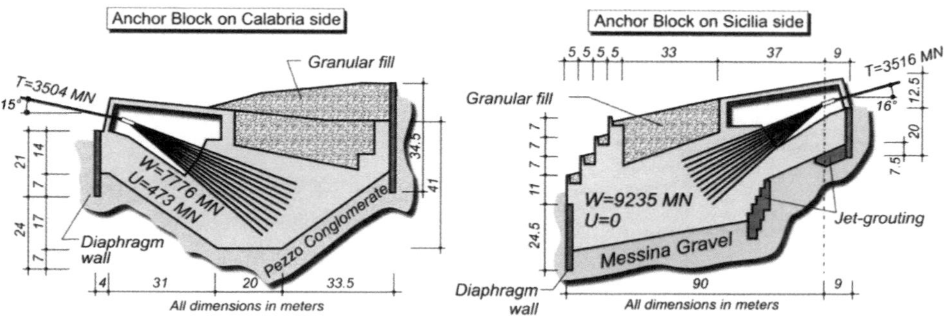

◄ Figure 7.26 Anchor blocks on Calabria and Sicily sides.

◄ Table 7.7 Salient anchor blocks features.

		Sicily	Calabria
Tension load			
Horizontal component	Th, Mn	3382	3382
Vertical component	Tv, Mn	959	918
Dead weight	W, Mn	9235	7776
Buoyancy force	U, Mn	0	473
Load component ratio	r, Mn	0.41	0.53

block is embedded in the Messina Gravel (MG) deposits. Their designs have been influenced by the following considerations:

$$\textit{Load component ratio}: r = \frac{I_h}{G - T_V - U}$$

- The narrow width of the land between the Strait and the Mediterranean Sea to receive the block on Sicilian shore.
- The temporary earth support of excavations more than 50 m deep required to accommodate the blocks in the ground.
- The need to avoid significant permanent displacements of the blocks after strong earthquakes.

The design was carried out following four different approaches, of which, for sake of simplicity, only the following two are briefly examined:

- 2-D coupled pseudostatic FE analysis using the Gefdyn computer code. In this analysis, owing the low correlation between horizontal (a_h) and vertical (a_v) PGA observed in the recorded strong motion time histories, only the $a_h = 0.32$ g was considered. FEM's depth and length were 200 and 300 metres respectively.
  The geotechnical parameters and stress-strain-time behaviour employed in such analyses were similar to those mentioned in connection with the tower design.
  Throughout the analysis, all the phases of the anchor block construction were modelled, from the excavation stage to the application, in a static mode, of the extreme earthquake loading (Kramer, 1996). For the Calabria anchor block, the analysis under earthquake loading yielded horizontal (s_h) and vertical (s_v) displacement of 60 and 40 millimetres respectively. Similar displacement values were obtained for the Sicily anchor block.
- With the aim of exploring the occurrence of permanent anchor block displacements during a third level earthquake, a simplified dynamic analysis of a block on an inclined plane was carried out so as to evaluate the critical seismic coefficient $k_c = a_c/g$. (Figure 7.27)

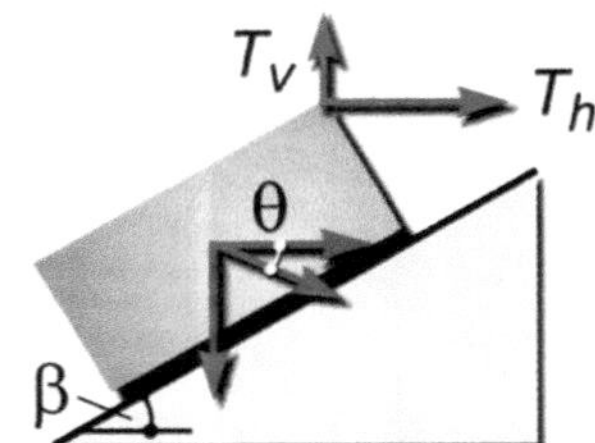

Figure 7.27 ▶ Model of block on inclined plane.

$$k_c = \frac{\sin\beta + \cos\beta\tan\delta + \frac{T_v}{W}(\sin\beta\tan\delta - \cos\beta) - \frac{T_v}{W}(\cos\beta\tan\delta + \sin\beta)}{\cos(\theta+\beta) - \sin(\theta+\beta)\tan\delta}$$

Value of θ for which k_c is minimum from $\frac{dk_c}{d\theta} = 0 \rightarrow \theta(k_c)_{min} = -(\delta+\beta)$

$k_c\, g$ = *critical ground acceleration;* $\theta = \arctan\left(\frac{a_v}{a_h}\right)$

$\tan\delta$ = *friction coefficient at block-inclined plane interface*

The input and the output data for the analysis are:

k_c = critical seismic coefficient;
g = acceleration of gravity, m/sec^2;
a_c = free field peak ground acceleration, m/sec^2;
φ' = peak angle of shearing resistance of the geomaterial in which the block is embedded;
φ'_{cv} = friction angle at critical state of the geomaterial in which the block is embedded;
β = sliding plane angle for which $k_c = k_{min}$.

The computations show that k_c is always higher than the largest design seismic coefficient k = 0.58. This holds even if one makes the conservative assumption that $\varphi' = \varphi'_{cv}$.

Therefore, owing to an overall conservative sizing of the anchor blocks, the simplified dynamic analysis indicates that they are not likely to suffer from any significant permanent displacements during a third level earthquake.

7.9 Operation & maintenance strategies

Since the very beginning of the design, in the late eighties and early nineties, it was clear that a particular strategy for optimizing management, operation and maintenance of the bridge would be a fundamental requirement, both in relation to the large scale and technical complexity of the structure and because of its economic importance as a "vital" link in the national transport system. Such a strategy was aimed at achieving best performances in terms of:

- Safety, durability and value of the structure; i.e. keeping the structure efficient and in good condition and maintaining its economic value for its entire life-cycle,
- Effective and economic maintenance; i.e. achieving the above target with minimum expense,
- Safety of users and maintenance personnel,
- Service continuity and quality; i.e. to minimise the need for traffic management during maintenance activities,
- Security against possible offences; i.e. preventing sabotage.

At that time, suitable provisions for inspection and maintenance were already considered essential for large suspension bridges. However, sophisticated control systems were not yet wide-ranging, and such that did exist were mostly focussed only on structural health monitoring.

Thus, in addition to extensive access provisions, a very advanced and innovative monitoring, control and management system was conceived for the Messina Bridge. (Figure 7.28) This is a fully integrated instrumentation and communication system involving the collection, distribution, processing

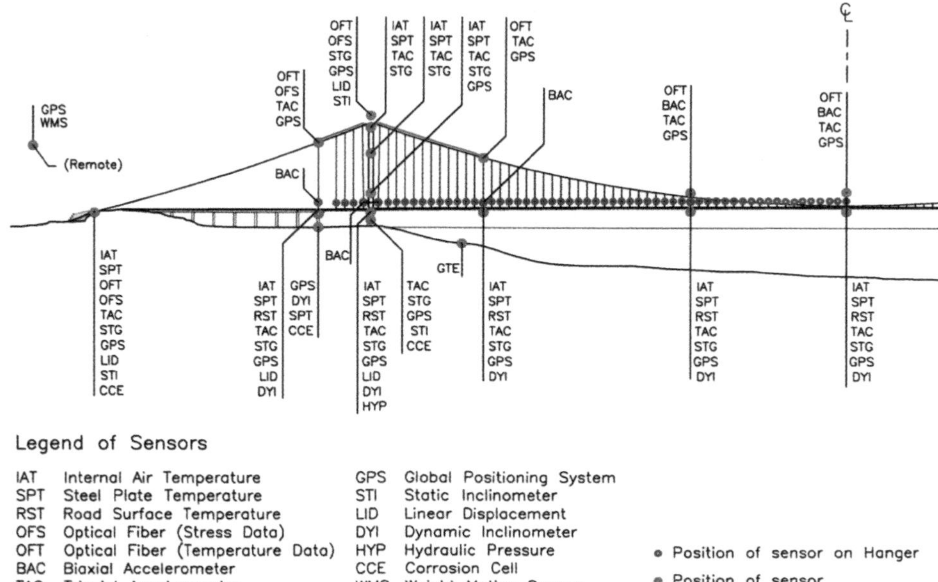

◄ Figure 7.28 Typical layout of structural instrumentation on the bridge.

and storage of all needed information, and including procedures and equipment to manage various events and scenarios.

The following paragraphs describe the main features of these provisions.

7.9.1 Access provisions for inspection & maintenance

The requirement to allow easy access to all parts of the deck, towers and cables is particularly demanding due to the exceptional length and height of the structure.

A cantilevered maintenance service lane is provided along each side of the deck and a continuous walkway runs alongside each side of the railway envelope. Four large under-deck travelling gantries are provided, spanning the full width between the fascia beams.These gantries need to hang low enough to pass under the crossbeams, so mobile scissor lifts mounted on the gantries give access to the underside of the longitudinal boxes at higher level.

A complete system of internal service walkways and utilities for inspection and maintenance (such as electricity, water, compressed air and communication facilities) is provided within the longitudinal boxes and crossbeams of the deck, as well as inside the towers. Inside the roadway boxes, a travelling maintenance trolley system runs the full length of the bridge passing through large openings in the transverse diaphragms. Each crossbeam will have access from the cantilevered maintenance service lane above. Due to the very slender cross section profile of the deck, the outer web of the crossbeam has a shallow depth which creates some difficulties of regular access to these areas that require relatively frequent close inspection. Special platforms are therefore suspended below the box at 360 m centres to give easier access into the crossbeams through the bottom flange.

Each tower leg is provided with large internal lifts and emergency stairs as well as service walkways and other utilities for inspection and maintenance. All outside surfaces of the towers can be reached using a system of permanent external suspended platforms.

A cable trolley which is supported on the handstrands provides an efficient system for access to all parts of the main cables, including the hanger top connections. Cable splay saddles where cables splay in a fan of wires (at anchor blocks) are housed in wide airtight chambers that have easy access. These rooms, as any other internal volume of deck and towers, are dehumidified to prevent corrosion.

All other important parts of the structure, such as bearings, joints, anchorages, etc. have similar provisions for easy and effective access.

7.9.2 Monitoring, control and managing system

The instrumentation and monitoring system will record and process any event influencing the use, safety, structural health and durability of the bridge, first during construction and then during operation. The system will constantly survey and record most significant physical parameters relevant to the environment and the structure, facilitating reliable modelling and reconstruction of the structural behaviour (strains, deformations, etc.). All data will be selectively recorded based on pre-determined criteria, such as periodic repetition or the exceedence of defined threshold values. In practice, the structural monitoring system will record all loading influences on the bridge by the physical environment and traffic, such as wind speeds and directions, temperatures, rainfall, ground accelerations, traffic loading etc., as well as the structural responses such as strains, displacements, accelerations, ground settlements, etc.

Processing this set of data will allow operators to:

- control and document the structural health of the bridge,
- verify the bridge performance and calibrate predictive analytical models for such structures,
- check components subject to wear or deterioration, directing attention to those areas most likely to require inspection,
- optimise safe and efficient operation of the link as well as efficient and cost-effective maintenance.

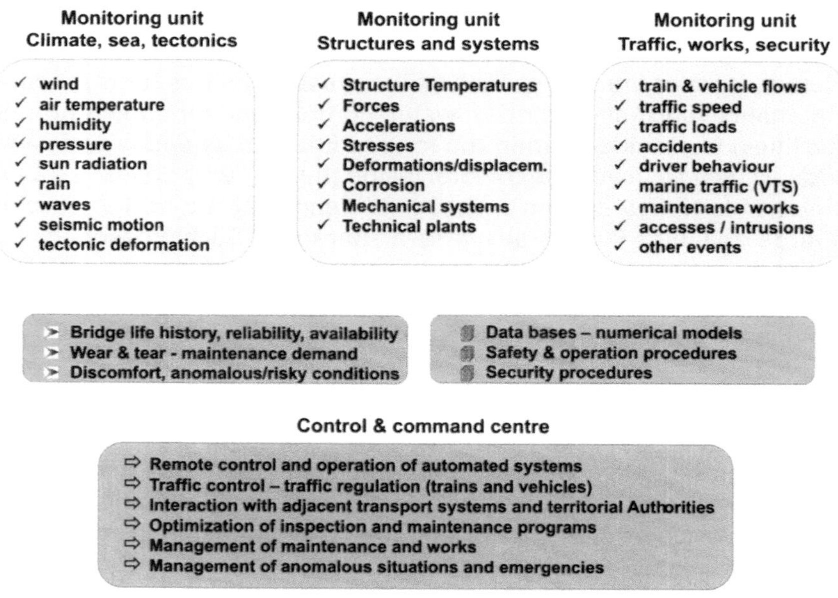

Figure 7.29 ▶ Flow diagram of the monitoring and control systems.

During the construction phase the system will record the structural history of each component from the very beginning, thus playing a fundamental role in documenting the ongoing condition of the structure.

Furthermore, other real-time information is collected by different sensor systems on the overall operation of the infrastructure, such as the type, speed and behaviour of traffic, the functioning of technical plant and equipment, access to restricted areas etc.

The entire flow of information is collected and processed in a control & command centre located in the Operation Centre near the Calabria anchor block. (Figure 7.29)

Great importance has been placed on the durability, reliability and redundancy of the instruments as well as the effectiveness and reliability of the overall communication systems, adopting the principles of "Network Centric Operation".

7.9.2.1 Operation of the infrastructure, safety and security

Staff at the Operation Centre will use the information obtained from the monitoring system to facilitate the management of the crossing, and in particular the traffic controls, also availing themselves of specialized teams, equipment and procedures. The management procedures will ensure, among other aspects:

- the safety of users and vehicles, as well as the best levels of service,
- the effective, efficient and economic operation of the crossing,
- coordination with the operators of adjacent transport systems (railway & highway networks, local roads, navigation etc.).

The required performance levels can thus be guaranteed in any normal, anomalous and emergency situation by the effective management of any event that could potentially affect the serviceability of the bridge, as well as by the proper handling of any possible emergency. The control & command systems will, for instance, enable the bridge operation to be regulated following any particularly severe meteo-climatic or seismo-tectonic event and also in the event of traffic anomalies (i.e. congestions, queues, accidents, or restrictions for maintenance works).

Due to the importance of the infrastructure, systems and procedures are also included which prevent, detect and respond to any specific terror, sabotage or vandalism actions.

During the construction phase the control system will also play a fundamental role in surveying the security, legality and efficiency of the work sites, monitoring personnel and vehicle access to site, and enabling efficient fleet management, etc.

7.9.3 LCC and RCM

The project has been promoted very much on the basis of minimising whole of life costs, and all aspects of the design are subject to review and checking in accordance with an established life cycle costing (LCC) model. Furthermore, reliability centred maintenance (RCM) strategies must be developed in conjunction with the LCC approach so that the detailed design reflects the most economic solution from the point of view of long term maintenance as well as initial construction. This philosophy affects the choice of materials as well as the detailed design of many components.

The same philosophy will also be fundamental during the operation phase.

7.10 Approach infrastructures

The project includes approach infrastructures connecting the suspension bridge with the existing road and railway network on the two sides.

In order to achieve the best relationship between the bridge project and the proposed future arrangement of road and railway networks, a "Framework Agreement" was set up between SdM and the regions of Sicily and Calabria, ANAS (Italian roadway Authority) and RFI (Italian railway Authority).

Among other aspects, this agreement included provision for the following main works which were deemed crucial for the whole project (Figure 7.30).

a. Modification and upgrading of the A3 highway (Salerno–Reggio Calabria) and relevant predisposition for connection ramps.
b. Extension of high speed/high capacity railway line to Reggio Calabria
c. Links to the existing railway running along Calabria coast.
d. New railway station in Messina that will be along the line (it is presently a terminal station).
e. Extension of Messina highway connection (from Annunziata to Giostra).

These works are outside SdM's scope and have been placed under the responsibility of other competent authorities, particularly ANAS and RFI, who have committed themselves to complete them in due time in respect to the bridge construction.

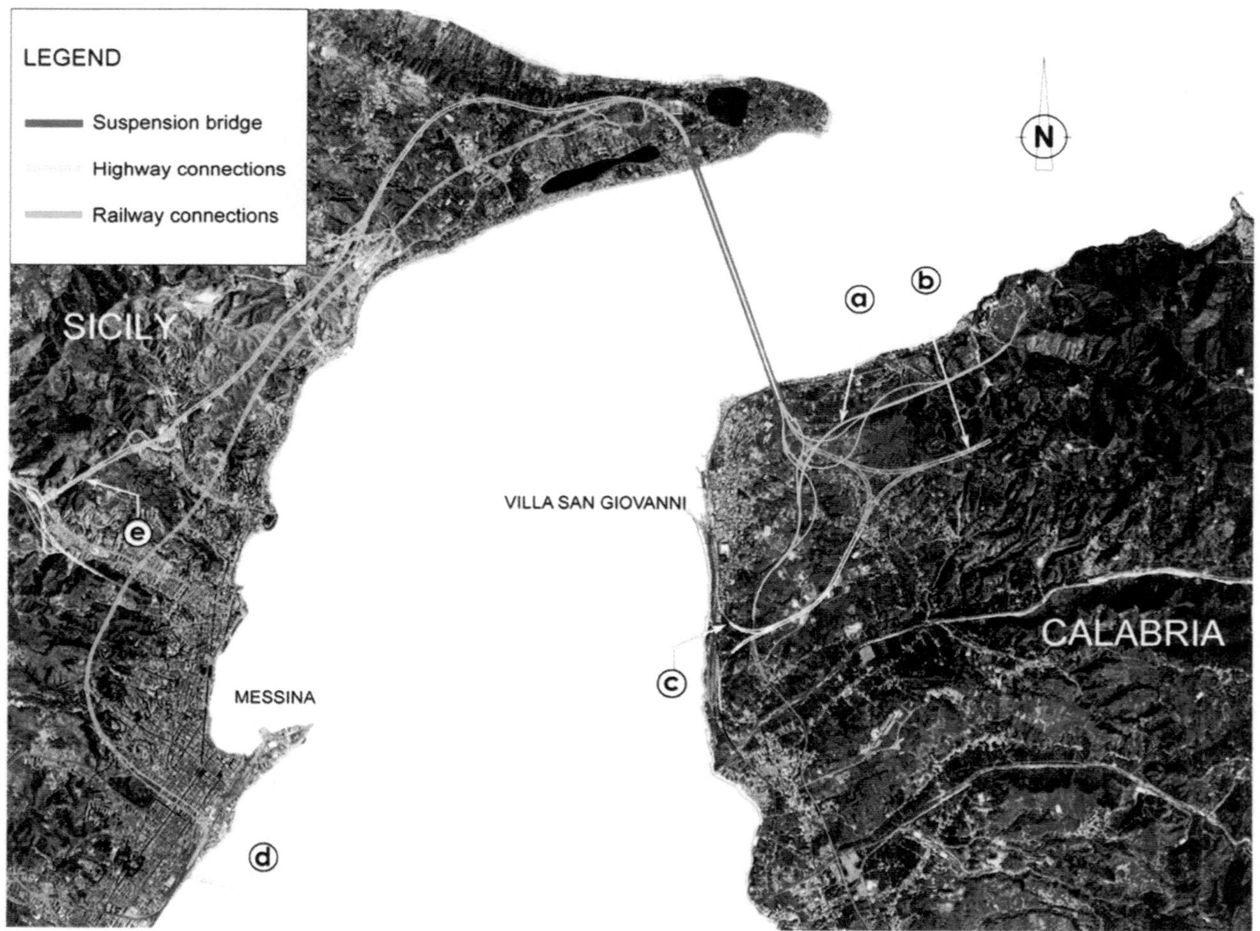

Figure 7.30 ▼ General layout of the bridge and its connections to existing roadway and railway networks. Letters a,b,c,d,e show works to be carried under separate contracts as mentioned in the text.

7.10.1 Railway connections

The railway connections are in total 19.8 km long (15.2 km in Sicily and 4.6 km in Calabria), most of which is in tunnel: 93% tunnels, 3% viaducts and 4% embankments. The new infrastructure will connect the bridge to the new Messina railway station in Sicily and the planned Naples-Reggio Calabria high speed, high capacity railway line on the Calabria side.

The whole railway infrastructure will comply with the European Standard for Railway Interoperability for the High Speed, High Capacity system (Category II).

7.10.1.1 Sicily side

The railway consists of a new 15.2 km line, from the bridge to Messina station, running in tunnel for 92% of its length.

Main technical requirements:

- 12.70 metre wide railway platform, with tracks at 4 metre centres and 0.5 metre emergency/maintenance pedestrian paths.
- Twin tunnels with internal diameter 8.20 metres, connected every 250 metres with smoke-proof bypasses incorporating shelters of adequate size.
- Access roads to the tunnels and appropriate areas for emergency parking, first aid/triage and helicopter rescue.
- Specific railway maintenance facilities.

7.10.1.2 Calabria side

The railway infrastructure consists of a new railway line, approximately 4.6 km long connecting the suspension bridge to the existing railway lines (Villa S. Giovanni station) and the proposed future extension of the high speed, high capacity line. The infrastructure runs in tunnel for 96% of its length.

The main technical requirements are the same as for the Sicily side.

7.10.2 Highway connections

The highway connections are 20.3 km long in total, 10.5 km in Sicily and 9.8 in Calabria, the majority of which is in tunnel: 63% tunnels, 22% viaducts and 15% embankments.

The new highways comply with the Italian standard for "A"-category highways[1] and consist of two carriageways, each having two 3.75 metre wide traffic lanes and one 3.0 metre emergency lane. The design speed ranges from 110 to 140 km/hour.

7.10.2.1 Sicily side

The new infrastructure will connect the bridge to the Messina-Catania and Messina-Palermo highway system via the above-mentioned extension (Annunziata to Giostra) of the highway junction into the centre of Messina, part of which is already under construction by the Messina Municipality.

The new highway comprises five tunnels (total length of 6.8 km), five viaducts (total length of 1.6 km) and embankments (total length of 2.1 km).

Main features:

- Dual carriageway approximately 10.5 km long;
- Toll station located approximately 2 km from the Sicily bridge tower;

[1] Ministerial Decree of November 5th 2001.

- Junction interchange with the ordinary road network at Curcuraci/Guardia.

7.10.2.2 Calabria side

The new infrastructure will connect the bridge to the modified and upgraded Salerno-Reggio Calabria highway through a system of ramps. Works relevant to this refurbishment are already in progress by ANAS.

The ramps consist in general of two traffic lanes and one emergency lane, whilst ramps to and from Reggio Calabria consist of a one-way single lane (3.75 m) plus emergency lane.

The highway connections consist of four tunnels (total length of 6.0 km), nine viaducts (total length of 2.9 km) and embankments (total length 0.9 km).

The building for the Operation Centre for the crossing is located near the Calabria anchor block and is linked to the infrastructure by connecting access, service and emergency roads.

7.11 Strategies and provisions for construction

The organization and logistical problems of construction will be unusually complex on this project, particularly in connection with the following requirements and constraints:

- Maximum efficiency and simultaneous working on several activities will be required to complete the complex construction programme within the required time constraints.
- Very large production volumes and speeds are required in a short time. This problem affects the whole work, including excavation and concreting. It is however particularly important for the steelwork (≈200,000 tonnes) and the main cables (≈170,000 tonnes) that also require very high level quality standards.
- The territory is very sensitive, local transport networks are weak and the land areas available for the construction sites are limited and constrained. It is therefore necessary to restrict the occupation of land areas, control the impact of construction activities and minimise the disruption due to transports to and from the sites.

These problems significantly affected the choice of construction processes as well as the provision of temporary works and the adoption of particular logistical approaches.

As far as the construction is concerned, the design envisages extensive use of pre-fabrication and pre-assembly in remote workshops and yards, without which the challenges of working at this scale could not be met economically and proper quality could not be achieved.

Production of basic materials, particularly high strength wires for the main cables, will have to be started early and will involve the combined output of several production units. The construction process will therefore involve several sites having different purposes and locations:

a. Remote workshops for production of prefabricated steel panels and sections, that can be located anywhere, depending only on the ability to economically produce high quality fabricated parts and deliver them to the assembly yards.
b. Nearby or distant sites and yards for assembling and/or stocking complete box and girder sections and semi-finished products ready for erec-

tion. These sites will be located in any case outside the immediate area of the Strait; only the reels of cable wire will need to be mostly stocked near to the site.

c. Main construction sites on both shores of the Strait, along the project alignment, for the erection of the steel structures and the execution of civil works.
d. Several sites for temporary stocking of spoil and final disposal.

Extensive use of marine transport is envisaged to and from the construction sites. Off-site prefabrication and marine transport are envisaged particularly for deck and tower construction, but also for other structures and components of the bridge. Sea-based activities and marine transport are also foreseen for civil works, including earthworks.

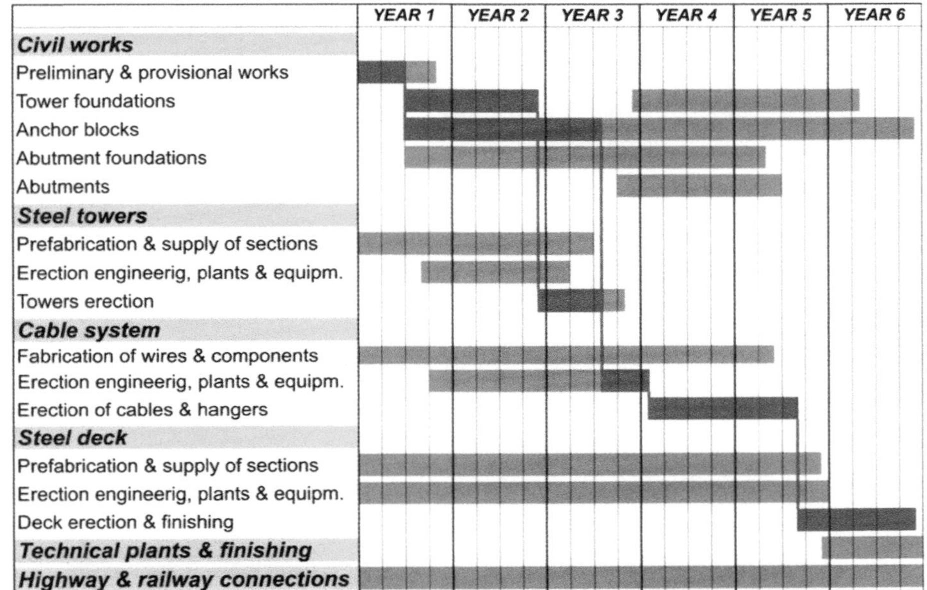

◄ Figure 7.31 Overall construction programme. (The red bars indicate the critical path).

◄ Table 7.8 Average production rates.

	Average production rates	
	Fabrication & prefabrication	**Erection — construction**
Suspension Bridge		
Deck steelwork	14,000–15,000 t/year	3 sections/week
Tower steelwork	60,000–75,000 t/year	1 section/week (per tower)
Cable system	50,000 t/year	126 wires/day (per cable)
Concrete structures	–	900–1,300 m³/day (per site)
Roadway & Highway connections		
Excavation	–	1,600 m³/day
Embankments	–	1,000 m³/day
Tunnelling – by traditional methods	–	2.5 m/day (per tube)
Tunnelling – by Tunnel Boring Machine	–	14 m/day (per tube)

The deck and towers will be prefabricated and assembled into complete sections, each one weighting more than 1,000 t that will be transported to the erection site by barges and lifted using special equipment.

Off-site pre-fabrication and assembly, and the use of marine transport, will strongly reduce the impact of construction activities on the territory and local infrastructure.

An overall construction programme of 6 years is envisaged – see Figure 7.31.

The main estimated average production rates that form the basis of the programme are given in Table 7.8.

8 – THE WAY AHEAD

Contents of this chapter:

- **State of the art of suspension bridges**
- **Future plans**
- **Conclusion**

BLOCCO DI ANCORAGGIO CALABRIA

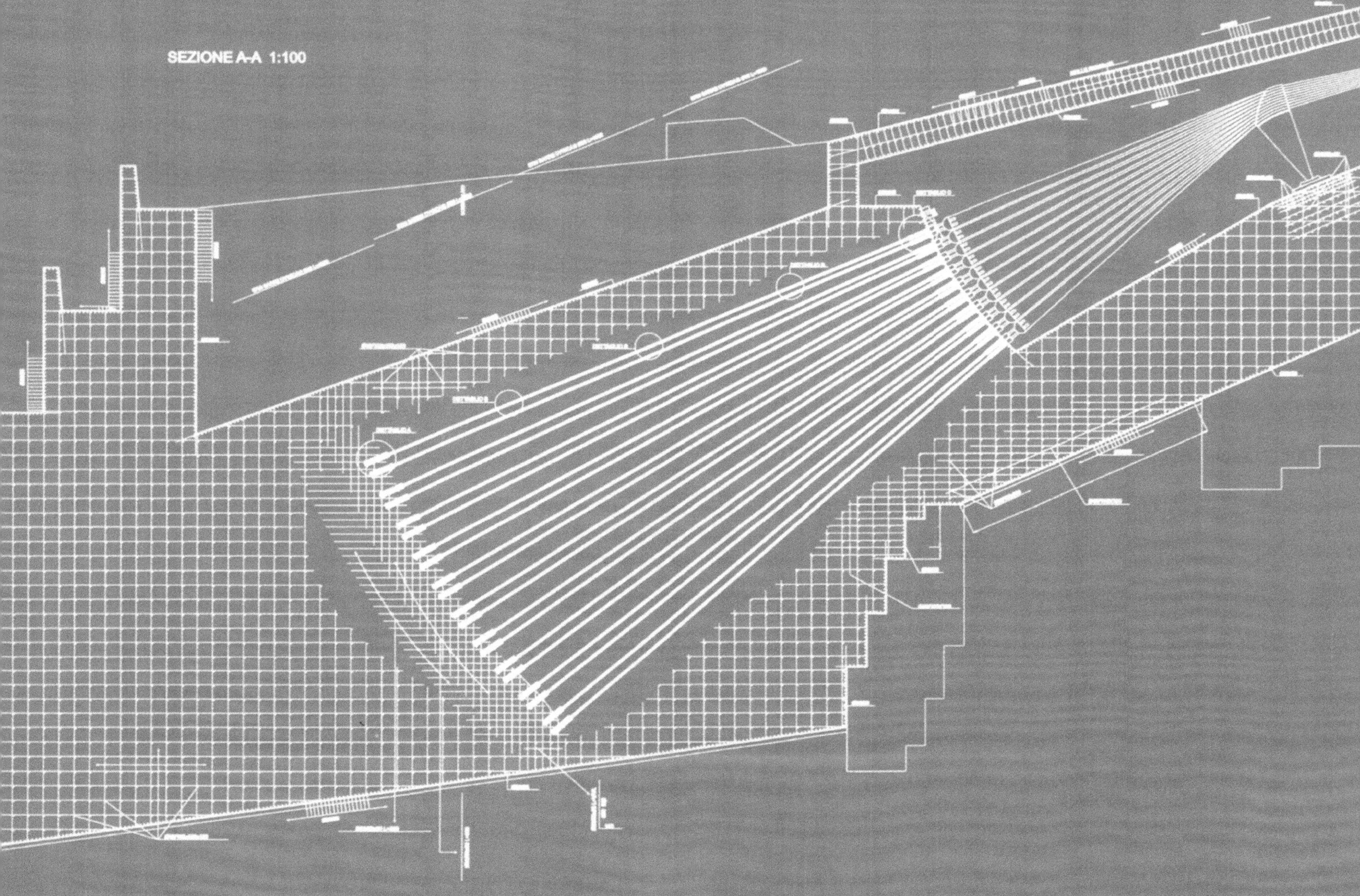

BLOCCO DI ANCORAGGIO SICILIA

This chapter reviews the state of the art in suspension bridge design, and looks ahead to the possible further developments that might occur in the design of the Messina Strait Bridge as it goes through the final detailed design and approaches the construction stage. By considering the experiences and innovations exhibited by other bridges around the world, some of which have been strongly influenced by the Messina design, the opportunities for some further design optimisations are explored.

8.1 State of the art of suspension bridges

The present design for the Messina Strait Bridge has been developed since the mid 1980s through a series of in-depth structural analyses and investigations.

At the time when the investigation started the required span of 3300 m for the Messina Strait Bridge was almost 2.5 times longer than the longest existing suspension bridge span in the Humber Bridge (1410 m). (Figure 8.1) Later, the completion of the Akashi Kaikyo Bridge in 1998 (Figure 8.2) moved the maximum span of suspension bridges up to 1991 m, but the span of the Messina Strait Bridge would still constitute a considerable jump in span length by more than 1300 m.

Early in the design process it became evident that the design could not be based on a simple extrapolation from the existing long span suspension bridge designs. If this extrapolation had been applied, the 14 m deep and 35 m

Figure 8.1 ▶ The Humber Bridge (Great Britain).

Courtesy of Neil McFadyen, Flint & Neill Partnership – Great Britain.

Figure 8.2 ▶ The Akashi Kaikyo Bridge (Japan).

◄ Figure 8.3
The Storebaelt East Bridge (Denmark).

wide truss of the Akashi Kaikyo Bridge would have become 23 m deep and 58 m wide if simply proportioned by the span ratio 3300/1991 = 1.66. So a deck with a truss configuration would be prohibitively heavy and be characterized by an unacceptably high drag.

Instead, the solution for Messina was to split the deck into three independent box girders each with an extreme slenderness. The depth of each of the three streamlined boxes is less than 1/1200 of the main span length. This can be compared to a depth-to-span ratio between 1/100 and 1/150 used in traditional American and Japanese suspension bridges with a truss-type deck and 1/300–1/400 used around the world in more recent suspension bridges with a mono-box deck.

The reference design prepared by SdM in the early 1990s was used (in a slightly modified form) as the basis for the bidding by the two prequalified contractor groups in 2005. During the preparation of their bids the two groups made only relatively minor adjustments to the reference design.

The present design for the Messina Strait Bridge is based on structural materials with well-known properties and proven construction procedures, and has been shown to be technically feasible. It is now ready to form the basis for the final detailed design and the subsequent construction.

It must also be appreciated that the analytical and experimental investigations that have been carried out during the preparation of the reference design are comprehensive and have addressed all the major issues related to extending the span of suspension bridges beyond previous limits. Consequently, it can be concluded that the bridge will have similar levels of safety and reliability as other suspension bridges built in recent years. In certain aspects the current design might even be slightly conservative, so it is likely that some savings can be achieved during the final design phase and still provide a structure with fully acceptable safety levels.

8.2 Future plans

8.2.1 Multi box decks

The design of the Messina Bridge deck has inspired the design of a number of recent long span bridges with two individual box girders separated by a relatively wide gap. A number of such bridges with dual box decks are currently under construction and these provide valuable information on fabrication and construction issues associated with multi box decks.

In the design competition for the Stonecutters Bridge in Hong Kong in 1999–2000, the winning design was a cable stayed bridge with a dual box deck. The dual box configuration for the Stonecutters Bridge was also influenced by the decision to form the two 300 m high towers as simple single vertical columns of conical shape on the bridge centreline. The width of the gap between the two boxes was consequently dictated by the tower diameter at deck level and not by aerodynamic considerations. (Figures 8.4 and 8.5) With a different tower design it would have been possible to arrive at a narrower gap or even to have a simple mono-box deck. This was actually later proved by the design of the Sutong Bridge where a mono-box deck was used for a span 70 m longer than that of the Stonecutters Bridge.

In the original competition design for the Stonecutters Bridge each of the two boxes had a continuously curved soffit but during detailed design the box shape was changed so that only the outer parts of the boxes were curved whereas the inner parts were plane. This design change simplified the joint between the longitudinal boxes and the cross beams, as well as improving the aerodynamic behaviour. With the sharp corners at the transition

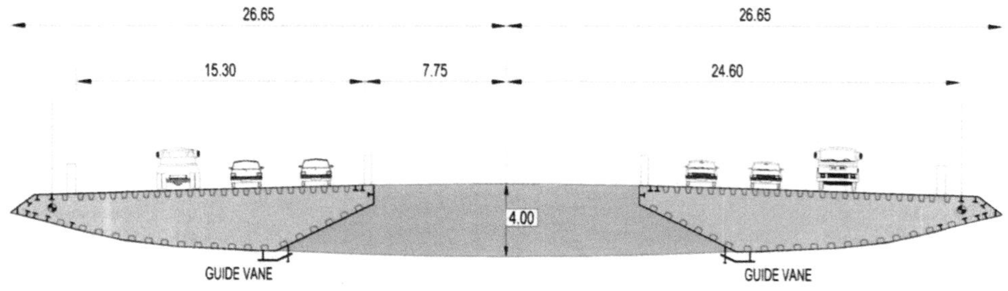

Figure 8.4 ► Stonecutters Bridge, Hong Kong; cross section of the bridge deck. The width of the centre gap is determined by the tower diameter.

Figure 8.5 ► Stonecutters Bridge; lifting of a twin box deck unit.

Courtesy of Ove Arup & Partners – Hong Kong Ltd.

between the curved and the plane soffit it proved advantageous to add guide vanes to suppress vortex shedding and avoid the associated vertical vibrations that could lead to user discomfort.

Coincident with the design of the Stonecutters Bridge, another major cable supported bridge was being designed in Hong Kong – theTsing Lung Bridge. With a span of 1450 m it would have had the third longest span of any suspension bridge at the time. (Figure 8.6) Due to its location close to Hong Kong's new international airport at Chep Lap Kok there was a height restriction for the towers. As a consequence, the main cable sag to span ratio had to be limited to 1/15 – a smaller value than found in any other major suspension bridge.

To achieve satisfactory aerodynamic behaviour the bridge deck had to be composed of two individual boxes separated by a 10 m wide gap. The shape of the two boxes was chosen to be almost identical to the two roadway boxes of the Messina Bridge design. The main reason for that was undoubtedly the influence of Dr. William C. Brown who had been a significant contributor to the development of the Messina Bridge design in the late 1980s and also played an active role during the conceptual design of the Tsing Lung Bridge.

Unfortunately the construction of the Tsing Lung Bridge did not proceed after the completion of the detailed design, and currently it is not expected that the project will be revived until at least some years after 2010.

At present most large suspension bridge activity is to be found in China, and among the Chinese suspension bridges under construction the most prominent example is the Xihoumen Bridge with a main span of 1650 m. (Figure 8.7) That will make it the second longest span in the world, just surpassing the 1624 m span of the Storebælt East Bridge in Denmark.

The deck of the Xihoumen Bridge consists of two box girders separated by a gap of 6 metres and interconnected by cross beams at the hanger positions. This design was shown to fulfill the requirement of a critical wind speed

◄ Figure 8.6 Tsing Lung Bridge project, Hong Kong, with twin box girder deck section (designer: Maunsell AECOM).

Courtesy of Hong Kong Highways Department.

Figure 8.7 ▶ The Xihoumen Bridge, China; note the twin box girder deck cross section.

Photo Alex Needham (www.wikipedia.org).

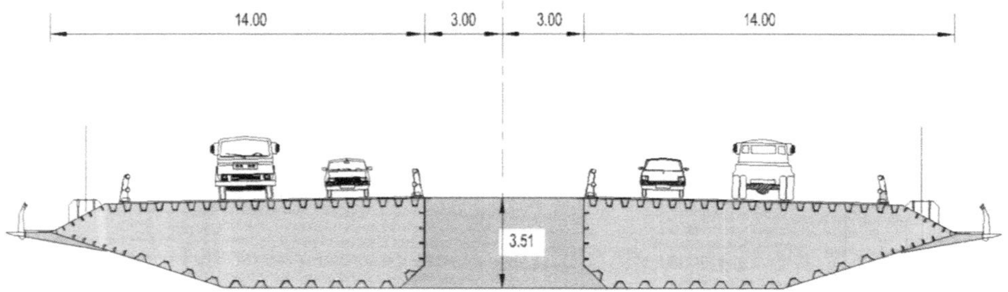

of more than 80 m/s, whereas tests on a single box girder deck had failed to demonstrate adequate aerodynamic stability. This had actually been expected because it was already known that the critical wind speed for the Storebælt East Bridge had been determined to be 72 m/s. However at Storebælt the requirement was only to keep the critical wind speed above 60 m/s so the single box deck was acceptable in that case.

Each of the two box girders forming the deck of the Xihoumen Bridge is composed of plane stiffened steel panels which simplifies the fabrication.

The design of the Xihoumen Bridge with a twin box deck (a "vented" deck) points towards an efficient solution for the design of future suspension bridges with long spans more than about 1600 to 2000 m.

The new San Francisco-Oakland Bay suspension bridge is due to be completed in 2012. (Figure 8.8) This will be the first dual box suspension bridge in

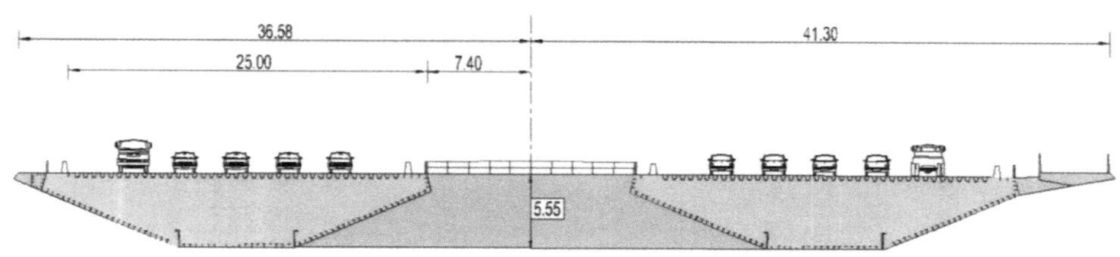

Figure 8.8 ▶ The new San Francisco-Oakland Bay Self-Anchored Suspension Bridge (USA): rendering and twin box deck cross section.

Courtesy of California Department of Transportation – USA.

the United States. Driven primarily by the desire to create a visually impressive structure, the bridge is quite unusual in its design, with one centrally located pylon and a self-anchored suspension system. What appears as a pair of cables is actually one continuous cable that loops around the west end of the girders and has both ends anchored at the east end.

The bridge is asymmetrical, with a main span of 385 m and one side span of 180 m. With such short spans and such a stiff structural system, the aerodynamic stability is of little concern. As for the Stonecutters Bridge, the central pylon forced the design to two boxes with a space of about 14 metres between them. The girders are connected together by 10 m wide transverse crossbeams spaced at 30 m. There is a combination cycleway/walkway cantilevered off of the eastbound girder, balanced by a counterweight in the outer edge of the westbound girder.

In the Stonecutters Bridge and the new Bay Bridge the width of the gap between the two boxes is determined by the width of the central tower at the deck level, but for the Messina Strait Bridge the two tower legs pass outside the edges of roadway boxes so the tower geometry does not dictate the width of the two gaps.

The width of the gaps was determined around 1990 when it was assumed that emergency lanes should be provided on grids along both the railway deck and the roadway deck. These emergency lanes have been omitted in the current design so as a consequence it may be possible to reduce the gap widths if permitted by aerodynamic constraints.

Most of the bridges with dual box decks currently under construction use flat bottom plates which are easier to fabricate than curved ones and are thus generally preferred. However, curved plates provide certain advantages in relation to aerodynamic response and out-of-plane buckling and can therefore be preferred when economically justifiable.

In the case of the Messina Strait Bridge the central railway box can undoubtedly be composed of flat panels, as has been proposed by the Impregilo JV in their successful tender design, and it might also be possible to use flat panels for the inner parts of the outer roadway boxes in a similar manner to the Stonecutters Bridge.

8.2.2 Main cable sag

In suspension bridges the ratio between the sag of the main cable and the span length is generally chosen to a value between 1/10 and 1/11. Outside this interval lie the Golden Gate Bridge from 1937 (Figure 8.9) with a sag ratio of 1/8.8, the Storebælt East Bridge with a sag ratio of 1/9 and at the other end some British suspension bridges with sag ratios of around 1/12.

The tension in the main cable varies with the sag ratio, and in cases where the dead load of the cable is modest compared to that of the deck plus the traffic load the cable tension will be approximately inversely proportional to the sag ratio. So with increasing sag the cable tension decreases, thus reducing its required cross section, and an overall saving in the quantity of cable steel will be achieved. This saving in cable steel is to some extent counteracted by the additional cost of the higher pylons required if the main cable sag is increased.

In recent suspension bridge designs a structural optimization has led to a trend towards larger sag ratios than the traditional value of between 1/10 and 1/11. Thus, for example, the Storebælt East Bridge in Denmark as well as two Korean suspension bridges under construction (the

Figure 8.9 ► Golden Gate Bridge, California (USA).

From http://www.pdphoto.org.

Jeokgeum-Yeongnam Bridge and the Myodo Bridge) all have main cable sag ratios of 1/9. These increased sag ratios were arrived at after a thorough optimization process, so in the light of this it would seem relevant to perform further optimization of the Messina Bridge main cable sag, e.g. within the interval from 300 m to 360 m. In the reference design, the cable sag was fixed at 300 m, corresponding to 1/11 of the span length, and the two bidders were not allowed to modify the sag ratio in the tender.

In bridges where the dead load of the cable constitutes a considerable part of the total load, the effect of increasing the sag ratio will be more pronounced as the cable tension, and thereby the cable cross section, decreases not only due to the increased sag but also due to the reduced cable dead load.

The stress in the cable due to its own weight will diminish with increasing sag so that a larger portion of the cable strength can be used for carrying other loads such as the dead load of the deck and the traffic load upon it. At the same time the cable tension arising due to loads on the deck will go down, so there are two contributions to reduction of the cable area with increasing sag. There is a limiting effect as the cables get longer and therefore heavier if the sag ratio reaches an unreasonable extreme.

The reduced self weight of the main cables will also have a positive effect on the aerodynamic stability as a reduced ratio between the mass in the cable planes and the distributed mass across the deck width will result in a more pronounced separation between the vertical and the torsional frequency.

For the Messina design with a main cable sag of 300 m the cost of the towers constitutes approximately 25% and the cost of the cable system accounts for approximately 35% of the total construction cost. For a sag of 330 m the quantity of the main cables would be reduced by a factor of approximately 0.85 but at the same time the steel tower height would have to be increased from 360 m to 390 m, i.e. by a factor of 1.083. Taking both effects into consideration it can be determined that a total potential saving of approximately 3% could be achieved. For a sag of 360 m the total saving (compared to a 300 m sag) would be around 5%.

8.2.3 Cross girder

In a number of recent suspension bridges the principal transverse elements inside the box girders have been changed from a plated diaphragm to a diagonally braced layout. (Figure 8.10)This could lead to a study of the adequacy of truss type cross girders in the Messina Strait Bridge.

Cross girders of a truss configuration might lead to weight savings and simplify the joint between the longitudinal deck boxes and the cross girders.

◄ Figure 8.10 Diagonally braced diaphragm inside the single cell box of the Storebaelt East Bridge (Denmark).

8.2.4 Analytical methods

The analytical methods available for the design of structures are constantly being improved and as a result it is now possible to base the dimensioning of the structural elements on realistic assumptions rather than to use conservative rules from standard specifications of the past.

In connection with the design and construction of major suspension bridges in seismic areas the analytical methods are being improved especially regarding soil-structure interaction. Such analytical improvements (and especially if linked to actual observations and measurements) will be of great interest for future analyses of the bridge.

8.2.5 Runnability

8.2.5.1 Vehicular traffic

Certain unbalanced load cases can produce rotations of the bridge deck that may affect the smooth operation of the crossing unless properly accounted for in the design. Of particular interest for suspension bridges is the case when half of the main span is loaded on one carriageway only.

The most realistic probable extreme loading, based on actual traffic measurements on Italian highways, has been estimated to produce a 4% twist of the bridge deck. With such a change in transverse gradient, the performance of vehicles traveling at high speed may be somewhat affected when changing lanes or making a sudden stop.

Since the accumulation of traffic necessary to create this highly unusual and specific condition would require an estimated 20 to 30 minutes, the incident should be detected soon enough to implement a speed restriction for all lanes of the bridge. Assuming that normal speed limits apply, it will be sufficient to apply a modest restriction on speed for traffic using the bridge until the incident has been cleared when traffic can then resume to normal. Bridge operators can implement such speed restrictions as soon as an incident potentially leading to such particular patterns queuing of traffic along the main span is detected.

The above is based on the present design geometry of the deck, in which the roadways are sloped towards the centre of the bridge. This aspect could alternatively be dealt with through minor modifications of deck configuration. If the deck design were to be changed so that the built-in cross-slope was towards the outside edges rather than inside, the above mentioned load condition would yield the maximum actual crossslope in the congested lanes where vehicle speed is necessarily low, so no further speed restriction would be required. The opposite roadway would rotate to a negative 2% slope, therefore never exceeding the built-in amount, and allowing it to operate without restriction.

Such a modification would require a design change for some details and in particular the drainage system. This in itself would not be a problem since outwardly-pitched drainage is the most common solution on other major bridges worldwide.

Such an alternative will be evaluated during the final design.

8.2.5.2 Railway runnability

Although the Design Basis for the bridge does not specify a maximum gradient or rate of change of slope for the purpose of controlling railway runnability, the maximum and minimum values using various combinations of rail and highway loadings compare favourably with specifications of the American Railway Engineers Association (AREA) and others such as Indian National Railways. These specifications require that the maximum change in longitudinal grade over a 100 metre length should not exceed 0.05% for sag vertical curves and 0.10% for summit vertical curves. The Messina Strait Bridge design does not exceed 0.012% for either case, and is therefore well within the requirements of these international standards.

8.2.5.2.1 Twist in the track due to rare loading events

Under rare live loading situations, there is a potential 4% twist of the bridge deck at the quarter point of the main span. This results in one rail being lower than the other by approximately 60 mm at the point of maximum twist. This situation, known as cant deficiency (negative super elevation), is covered by AREA, Indian Railways, and various European specifications, which all permit cant deficiency of 75 mm in normal cases and 100 mm in extreme cases. Again, the Messina Strait Bridge will perform within these accepted standards.

8.2.5.2.2 Rate of change of twist

The 4% twist, which is equivalent to 60 mm cant deficiency, occurs gradually over a length of approximately 1600 m. A train moving at a speed of 160 km/h (44 m/second) experiences a rate of change of twist of 60 mm in

36 seconds, equal to 1.7 mm/sec. International railway practice, including high-speed train operations, generally permits a rate of change of twist of 30 to 35 mm/sec. The Messina Strait Bridge design falls well within this allowable limit.

With the increase of high-speed railway construction, the analytical methods available for determining train runnability on bridges are constantly being improved. Therefore, it might be possible to refine (and verify) the analytical results obtained in earlier analytical investigations and improve the designer's understanding of predicted bridge performance in the final design.

8.2.6 Roadway surfacing

The roadway surfacing on orthotropic steel decks are generally composed of asphaltic layers with a total thickness of between about 40 and 60 mm. Thus, for example, in the Storebælt East Bridge the surfacing under the traffic lanes consists of (from bottom to top):

- 4 mm mastic.
- 25 mm intermediate layer of mastic asphalt.
- 30 mm wearing layer of mastic asphalt.

This type of surfacing has been used on steel decks in Denmark since 1970 and has proved to be very durable. Other systems have been used elsewhere, and all generally provide satisfactory performance.

With the record long span of the Messina Strait Bridge it becomes attractive to reduce dead load as much as possible, and this is the reason why the successful Impregilo tender design proposes a very thin (10–12 mm thick) polymeric resin layer on top of the steel deck. This type of road surfacing has already been used on several movable bridges but never before on major fixed bridge spans. From the point of view of weight reduction a thin road surfacing is attractive, but it will require that very tight geometric tolerances are achieved during fabrication of the box girders to arrive at an almost perfectly plane deck plate. Furthermore, the orthotropic deck plate will be subjected to a more serious fatigue loading. Recognising this, the Impregilo proposal also introduced a number of measures to improve the fatigue strength of the orthotropic deck. For example, under the traffic lanes the deck plate thickness was increased to 16 mm, and the depth of the trapezoidal trough stiffeners was increased and their transverse spacing reduced.

The saving in total deck dead load due to a thin roadway surfacing will be around 10%, but at the same time the fabrication of the orthotropic deck plate will be more complicated and costly.

To determine the adequacy of a thin polymeric resin surfacing on the deck of the Messina Strait Bridge it will be relevant to study the behaviour and durability of this type of road surfacing on existing bridges. It will also be important to develop suitable procedures for removal and renewal of the surfacing and to test these to demonstrate that future re-surfacing of the bridge deck can be carried out without damage to the steel deck plate.

8.2.7 Construction and maintenance of main suspension cables

The main suspension cables of the Messina Strait Bridge will be very much larger than any previously constructed. Over 160,000 tonnes of high strength wire will need to be used to form its four 1.24 metre diameter cables.

By comparison, the two cables of the Golden Gate Bridge weigh a total of 22,000 tonnes and the cables of the Storebaelt East Bridge total slightly less than 20,000 tonnes.

There are two conventional ways of constructing parallel wire suspension bridge cables; either aerial spinning of individual wires or using shop fabricated parallel wire strands (see Appendix 2). The first method, perfected by John Roebling in the middle of the 19th century, employs a tramway extending from one anchorage to the other. The tramway is used to pull a "spinning wheel" carrying several loops of wire at a time from one end to the other. The loops are secured at each anchorage and the wheel is reloaded for another trip. The Golden Gate Bridge cables were spun with up to six loops of wire per trip, the most ever. Using a similar system for the Messina Bridge it would take almost 3,700 trips of the spinning wheel to place all 44,352 wires for one cable. This might take up to 3 years.

Instead, it has been proposed to erect the cable using the alternative prefabricated parallel wire strand (PPWS) method. In this system, a number of wires are bundled together into a "strand". The strand is socketed at both ends and hauled across the bridge from one anchorage to the other, and secured to the anchorage at both ends. To date, the largest strands have comprised 127 wires. For the massive cables of the Messina Strait Bridge, it may be necessary to use larger strands, and this would require the development of new manufacturing and handling techniques.

The usual protection system for suspension bridge main cables has been to compact the cable into a circular cross section and tightly wrap the entire length with soft wire. Various kinds of waterproofing pastes have been used under the wrapping wire and various paint systems have been used to coat the outside surface.

In the late 1990's, the Honshu-Shikoku Bridge Authority developed an innovative protection system for the cables of their suspension bridges. The cables are wrapped with an airtight elastomeric material, similar to synthetic rubber, and dry air is injected into the cables at discrete locations. The injection of dry air effectively prevents moisture from entering the cables which makes corrosion of the wires virtually impossible. This is a very valuable innovation in cable protection as recent experience in the United States and Europe has shown that the traditional system is not always reliable and many suspension bridges have suffered corrosion of their main cables.

Recently, continuous monitoring methods have also been developed to indicate the conditions inside the main cables. These include acoustic monitoring to detect breaking wires, and sensors that measure humidity, temperature and chemical activity.

It is anticipated that all of these innovations will be applied to the main cables of Messina Strait Bridge. Nevertheless, it will be necessary to develop a detailed inspection and maintenance program to ensure that the main cables remain well protected and capable of supporting the entire weight of the roadways and railways for at least two hundred years.

A more detailed discussion covering this critically important subject of main cable construction and maintenance is included in Appendix 2.

8.2.8 New materials

The substantial contribution of the main cable self weight to the total cable tension means that it would be especially attractive if materials with higher strength-to-weight ratios could be used in place of conventional high strength steel wires. It has, therefore, been suggested on several occasions that the use of carbon fibre rods for the main cables of super-long span bridges could be advantageous as these can have at least the same strength as steel wires but with a weight of less than one quarter of the equivalent steel design.

However, the present cost of carbon fibre rods is very high, so the benefits of a reduced self weight of the cables are far outweighed by the extra cost, so a net saving will not be achieved. It must also be remembered that there is so far no experience on construction procedures and long term behaviour of carbon fibre cables in major suspension bridges. Therefore, it is unlikely that it will be realistic to consider carbon fibre cables for the Messina Bridge.

Due to the significant contribution of the self weight to the dimensions of the main load carrying elements in a long span bridge, it would also seem attractive to use lightweight materials in the deck. However, the lack of experience on long term behaviour and fatigue performance also makes it unattractive to consider materials such as aluminum or fibre composites in the main deck structure of such a major piece of infrastructure without first gaining real experience on smaller bridges. However, in secondary elements such as wind screens, railings, lamp posts, etc. the design might benefit from a weight saving through the use of lightweight materials especially if these elements are replaceable.

The use of other structural materials such as stainless steel and titanium would be interesting from the point of view of improved durability, but the extra cost of such materials is unlikely to be attractive in the current market.

8.3 Conclusion

The design of the Messina Strait Bridge as developed in the early 1990's was based on several innovative ideas for the construction of suspension bridges with very long spans exceeding the spans of existing suspension bridges by a large margin.

In spite of the fact that the Messina Strait Bridge is not yet under construction its design has already inspired the design of many other long span bridges, including some presently under construction.

With the additional experience gained since the early 1990's in relation to topics such as erection techniques, damping measures and durability, it is evident that the Messina Strait Bridge today can be built with even greater confidence than could be achieved before.

APPENDIX 1: THE AEROELASTIC THEORY

Contents of this appendix:

- **Introduction**
- **Different types of models**

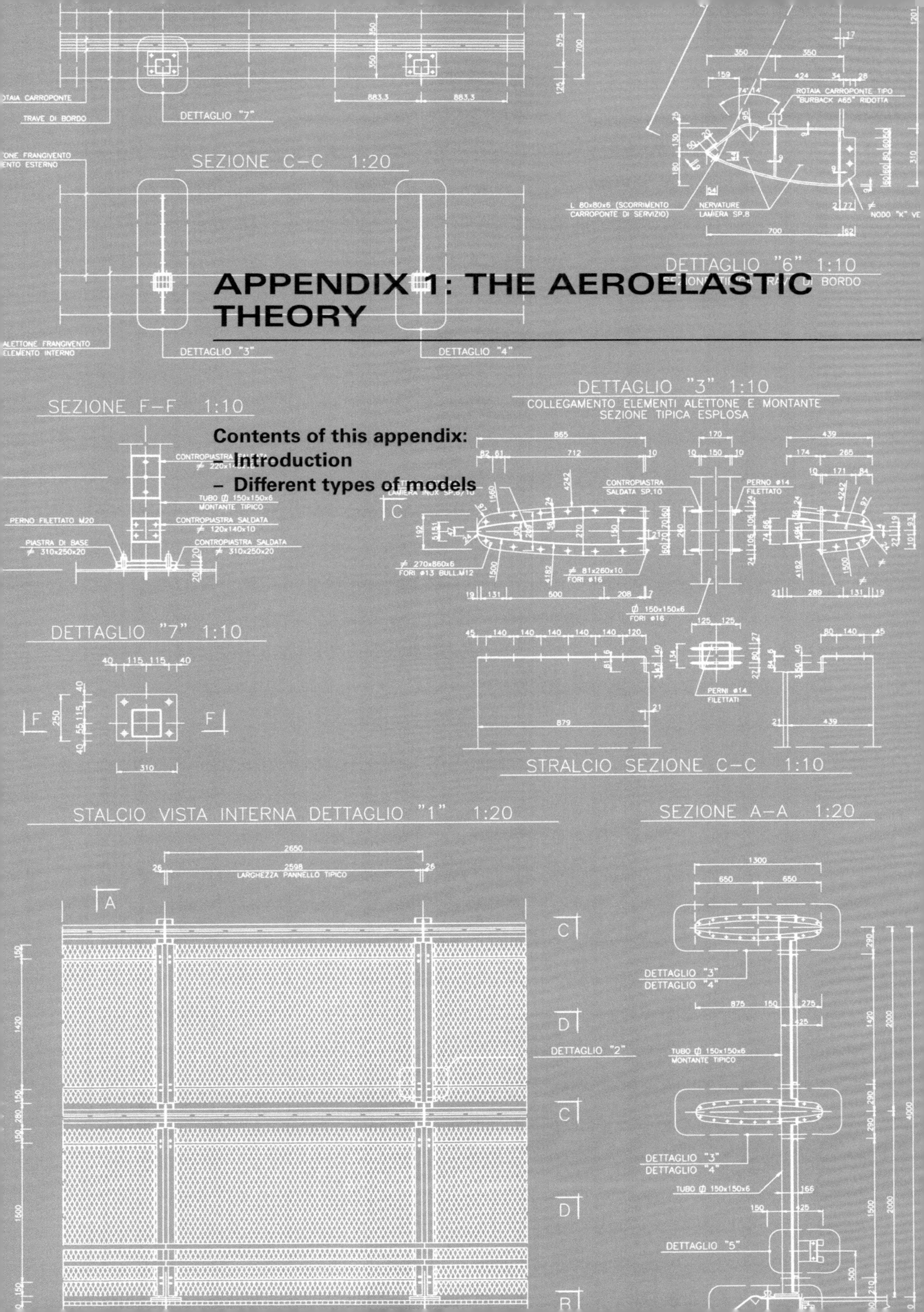

As pointed out in Chapter 6, the wind is the major problem affecting the feasibility of a very long suspension bridge. More precisely the two degree of freedom "flutter type" instability is the major problem conditioning the structure's stability and, as a consequence, the feasibility of the bridge.

A1.1 Introduction

As already explained, the flutter problem is an aeroelastic problem defined by the interaction between the structure's elastic and inertial characteristics and the aerodynamic forces.

In order to better understand these problems, it is useful to describe here the different analytical tools developed for understanding and controlling these aeroelastic problems.

This appendix is organised as follows:

1. Introduction
2. Different types of models
   - 2.1 Quasi Steady Theory
   - 2.2 Linearised theory
   - 2.3 Non-linear approaches

A1.2 Different types of models

Among the different analytical approaches, the QST (Quasi Steady Theory) is the easiest to understand. For this reason it will be explained in detail in the first section below.

As we will see, the aerodynamic forces are functions of the motion and the incoming turbulence in a non-linear way.

The linearization of the aeroelastic forces is widely used for computing the stability of the bridge and the response to turbulent wind using the well known "flutter derivative" coefficients and the aerodynamic admittance functions. This part will be described in the second section.

Beside the QST, other approaches are used to reproduce the non-linear behaviour of the aeroelastic forces. A short description of these approaches will be reported in the last section.

A1.2.1 Quasi Steady Theory (QST)

The QST is well suited to understand the physics of the aeroelastic response of the bridge to turbulent wind. The QST supposes that the aeroelastic forces, acting on the bridge deck, are the same static forces that are measured in the wind tunnel on a sectional model, expressed as:

$$F_y = \frac{1}{2}\rho V_{Rel}^2 BLC_D(\alpha)$$

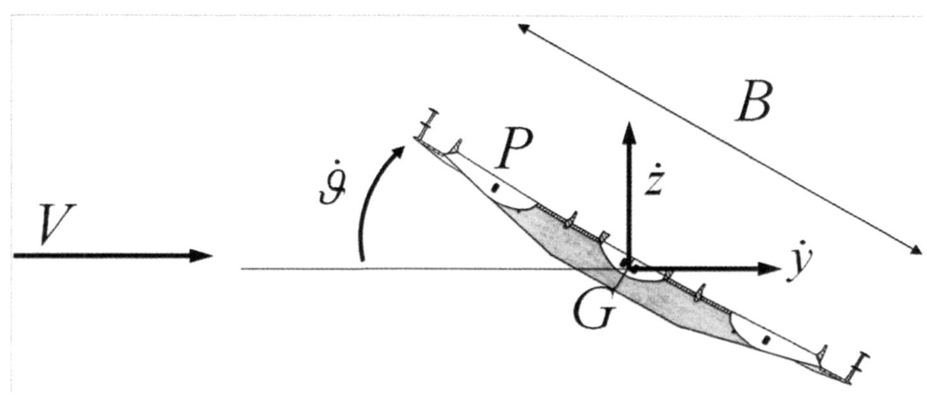

◄ Figure A1.1 Deck velocity components.

$$F_z = \frac{1}{2}\rho V_{\text{Rel}}^2 BLC_L(\alpha)$$

$$F_\vartheta = \frac{1}{2}\rho V_{\text{Rel}}^2 B^2 LC_M(\alpha)$$

where V_{Rel}^2 represents the square of the wind velocity, relative to the deck, that can be computed taking into account the deck velocity component while α represents the angle between the wind velocity and the deck during its motion.

Under these hypotheses, calling y and z respectively the horizontal and vertical displacements of the bridge deck motion and ϑ its rotation and $\dot{y}$, $\dot{z}$ and $\dot{\vartheta}$ the corresponding velocities, the relative wind speed is easily expressed. Actually, as shown in Figure A1.1, the velocity of the deck point P has a horizontal component equal to:

$$V_{P,y} = \dot{y},$$

and a vertical component equal to:

$$V_{P,z} = \dot{z} + \overline{PG}\dot{\vartheta}$$

As a consequence, according to the choice of the position of the considered point P a different relative velocity may be defined

We decide to assume, as reference point, the deck leading edge positioned $B/2$ upwind the deck centre G. The relative velocity, defined as the velocity seen by an observer that is moving together with the deck leading edge, may therefore be defined by its components along the y and z direction:

$$V_{\text{Rel},y} = V_y - \dot{y}$$

$$V_{\text{Rel},z} = V_z - \dot{z} - \frac{B}{2}\dot{\vartheta}$$

where V_y and V_z are the horizontal and vertical components of the incoming wind.

If the wind is turbulent, it is possible to define a mean value of the wind speed V_m, a horizontal turbulent component v and a vertical turbulent component w.

The vector of the relative velocity is shown in Figure A1.2, taking into account also the turbulent wind components.

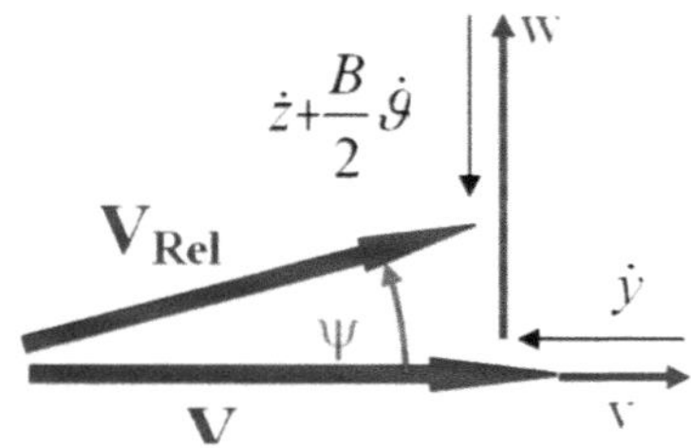

Figure A1.2 ▶ Relative wind vector definition.

The square of the modulus of the relative velocity can be expressed as:

$$V_{Rel}^2 = (V_m + v - \dot{y})^2 + \left(w - \dot{z} - \frac{B}{2}\dot{\vartheta}\right)^2$$

while ψ represents the angle of attack of the relative wind speed:

$$\psi = arctg\frac{w - \dot{z} - \frac{B}{2}\dot{\vartheta}}{V_m + v - \dot{y}}$$

The aerodynamic forces acting on the bridge deck according to the QST are:

$$F_y = \frac{1}{2}\rho V_{Rel}^2 BLC_D(\alpha)$$

$$F_z = \frac{1}{2}\rho V_{Rel}^2 BLC_L(\alpha)$$

$$F_\vartheta = \frac{1}{2}\rho V_{Rel}^2 B^2 LC_M(\alpha)$$

where α is the angle of attack between the relative velocity and the deck:

$$\alpha = \psi + \vartheta$$

and the sign convention is as shown in Figure A1.3.

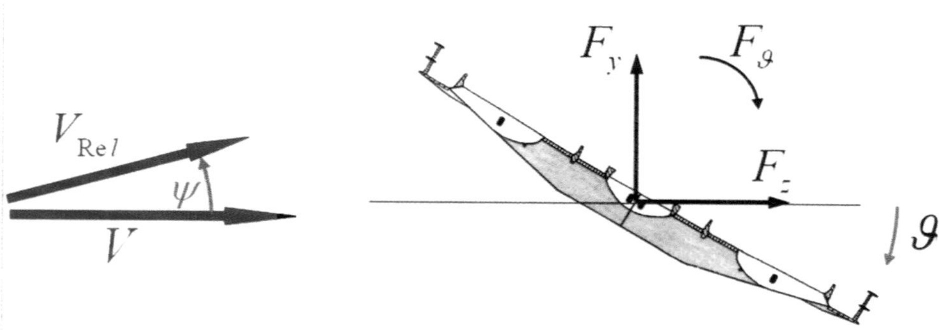

Figure A1.3 ▶ Aerodynamic forces: sign convention.

The bridge has natural modes of vibration described as vertical, horizontal and torsional motions. To understand the bridge behaviour under turbulent wind conditions and flutter instability it is usually sufficient to consider the first vertical and torsional modes.

The problem can be therefore modelled as in Figure A1.4.

where y and z stand for the horizontal and vertical displacement components and ϑ is the rotation. The elastic links are chosen so that the

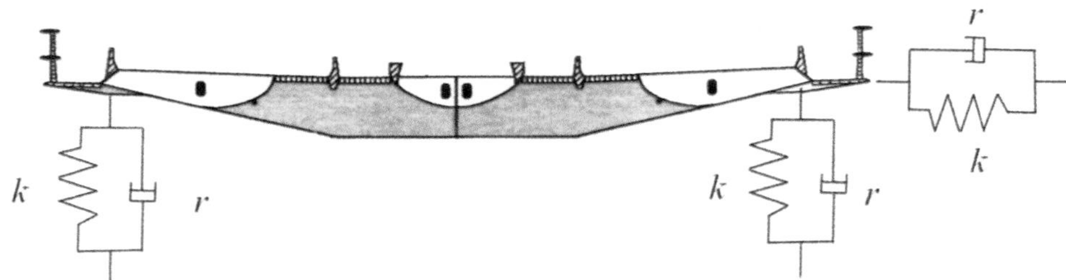

◄ Figure A1.4 Elastically suspended deck section.

model has the same natural frequencies as the real structure for the first modes.

The aerodynamic forces acting on the bridge deck are F_y, F_z and F_ϑ for a given incoming wind.

The equations of motion of the system are:

$$\begin{aligned} m\ddot{y} + r\dot{y} + ky &= F_y \cos\psi - F_z \sin\psi \\ m\ddot{z} + r\dot{z} + kz &= F_y \sin\psi + F_z \cos\psi \\ J_G\ddot{\vartheta} + rl^2\dot{\vartheta} + kl^2\vartheta &= F_\vartheta \end{aligned}$$

where m and J_G are the modal masses, r the damping, and k the stiffness of the elastic links providing the correct modal parameters.

By substituting the previously presented non-linear relations in the expressions for F_y, F_z, F_ϑ the equations become:

$$m\ddot{y} + r\dot{y} + ky = \frac{1}{2}\rho S\left((V_m + v - \dot{y})^2 + \left(w - \dot{\vartheta}\frac{B}{2} - \dot{z}\right)^2\right)(C_D(\alpha)\cos\psi - C_L(\alpha)\sin\psi)$$

$$m\ddot{z} + r\dot{z} + kz = \frac{1}{2}\rho S\left((V_m + v - \dot{y})^2 + \left(w - \dot{\vartheta}\frac{B}{2} - \dot{z}\right)^2\right)(C_D(\alpha)\sin\psi + C_L(\alpha)\cos\psi)$$

$$J_G\ddot{\vartheta} + rl^2\dot{\vartheta} + kl^2\vartheta = \frac{1}{2}\rho S\left((V_m + v - \dot{y})^2 + \left(w - \dot{\vartheta}\frac{B}{2} - \dot{z}\right)^2\right)C_M(\alpha)$$

where:

$$\alpha = \theta + \psi = \theta + \operatorname{arctg}\frac{w - \frac{B}{2}\dot{\theta} - \dot{z}}{V_m + v - \dot{y}}$$

Using the displacement vector:

$$\underline{X} = \begin{Bmatrix} y \\ z \\ \vartheta \end{Bmatrix}$$

the equation of motion system may be written using a matrix formulation as:

$$M\underline{\ddot{X}} + R\underline{\dot{X}} + K\underline{X} = \underline{F}_A\left(\underline{X}, \underline{\dot{X}}, V_m, v(t), w(t)\right)$$

A1.2.2 Linearised theory

A simplified approach, for the study of the aeroelastic problem, may be obtained by considering the linear formulation of the aerodynamic forces, acting on a generic section of a bridge deck. The aerodynamic forces, as we have seen in the previous section, are generally non-linear functions of the bridge motion, the wind speed and the turbulent wind velocity components:

$$\underline{F}_A = \underline{F}_A\left(\underline{X},\underline{\dot{X}},V_m,v(t),w(t)\right) = \underline{F}_A\left(\underline{X},\underline{\dot{X}},V_m,\underline{b}\right)$$

using a matrix formulation where:

$$\underline{F_A} = \begin{Bmatrix} F_y \\ F_z \\ F_\vartheta \end{Bmatrix};\quad \underline{X} = \begin{Bmatrix} y \\ z \\ \vartheta \end{Bmatrix};\quad \underline{\dot{X}} = \begin{Bmatrix} \dot{y} \\ \dot{z} \\ \dot{\vartheta} \end{Bmatrix};\quad \underline{b} = \begin{Bmatrix} v \\ w \end{Bmatrix}$$

A linear formulation of the aerodynamic forces may be considered, if the hypothesis of small variations of $\underline{X}$, $\underline{\dot{X}}$ and $\underline{b}$ is assumed.

Starting from the static deformation, reached under mean wind speed conditions (equilibrium position):

$$X_o = \begin{Bmatrix} y_0 \\ z_0 \\ \vartheta_o \end{Bmatrix}\quad \underline{\dot{X}}_o = \begin{Bmatrix} 0 \\ 0 \\ 0 \end{Bmatrix}\quad \underline{b}_0 = \begin{Bmatrix} 0 \\ 0 \end{Bmatrix}$$

under the hypothesis of small variation of deck motion and turbulent wind fluctuations close to the equilibrium values, the linear expression of the aerodynamic forces is:

$$\underline{F}_A\left(\underline{X};\underline{\dot{X}};\underline{b}\right) = \underline{F}_A\left(\underline{X_o};\underline{\dot{X}_o};\underline{b_o}\right) + \left.\frac{\partial \underline{F}_A}{\partial \underline{X}}\right|_o \underline{dX} + \left.\frac{\partial \underline{F}_A}{\partial \underline{\dot{X}}}\right|_o \underline{d\dot{X}} + \left.\frac{\partial \underline{F}_A}{\partial \underline{b}}\right|_o \underline{db}$$

where $\underline{dX}$, $\underline{d\dot{X}}$ and $\underline{db}$ represent the variations of the corresponding quantities, starting from the values assumed at the equilibrium position:

$$\underline{dX} = \begin{Bmatrix} y(t)-y_0 \\ z(t)-z_0 \\ \vartheta(t)-\vartheta_0 \end{Bmatrix};\quad \underline{d\dot{X}} = \begin{Bmatrix} \dot{y}(t) \\ \dot{z}(t) \\ \dot{\vartheta}(t) \end{Bmatrix};\quad \partial\underline{b} = \begin{Bmatrix} v(t) \\ w(t) \end{Bmatrix}$$

Considering the QST formulation of the aerodynamic forces:

$$F_y = \frac{1}{2}\rho V_{Rel}^2 BL\left(C_D(\alpha)\cos\psi - C_L(\alpha)\sin\psi\right)$$

$$F_z = \frac{1}{2}\rho V_{Rel}^2 BL\left(C_D(\alpha)\sin\psi + C_L(\alpha)\cos\psi\right)$$

$$F_\vartheta = \frac{1}{2}\rho V_{Rel}^2 B^2 LC_M(\alpha)$$

according to the sign convention shown in Figure A1.5

with:

$$\psi = arctg\frac{w-\left(B_1\dot{\vartheta}+\dot{z}\right)}{V+v-\dot{y}},\quad V_{Rel}^2 = \left(V+v-\dot{y}\right)^2 + \left(w-B_1\dot{\vartheta}-\dot{z}\right)^2,$$

and

$$\alpha(t) = \vartheta(t) + \psi(t)$$

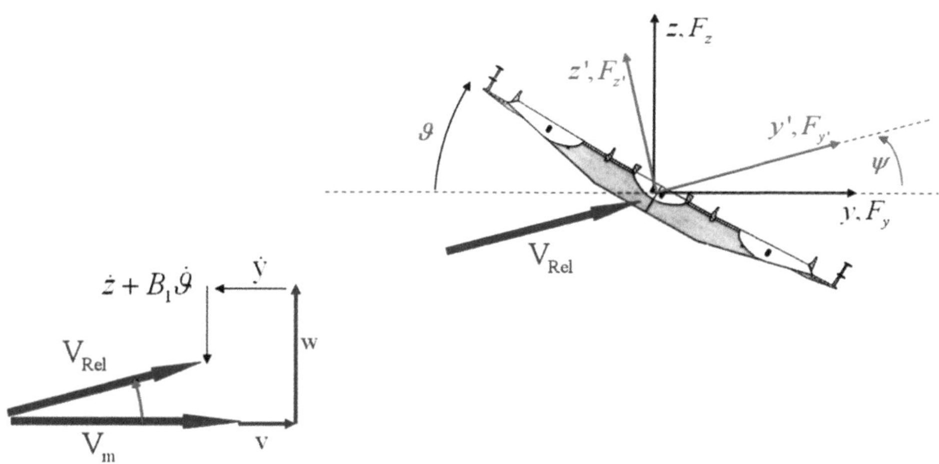

◄ Figure A1.5 Absolute and apparent reference frame.

the 1st order approximation of the aerodynamic forces consists of the following components:

a) a static component:

$$\underline{F}_A\left(\underline{X}_o;\underline{\dot{X}}_o;\underline{T}_o\right) = \frac{1}{2}\rho V_m^2 BL \begin{Bmatrix} C_D(\vartheta_o) \\ C_L(\vartheta_o) \\ BC_M(\vartheta_o) \end{Bmatrix} = \underline{F}_{A,St}$$

b) terms depending on the deck displacement that act as equivalent aerodynamic stiffness:

$$\left.\frac{\partial \underline{F}_A}{\partial \underline{X}}\right|_o \underline{dX} = \underbrace{\frac{1}{2}\rho V_m^2 BL \begin{bmatrix} 0 & 0 & K_{D0} \\ 0 & 0 & K_{L0} \\ 0 & 0 & BK_{M0} \end{bmatrix}}_{-[K_A]_0} \begin{Bmatrix} y(t)-y_0 \\ z(t)-z_0 \\ \vartheta(t)-\vartheta_0 \end{Bmatrix} = -[K_A]_0\,\underline{dX}$$

c) terms depending on the deck motion velocity that act as equivalent aerodynamic damping:

$$\left.\frac{\partial \underline{F}_A}{\partial \underline{\dot{X}}}\right|_o \underline{d\dot{X}} = \underbrace{-\frac{1}{2}\rho V_m^2 BL \begin{bmatrix} 2C_{D0} & (K_{D0}-C_{L0}) & (K_{D0}-C_{L0}) \\ 2C_{L0} & (C_{D0}+K_{L0}) & (C_{D0}+K_{L0}) \\ B_1 2C_{M0} & B_1 K_{M0} & B_1 K_{M0} \end{bmatrix} \frac{1}{V_m}}_{-[R_A]_0} \begin{Bmatrix} \dot{y}(t) \\ \dot{z}(t) \\ \dot{\vartheta}(t) \end{Bmatrix} = -[R_A]_0\,\underline{d\dot{X}}$$

d) terms depending on the turbulent wind velocity components called buffeting terms:

$$\left.\frac{\partial \underline{F}_A}{\partial \underline{T}}\right|_o \underline{db} = \frac{1}{2}\rho V_m^2 BL \begin{bmatrix} 2C_{Do} & (K_{Do}-C_{Lo}) \\ 2C_{Lo} & (C_{Do}+K_{Lo}) \\ B2C_{Mo} & BK_{Mo} \end{bmatrix} \begin{Bmatrix} \dfrac{v(t)}{V_m} \\ \dfrac{w(t)}{V_m} \end{Bmatrix} = \underline{F}_{A,Buff}$$

The equation of motion becomes:

$$[M_S]\underline{d\ddot{X}} + \underbrace{\left[[R_S]+[R_A]_o\right]}_{[R_{Tot}]}\underline{d\dot{X}} + \underbrace{\left[[K_S]+[K_A]_o\right]}_{[K_{Tot}]}\underline{dX} = F_{A,Buff}$$

with

$$[K_S]\underline{X}_0 = F_{A,St}$$

The equivalent aerodynamic stiffness and damping matrices modify the structural ones generating the aeroelastic coupling between the structural and the aerodynamic aspects of the problem.

In fact, starting from a completely uncoupled system and considering only the vertical and torsional degrees of freedom with structural matrixes

$$[M_S] = \begin{bmatrix} m_z & 0 \\ 0 & m_\vartheta \end{bmatrix};\ [R_S] = \begin{bmatrix} r_{z,S} & 0 \\ 0 & r_{\vartheta,S} \end{bmatrix};\ [K_S] = \begin{bmatrix} k_{z,S} & 0 \\ 0 & k_{\vartheta,S} \end{bmatrix}$$

the total stiffness and damping matrixes summing the structural and aerodynamic contributions become:

$$[K_{Tot}] = \begin{bmatrix} k_{z,S} & -qBLK_{Lo} \\ 0 & k_{\vartheta,S} - qB^2LK_{Mo} \end{bmatrix}$$

$$[R_{Tot}] = \begin{bmatrix} r_{z,S} + \dfrac{qBL(C_{Do} + K_{Lo})}{V} & \dfrac{qBLB_1(C_{Do} + K_{Lo})}{V} \\ \dfrac{qB^2LK_{Mo}}{V} & r_{\vartheta,S} + \dfrac{qB^2LB_1K_{Mo}}{V} \end{bmatrix}$$

where q is the dynamic pressure:

$$q = \frac{1}{2}\rho V^2$$

The matrices are therefore dependent on the value of q and as a consequence the static and dynamic behaviour of the bridge depends on the value of the wind speed.

Considering the direct terms of the damping matrix for the vertical degree of freedom, if

$$(C_{D0} + K_{L0}) < -r_{zS}\frac{1}{qBL\frac{1}{V}} = -\frac{2r_{zS}}{\rho VBL}$$

a negative value of damping arises and dynamic instability occurs. Since the C_{D0} coefficient is always positive, the instability condition only arises if the slope of the lift coefficient K_{L0}, considered at the static position, has a negative value with respect to the considered sign convention.

A positive lift coefficient slope is therefore a mandatory requirement in the aerodynamic design of a bridge.

In the same way, considering the direct terms of the damping matrix for the torsional degree of freedom, if

$$K_{Mo} < -r_{\vartheta S}\frac{1}{qB^2L\frac{B_1}{V}} = \frac{2r_{\vartheta S}}{\rho VB^2LB_1}$$

a negative value of the damping arises and dynamic instability occurs. The instability condition arises only if the slope of the moment coefficient K_{M0}, considered at the static position, assumes a negative value with respect to the considered sign convention.

Thus, a positive moment coefficient slope is therefore also a mandatory requirement in the aerodynamic design of a bridge.

Considering the direct terms of the stiffness matrix for the torsional degree of freedom, if

$$K_{M0} > k_{\vartheta,S} \frac{1}{qB^2L} = \frac{2k_{\vartheta,S}}{\rho V^2 B^2 L}$$

then a negative value of the torsional stiffness occurs and static instability appears. The slope of the aerodynamic moment coefficient must be therefore positive but with small values.

Analysing the total matrices, it can be seen that even though the starting structural scheme was completely uncoupled, the aerodynamic terms produce a stiffness and damping coupling.

The coupling is stronger if the aerodynamic terms assume values similar to the structural ones. Usually this condition appears at high wind speed.

In the presence of a positive value of K_{M0}, with increasing wind speed the total torsional stiffness $k_{\vartheta,S} - qB^2LK_{Mo}$ decreases and the natural frequency of the torsional modes tends also to decrease. Under flutter instability conditions the torsional frequency becomes close to the vertical natural frequency and a completely coupled vibration occurs. The indirect terms of the total damping matrix contribute to add energy into the system instead of damping the oscillation, taking advantage from the phase shift between the vertical and the torsional components of the new coupled vibration mode.

Even if positive values of the K_{L0} and K_{M0} coefficients are used, the flutter instability may occur anyway. The importance of the aeroelastic coupling effects and of the related flutter instability becomes more significant as the aerodynamic terms become larger compared to the structural ones, and the closer together are the torsional and the vertical natural frequencies. (Figure A1.6)

The presented linear approximation of the QST is valid only if the variation of the deck motion terms and the turbulent wind terms are slow in

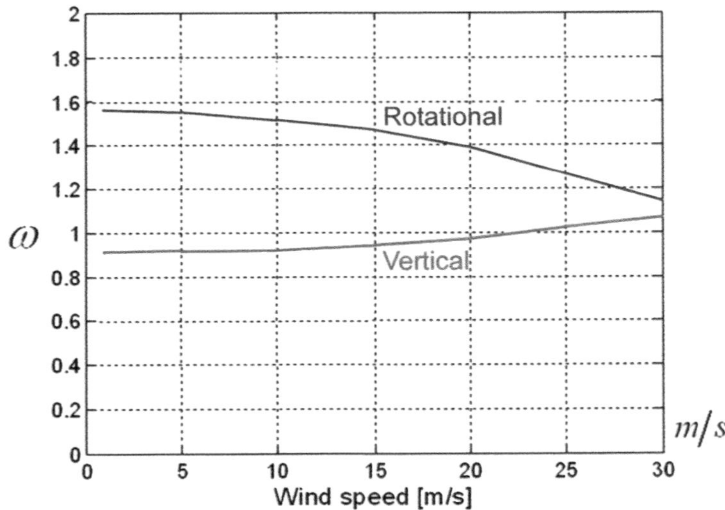

◄ Figure A1.6 Torsional and vertical natural frequency trend versus speed.

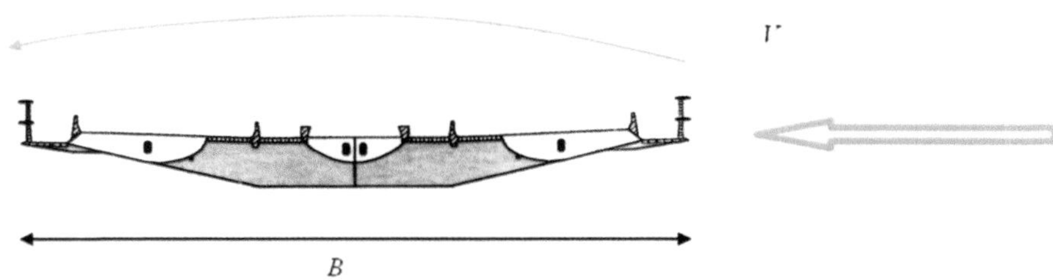

Figure A1.7 ► Fluid particle crossing the deck chord.

comparison to a characteristic aerodynamic time scale defined as the time taken for a particle of fluid to completely cross the width of the deck:

$$T_{Aer} = \frac{B}{V}$$

The ratio between the period of fluctuation T of the deck motion or of the turbulent wind velocity component and the characteristic aerodynamic time defines the reduced velocity parameter V^*, which is used in bridge aerodynamics to define the dependency of the aerodynamic forces on frequency:

$$V^* = \frac{T}{T_{Aer}} = \frac{V}{fB}$$

where f represents the fluctuation frequency.

In order to take into account the frequency dependency with the linear approach, aerodynamic transfer functions between the aerodynamic forces and the deck motion and between the aerodynamic forces and the turbulent wind velocity components are experimentally measured through wind tunnel tests using deck sectional models.

The aerodynamic transfer functions are measured by reproducing fluctuations of the deck motion components and the wind velocity components with small amplitudes in the wind tunnel, and measuring the corresponding aerodynamic forces and considering different mean angles of attack.

In the experimental procedures followed for the Messina Bridge aerodynamic characterization, the deck sectional model was moved by means of oil dynamic actuators according to a sinusoidal motion law along one single degree of freedom (horizontal or vertical or torsional) under mean wind conditions. (Figure A1.8) The reduced velocity parameter may be varied by changing the motion frequency or the mean wind velocity [1].

The definition of the aerodynamic transfer function related to the deck motion oscillation, according to the Scanlan formulation and the sign convention of Figure A1.3 are:

$$\underline{F_y} = \frac{1}{2}\rho V^2 BL \frac{1}{V^{*2}} \left\{ \left(P_4^* + iP_1^*\right)\frac{\underline{z}}{B} + \left(P_3^* + iP_2^*\right)\underline{\vartheta} + \left(P_6^* + iP_5^*\right)\frac{\underline{y}}{B} \right\}$$

$$\underline{F_z} = \frac{1}{2}\rho V^2 BL \frac{1}{V^{*2}} \left\{ \left(H_4^* + iH_1^*\right)\frac{\underline{z}}{B} + \left(H_3^* + iH_2^*\right)\underline{\vartheta} + \left(H_6^* + iH_5^*\right)\frac{\underline{y}}{B} \right\}$$

$$\underline{F_\vartheta} = \frac{1}{2}\rho V^2 B^2 L \frac{1}{V^{*2}} \left\{ \left(A_4^* + iA_1^*\right)\frac{\underline{z}}{B} + \left(A_3^* + iA_2^*\right)\underline{\vartheta} + \left(A_6^* + iA_5^*\right)\frac{\underline{y}}{B} \right\}$$

◄ Figure A1.8
Deck sectional model in the wind tunnel on oil dynamic actuators.

The non dimensional coefficients P_i^*, H_i^* and A_i^* (with $i = 1$ to 6), representing the real and the imaginary parts of the aerodynamic admittance functions, defined with the deck motion components, are called the "Flutter Derivatives".

A different non-dimensional formulation of the flutter derivatives is proposed by Zasso [2] and offers the advantage to show a more meaningful asymptotic behaviour to the linearized QST values at high reduced velocity values:

$$\underline{F}_y = \frac{1}{2}\rho V^2 BL\left(-p_1^*\frac{i\omega z}{V} - p_2^*\frac{i\omega B\vartheta}{V} + p_3^*\vartheta + p_4^*\frac{\pi}{2V_\omega^{*2}}\frac{z}{B} - p_5^*\frac{i\omega y}{V} + p_6^*\frac{\pi}{2V_\omega^{*2}}\frac{y}{B}\right)$$

$$\underline{F}_z = \frac{1}{2}\rho V^2 BL\left(-h_1^*\frac{i\omega z}{V} - h_2^*\frac{i\omega B\vartheta}{V} + h_3^*\vartheta + h_4^*\frac{\pi}{2V_\omega^{*2}}\frac{z}{B} - h_5^*\frac{i\omega y}{V} + h_6^*\frac{\pi}{2V_\omega^{*2}}\frac{y}{B}\right)$$

$$\underline{F}_\vartheta = \frac{1}{2}\rho V^2 B^2 L\left(-a_1^*\frac{i\omega z}{V} - a_2^*\frac{i\omega B\vartheta}{V} + a_3^*\vartheta + a_4^*\frac{\pi}{2V_\omega^{*2}}\frac{z}{B} - a_5^*\frac{i\omega y}{V} + a_6^*\frac{\pi}{2V_\omega^{*2}}\frac{y}{B}\right)$$

The adoption of the flutter derivative coefficients allows the definition of the aerodynamic equivalent stiffness and damping matrices for different values of reduced velocity and different mean angles of attack:

$$\begin{Bmatrix} F_y \\ F_z \\ F_\vartheta \end{Bmatrix} = \underbrace{\frac{1}{2}\rho V^2 BL\begin{bmatrix} p_6^*\frac{\pi}{2V_\omega^{*2}B} & p_4^*\frac{\pi}{2V_\omega^{*2}B} & p_3^* \\ h_6^*\frac{\pi}{2V_\omega^{*2}B} & h_4^*\frac{\pi}{2V_\omega^{*2}B} & h_3^* \\ a_6^*\frac{\pi}{2V_\omega^{*2}} & a_4^*\frac{\pi}{2V_\omega^{*2}} & a_3^*B \end{bmatrix}}_{-[K_A(V^*,\alpha)]_0}\begin{Bmatrix} y \\ z \\ \vartheta \end{Bmatrix} \underbrace{-\frac{1}{2}\rho V^2 BL\begin{bmatrix} p_5^* & p_1^* & p_2^*B \\ h_5^* & h_1^* & h_2^*B \\ a_5^*B & a_1^*B & a_2^*B^2 \end{bmatrix}\frac{1}{V}}_{-[R_A(V^*,\alpha)]_0}\begin{Bmatrix} \dot{y} \\ \dot{z} \\ \dot{\vartheta} \end{Bmatrix}$$

It is easy to perform a comparison with the linearized QST formulation of the aerodynamic equivalent stiffness and damping matrices. This comparison allows a check on the asymptotic values assumed by the flutter derivatives at high reduced velocity values.

As an example of the dependence of the flutter derivatives on the reduced velocity, the trend of the h_1^* coefficient versus the reduced velocity is shown in Figure A1.9. It is possible to appreciate that the asymptotic values reached by the coefficient at high reduced velocity for different angles of attack

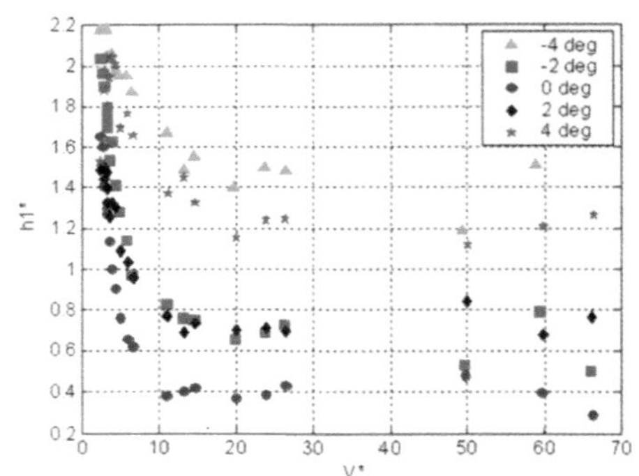

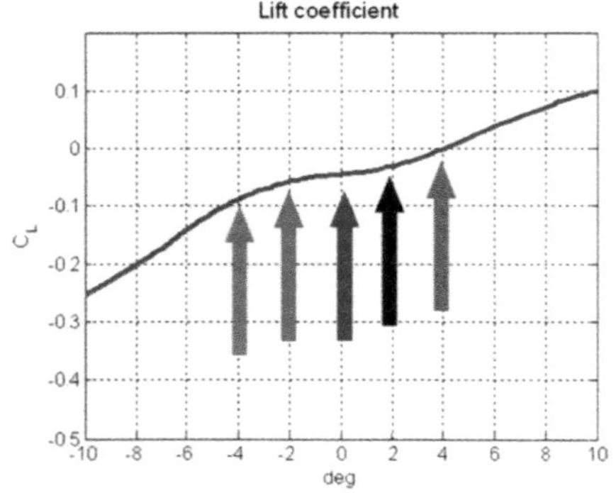

Figure A1.9 ▶ h_1^* flutter coefficient versus reduced velocity at different mean angles of attack. Comparison with the static coefficient slope.

Figure A1.10 ▶ Active turbulence generator and flow visualization.

follow the trend of the slope of the C_L static coefficient as expected by the linearization of the QST approach.

A strong dependence on the reduced velocity appears at V* < 10 for the case of the Messina Bridge multi box deck section.

Similar considerations may be drawn for the aerodynamic transfer function defined using the turbulent wind velocity components as input.

For the Messina Bridge, an active turbulence generator was used to produce a highly correlated wind with a vertical velocity component fluctuating with a single harmonic law [1, 3]. Figure A1.10 shows a visualization of the wave generated in the wind tunnel together with a sketch reporting the sign convention.

The following formulation reports the aerodynamic transfer function related to the turbulent wind components:

$$\underline{F}_{A,Buff} = \begin{Bmatrix} |F_y| \\ |F_z| \\ |F_\vartheta| \end{Bmatrix} = \frac{1}{2}\rho V^2 BL \begin{bmatrix} \chi_{vy} & \chi_{wy} \\ \chi_{vz} & \chi_{wz} \\ \chi_{v\vartheta} B & \chi_{w\vartheta} B \end{bmatrix} \begin{Bmatrix} \dfrac{v}{V} \\ \dfrac{w}{V} \end{Bmatrix}$$

where the coefficients χ_{ij}, $i = v, w$; $j = y,z,\vartheta$ are called "aerodynamic admittance functions".

The admittance function coefficient values at high reduced velocity may be compared to the values of the linearized QST. The trend of the χ_{wz} coefficient versus the reduced velocity is shown in Figure A1.11.

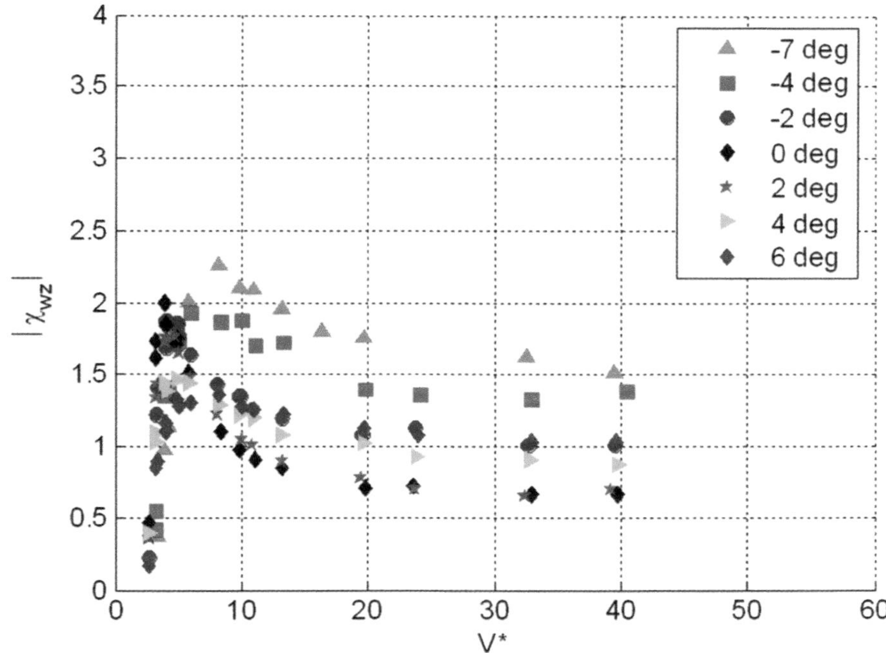

◀ Figure A1.11 Aerodynamic admittance function versus reduced velocity.

A complete linear formulation of the aeroelastic problem can therefore be written using the flutter derivatives and the aerodynamic admittance function coefficients.

$$[M_S]\underline{d\ddot{X}} + \underbrace{\left[[R_S]+\left[R_A\left(V^*,\alpha\right)\right]\Big|_0\right]}_{[R_{Tot}]}\underline{d\dot{X}} + \underbrace{\left[[K_S]+\left[K_A\left(V^*,\alpha\right)\right]\Big|_0\right]}_{[K_{Tot}]}\underline{dX} = F_{A,Buff}$$

A1.2.3 Non linear approaches

Even if linear approaches are the widest adopted method to study bridge aerodynamic stability and the response of the bridge to turbulent wind, they are not able to deal with the important non-linear effects intrinsically present in the aeroelastic problem. As already presented, the dependence of the aerodynamic forces from the deck motion and from the turbulent wind velocity components is fully non-linear.

$$+M\underline{\ddot{X}} + R\underline{\dot{X}} + K\underline{X} = \underline{F}_A\left(\underline{X},\underline{\dot{X}},V_m,v(t),w(t)\right)$$

The two main causes of non-linearities are the dependence of the aerodynamic forces on the angle of attack and on the reduced velocity. The QST is able to take into account the dependence on the mean angle of attack in a non-linear way, but it is valid only at high reduced velocities. On the contrary, the linear approaches are able to take into account the dependence on the reduced velocity through the flutter derivatives and the admittance function coefficients but they represent a solution around a fixed mean angle of attack. Measurements performed on full scale bridges highlighted that during the common operating condition the relative angle of attack between the bridge deck and the wind direction may reach large values that make the linear hypothesis not applicable. Some non-linear approaches have been developed [3, 4, 5] to deal with the fully non-linear aeroelastic problem. In this section we can only introduce the basic ideas of the "Band Superposition" approach and of the "rheological model" approach, and refer the reader to the literature for a more exhaustive description.

The "Band Superposition" approach relies on the possibility of separating the low frequency and high frequency contributions of the wind spectrum. The idea is that the low frequency part of the wind spectrum is responsible for the larger fluctuations of the angle of attack and it acts at high reduced velocity. A corrected QST is therefore proposed to account for this part by taking into account the non-linear dependence on the angle of attack. The part of the wind spectrum at high frequency is characterised by smaller amplitudes and lower reduced velocity. To consider the effects induced by this part of the wind spectrum, the aeroelastic forces are linearized around the angle of attack that is instantaneously computed by solving the low frequency part of the problem.

An alternative approach is to consider the whole wind spectrum effects in a single time domain model as represented by the "rheological model". This approach is based on a numerical model that is able to consider both the angle of attack and the reduced velocity dependence in a single time domain approach. To identify the numerical model parameters, aerodynamic hysteresis loops are measured on deck sectional models in the wind tunnel by changing the instantaneous angle of attack. The variation of the angle of attack is produced by moving the sectional model or by using the active turbulence generator with high amplitudes at different frequencies. (See [4], [5] for more details.)

References and Further Readings

[1] G. Diana, F. Resta, A. Zasso, M. Belloli, and D. Rocchi (2004), *Forced motion and free motion aeroelastic tests on a new concept dynamometric section model of the Messina suspension bridge*, J. Wind Eng & Ind. Aerodyn., 92, 441-462.

[2] A. Zasso (1996), *Flutter derivatives: advantages of a new representation convention*, J. Wind Eng & Ind. Aerodyn., 60, 35-47.

[3] G. Diana, S. Bruni, and D. Rocchi, *A numerical and experimental investigation on aerodynamic non linearities in bridge response to turbulent wind*, EACWE 4, 11-15 July 2005 Prague.

[4] Diana, G., Resta, F. and Rocchi, D. (2007), *"A new approach to model the aeroelastic response of bridges in time domain by means of a rheological model"*, Proc. of the ICWE 12, Cairns Australia.

[5] Diana, G., Resta, F. and Rocchi, D. (2008), *"A new numerical approach to reproduce bridge aerodynamic non-linearities in time domain"*, accepted for publishing on J. Wind Eng & Ind. Aerodyn.

[6] M. Bocciolone, F. Cheli, A. Curami, and A. Zasso (1992), *Wind measurements on the Humber bridge and numerical simulations*, J. Wind Eng & Ind. Aerodyn., 41-44, 165-173.

[7] Jain, A., Jones, N.P. and Scanlan, R.H. (1996), *"Coupled aeroelastic and aerodynamic response of long-span bridges"*, J. Wind Eng & Ind. Aerodyn., 89, 1335-1350.

[8] Caracoglia, L. and Jones, N.P. (2003), *"Time domain vs. frequency domain characterization of aeroelastic forces for bridge deck sections",* J. Wind Eng & Ind. Aerodyn., 91, 371-402.

[9] Chen, X., Matsumoto, M. and Kareem, A. (2000), *"Time domain flutter and buffeting response analysis of bridges"*, J. of Eng Mechanics., 126, 7-16.

[10] N.N. Minh, T. Miyata, H. Yamada, and Y. Sanada (1999), *Numerical simulation of wind turbulence and buffeting analysis of long span bridges*, J. Wind Eng & Ind. Aerodyn., 83, 301-315.

APPENDIX 2: CABLE CONSTRUCTION AND DURABILITY

Contents of this appendix:

- Cable construction
- Cable corrosion & traditional techniques for protection
- Protection of the unwrapped portions of the main cable
- Conventional and modern main cable protection systems
- Waterproofing zinc paste
- S-shaped wrapping wire
- Elastomeric wrapping
- Elastic coatings
- Dry air injection – background and system description
- Inspection and evaluation of main suspension cables
- Continuous acoustic monitoring
- Hanger cables

On a conventional suspension bridge, all of the "heavy lifting" is done by the towers, anchorages and main cables. These elements bear the entire load of the bridge girders, roadways, rail tracks and all traffic that crosses the bridge. Without question, the main cables are the most critical structural element, as they are most susceptible to damage from corrosion. They are extremely difficult and expensive to replace or repair. This Appendix addresses these critical issues.

A2.1 Cable construction

Since the Brooklyn Bridge was built in New York City in the late 1800's (Figure A2.1) nearly all main cables of major suspension bridges have been constructed of high strength galvanized steel wire. The wires are typically made of high quality steel with a very carefully controlled carbon content (0.8%) that ensures high tensile strength with sufficient ductility. The wire is manufactured by drawing rolled rods through successively smaller dies until the specified diameter has been obtained. This process of drawing the wires "cold works" the material resulting in an increase in tensile strength. After the drawing process, the wires are dipped in a bath of molten zinc, a treatment known as hot-dip galvanizing. The molten zinc actually forms an alloy at the interface with the steel, thus creating an impervious protective layer. The high temperature of the zinc bath also lowers the strength of the wire slightly and improves its ductility (the ability to deform without fracturing).

Modern bridge wire typically has a diameter of about 5 to 5.4 mm and an ultimate tensile strength of between 1570 N/mm^2 and 1850 N/mm^2. Bridge wire must be manufactured under strict quality control. The ability to draw wire of the proper strength and ductility is highly dependent upon the chemistry of the steel, particularly with respect to carbon content and freedom from impurities. The physical properties of the wire must be uniformly and regularly confirmed through testing of the specified sampling lots. It is

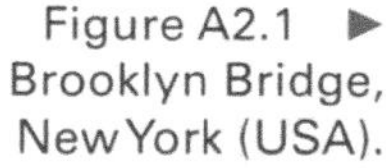
Figure A2.1 ► Brooklyn Bridge, New York (USA).

essential that all manufacturers perform the physical and chemical tests in an identical manner to assure uniformity throughout the cables.

Each suspension bridge cable is composed of thousands of wires laid parallel to each other. The wires are separated into bundles of several hundred wires, called "strands." The individual strands allow better adjustment of the wires to ensure that each carries its share of the load, but even more important, the individual strands are readily connected to the cable anchorages whereas one huge bundle of wires would be impossible to connect as a whole.

There are two ways that the main cables can be assembled: by aerial spinning and using prefabricated strands. In the aerial spinning method, individual wires are pulled across the bridge from one anchorage to the other using a spinning wheel (Figure A2.2), possibly two to twelve wires at a time. The individual wires of a strand are looped around a shoe at each end that is anchored into the anchor block.

In the prefabricated strand method (prefabricated parallel wire strands are known as PPWS), a number of wires are arranged in a parallel bundle that is held together by special bindings and socketed at both ends. The sockets are typically cast steel blocks with a conical hollow into which the wires are inserted and bonded using a polymeric resin or a molten metal alloy. The prefabricated strands are delivered to site on large drums and pulled across from one end to the other one at a time until the full number of strands in the cable have been erected and attached to the anchorages at each end. (Figure A2.3) The prefabricated strands must be made to the full length of the finished cable, so for longer spans, there are practical weight and size limitations to the number of wires that can be built into one strand. The first bridge to be constructed using PPWS was the Newport-Pell Bridge in the United States. (Figure A2.4) To date, the largest prefabricated strands have contained 127 wires. This limitation will no doubt be improved upon as more bridges are built using this method. The largest bridge built using this method of cable construction is the Akashi Kaikyo Bridge near Kobe, Japan, which has a main span of 1991 m. Each main cable comprises 290 strands of 127 wires each, making a total of 36830 wires of 5.23 mm diameter. The completed compacted cables are 1.12 m in diameter.

◄ Figure A2.2 Aerial spinning of the main cable on Storebaelt East Bridge (Denmark).

Courtesy of COINFRA S.p.A. – Italy.

Figure A2.3 ▶ Prefabricated parallel wire strand (PPWS) erection on Kurushima Kaikyo Bridge, Japan.

Courtesy of Honshu-Shikoku Bridge Expressway Company Limited – Japan.

Figure A2.4 ▶ Newport-Pell Bridge, Rhode Island, USA. Its cables were the first to use PPWS and are wrapped in fibreglass reinforced acrylic resin.

A2.2 Cable corrosion & traditional techniques for protection

Since the mid-19th century, when John A. Roebling pioneered the art of suspension bridge design, the main cables of suspension bridges have typically been protected by a tight covering of soft wire wrapping bedded in a sealing paste, usually red-lead (Pb_3O_4) in linseed oil, and overcoated with paint. Some exceptions are notable, such as the Newport-Pell Bridge (Rhode Island, USA, 1969) and Bidwell Bar Bridge (Oroville, California, USA, 1965) where glass-reinforced acrylic was used, and the William Preston Lane Bridge (Maryland, USA, 1973) where neoprene sheet was used. In 2007 the first internal inspection of the main cable wires of the Newport Bridge indicated that the acrylic wrapping system had performed extremely well, especially when compared to the very variable results observed in wire-wrapped cables inspected throughout the United States. Recognizing the advantage of using an impervious covering on the cables, a number of US suspension bridges have been retrofitted with elastomeric coverings placed over the existing wrapping wire. There are now a number of bridges in Europe and Japan that use dehumidification for corrosion protection – a dry-air injection system in conjunction with an elastomeric wrapping to ensure that no moisture can enter the cables.

Proper corrosion protection of the main cables is of extraordinary importance on any suspension bridge, and particularly on the Messina Strait Bridge due to the following reasons:

1. The extremely long design life (200 years).
2. The difficulty of replacing the main cables, especially due to the very long span. Although replacement of the main cables of suspension bridges has been undertaken in recent years (for example for the Tancarville Bridge in France), the sheer length of the Messina cables and their large diameter would make their replacement a very difficult task.
3. The relatively harsh marine environment.
4. The high stress levels in the wires.

A2.3 Protection of the unwrapped portions of the main cable

The protection of the most sensitive portions of the cables, at the tower top saddles and inside the anchorages, where the individual wires are exposed, relies on the hot-dip zinc coating of the wires. It has become common to install dehumidification units for the air inside the saddle and anchorage chambers. These two means are sufficient to ensure the long life of the cable wires in these critical areas.

A2.4 Conventional and modern main cable protection systems

Most new suspension bridges around the world still use the basic protection system developed by John Roebling for the Brooklyn Bridge. The parallel galvanized wires are firmly compacted, coated with a sealing paste, tightly wrapped with soft zinc coated wire, and painted. (See Figure A2.5)

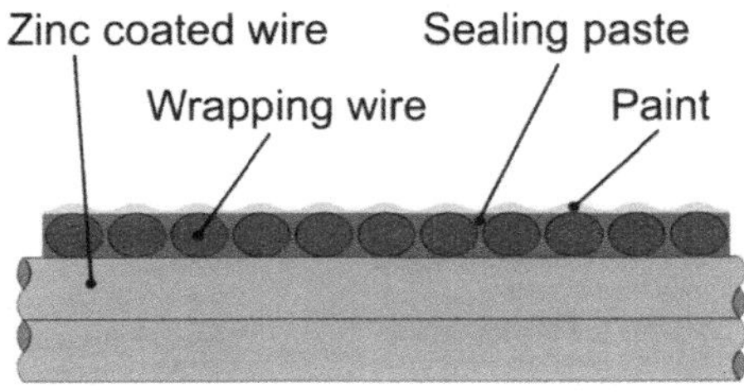

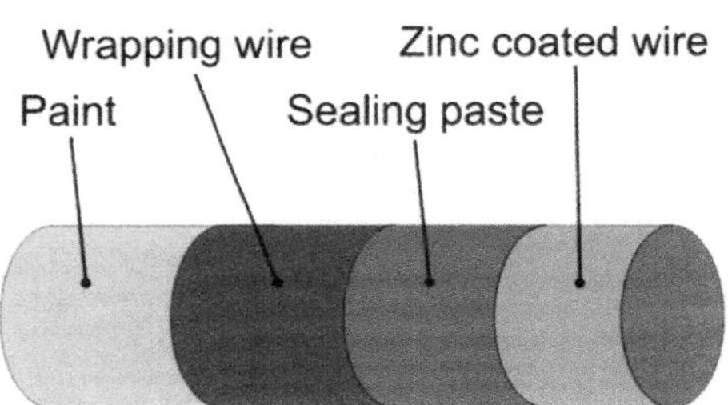

◄ Figure A2.5 Conventional cable corrosion protection system.

The in-depth inspection of the main cables of several bridges in the United States, performed over the last two decades, has revealed various degrees of corrosion and prompted different solutions for the protection of the cables, both for new construction and for cable rehabilitation projects. Similar inspections have recently been performed on bridges in England and Scotland. The information acquired through those cable investigations and testing of their wires has been very useful in the development of new methods for the protection of main cables from corrosion.

A2.5 Waterproofing zinc paste

Due to health and environmental concerns with the handling and disposal of lead-based materials such as the traditional waterproofing red lead paste, oil-based zinc paste has been specified for the protection of main cables in the last few decades.

More recently, an Italian firm specifically developed the proprietary elastomeric paste with high metallic zinc content (Elettrometall 8870) for use on the main cables of the Storebælt East Bridge in Denmark, completed in 1998. (Figure A2.6) The basic composition of this material has a proven track record of over 20 years in dam waterproofing applications. The mix was somewhat modified to adapt it to the particular application on suspension cables, and to ensure the long-term durability of all components and materials called for by the Storebælt design criteria. The paste was subject to extensive testing and scrutiny by the bridge owner prior to its acceptance. (Figure A2.7) Elettrometall 8870 consists of a moisture curing liquid polyurethane elastomer mixed just prior to application with metallic zinc powder. Metallic zinc accounts for 95% of the weight of the paste, thereby ensuring optimal galvanic protection of the wires. The paste cures to a sheath-like permanent elastic continuous coat, which fills all the voids between the wrapping and main cable wires, and also prevents any water penetration from the cable exterior.

The paste was recently used for the protection of the cables on the New Carquinez Straits Bridge in California, USA, as well as for test panel applications on the George Washington Bridge in New York and the Golden Gate Bridge in San Francisco, California. This elastomeric paste has excellent properties

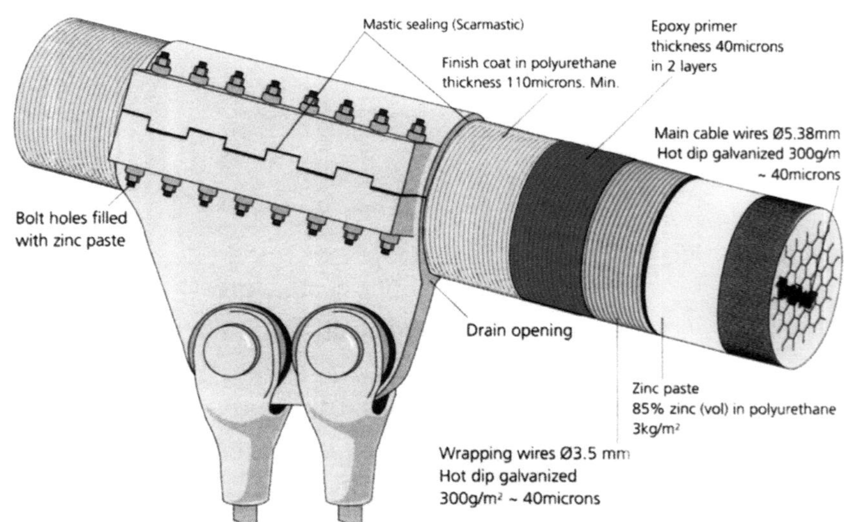

Figure A2.6 ► Protective layers on the main cables of the Storebælt East Bridge (Denmark).

From "The Storebælt Publications – East Bridge", courtesy of Sund & Bælt Holding A/S – Denmark.

◄ Figure A2.7 Elettrometall Paste applied to the main cable of the Storebælt East Bridge (Denmark).

◄ Figure A2.8 Grikote–Z Paste applied on Bear Mountain Bridge cable (USA).

with regard to cable protection, and its use will provide a much higher degree of protection than the traditional red lead paste.

Another paste recently developed is grease-based with powdered zinc and zinc oxide solids and corrosion inhibitors added. This is a proprietary material known as "Grikote-Z", developed in the United States for use in cable rehabilitation projects. In 2000–2001, it was used under the new wrapping wire on the Bear Mountain Bridge, which crosses the Hudson River in New York State. (Figure A2.8) The cables were reopened in the summer and fall of 2007, and it was found that the paste had performed very well, remaining soft and continuous over the surface of the cable, thereby precluding the entrance of water through the wrapping. This same material has been specified for the wrapping system of the new San Francisco-Oakland Bay Bridge, which is currently under construction.

A2.6 S-shaped wrapping wire

Typically, round wire has been used for wrapping the main cables. Due to the constant expansion and contraction of the cable under load, it is possible for the paint on the wrapping wire to crack and allow water penetration through the gaps between adjacent wire wrapping turns, especially when conventional non-flexible paint systems are used.

Consequently, a new type of S-shaped wrapping wire was developed in Japan, in an attempt to improve on the water tightness of the wire wrapping. This wire has an interlocking cross section, as shown in Figure A2.9, which does not leave gaps between consecutive turns, and which presents a smoother outer surface that results in improved paint durability.

S-shaped wire wrapping has been used on the Hakucyo and the Kurushima-Kaikyo bridges, completed in 1999 in Japan. It has not been used elsewhere in the world because it remains a proprietary item and requires special equipment to install it properly, however, this wrapping wire has been specified for the new San Francisco-Oakland Bay Bridge currently under construction in California.

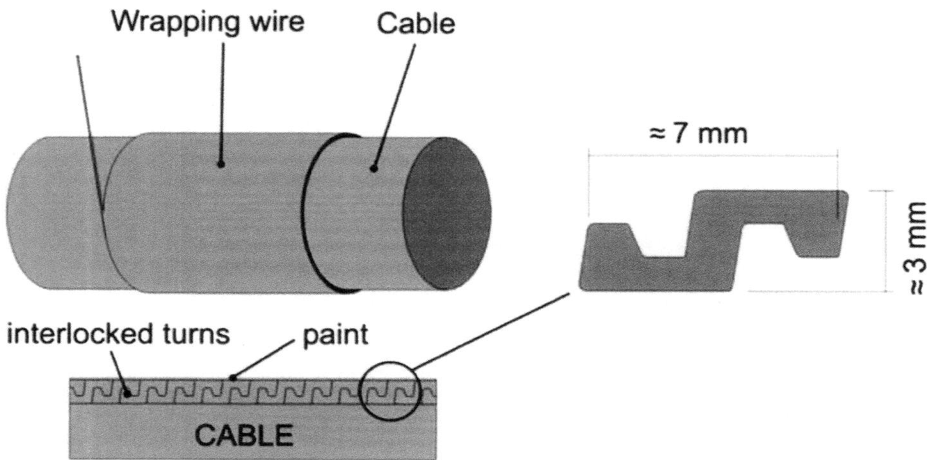

Figure A2.9 ► S-shaped wrapping wire details.

A2.7 Elastomeric wrapping

Several new wrapping materials have been used in recent years, sometimes as a replacement for wire wrapping, as in the second Chesapeake Bay Bridge in Maryland (William Preston Lane Bridge), and in other cases to add an additional level of protection for the wire wrapped cable. (Figure A2.10) These systems were developed in response to the need for the retrofit of existing bridges that show signs of corrosion of the main cable wires. Among the materials that have performed well is the neoprene wrapping system.

The neoprene wrapping system provides a watertight sheath against atmospheric attacks. The system consists of a liquid air-curing neoprene applied by brush to the surface of the wire wrapped cable or directly on the main cable wires. Then a 150 mm wide uncured neoprene strip is spirally wrapped around the main cable with a 50% overlap to create a "shingle" effect that

◄ Figure A2.10 William Preston Lane Bridge- Maryland, (USA).

Courtesy of Ammann & Whitney Consulting Engineers – New York.

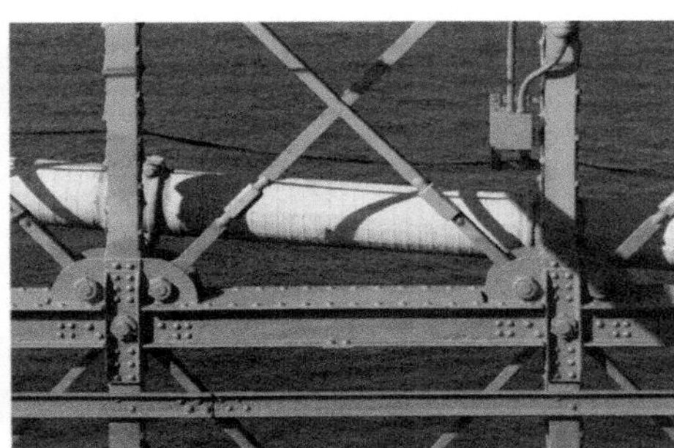

◄ Figure A2.11 Neoprene wrap on cables of Williamsburg Bridge (left) and Brooklyn Bridge (right) – New York (USA).

Courtesy of Ammann & Whitney Consulting Engineers – New York.

prevents the entry of water through the spiral joint. The cable band (or cable clamp) edges are caulked using a polyurethane sealant. An air curing Hypalon coating is then applied on the neoprene wrapped surface. Ground walnut shells or other gritty material sprinkled over the top surface of the cable, prior to the final coating, provide an anti-slip surface to allow maintenance and inspection.

The first application of this wrapping system over wire wrapping was on the cable rehabilitation of the first Chesapeake Bay Bridge in 1992, and it was used on a large scale on the rehabilitation of the Williamsburg Bridge in New York, completed in 1994. (Figure A2.11) The neoprene wrapping system was also used on the Akashi bridge, as part of a cable protection system that also includes a dehumidification system. This system is described in more detail in the next section.

The neoprene wrapping system, when properly installed over wire wrapping, has been quite effective in preventing the entry of water in the cables.

An alternative material that was recently used on the Paseo Bridge in Kansas City, Missouri is a neoprene substitute with similar properties, EPDM (Ethylene Propylene Diene Monomer), supplied by Carlisle-Syntech, Inc. This is a very widely used material in commercial roofing applications and may be less expensive than the neoprene material. Installation is virtually identical for either material.

Figure A2.12 ► Heat-curing of Cableguard wrapping – Hoga Kusten Bridge, Sweden.

Courtesy of Ove Sørensen – COWI A/S – Denmark.

The third material that has become popular when coupled with dry air injection (see Figure A2.12) is a proprietary system called "Cableguard," marketed by the D.S. Brown Company of the United States. This system uses chlorosulfonated polyethelyne, installed in a spiral wrap and heat-cured in place using specially constructed heating blankets. The heating tightens and welds the material to itself, creating a very reliable watertight covering.

A2.8 Elastic coatings

Traditionally suspension bridge cables were protected with the same paint system used for the steel structure.

More recently water-based acrylic coatings that contain highly elastic polymers and cure to a rubbery coating have been used for painting suspension

◄ Figure A2.13 Noxyde paint is used on the main cables of the Bear Mountain Bridge, Fort Montgomery, NY – USA (left) and Tagus River Bridge, Lisbon, Portugal (right).

cables. Because of their ability to sustain up to 200% elongation without cracking or peeling, they have been successfully used for the maintenance painting of wire wrapped cables on many existing suspension bridges (Bear Mountain Bridge, Mid-Hudson Bridge, Tagus River Bridge), and new bridges such as Carquinez Strait Bridge. In addition, these coatings have proved to have a long life in other applications, especially in environments where superior salt water and chemical resistance are required. One of such materials is the proprietary coating Noxyde, manufactured originally in Belgium and now licensed for manufacture in other countries. (Figure A2.13)

In Japan, a similar coating based on soft fluororesins has been used for the same purpose.

A2.9 Dry air injection – background and system description

A comprehensive study on the corrosion protection of existing suspension bridge cables was performed during the selection of the high strength wires for the Akashi Kaikyo Bridge. Starting in 1988, the interiors of the main cables of several existing bridges on the Honshu-Shikoku crossings were inspected, including the area under the cable bands, and some cables were found to have corroded surfaces after only a few years in service. This prompted the Honshu-Shikoku Bridge Authority to begin a probe of the causes and possible remedies for the unexpected deterioration. As part of the study, New York City based Ammann & Whitney Consulting Engineers were engaged by the Authority to advise on the mechanics of cable corrosion and protection.

The experience in the United States, where many older suspension bridges had previously been investigated, was found to be representative of suspension bridges elsewhere in the world. The conventional corrosion protection system was not always sufficient to prevent water intrusion. Water can enter the cable through discontinuities in the outer wrapping, or through water vapour carried by the air passing in and out of the cable as the atmospheric pressure varies, for example, through openings at the joints of the cable bands. Air entrapped in the voids between the wires inside the cables contains water that evaporates as the temperature rises, and condenses when the temperature falls. Tests performed during this study indicated that internal surfaces tend to remain wet, increasing from sides to the bottom, and decreasing toward the top. Typically, measurements of relative humidity inside cables showed that in most parts the humidity was always high, regardless of outside temperature, except at the cable bands, where the tight compaction was effective in limiting the movement of moisture and the level of humidity was found to be similar to the outside.

Further testing was done to verify the critical humidity for the corrosion of galvanized wires used on an existing bridge, where zinc corrosion, ferrous corrosion, and deteriorated paste were present with airborne salt in the injected air. This showed that when relative humidity is held below 60%, almost no corrosion occurred even without galvanization and in a salty atmosphere. Removal of salt from the air further improved the results.

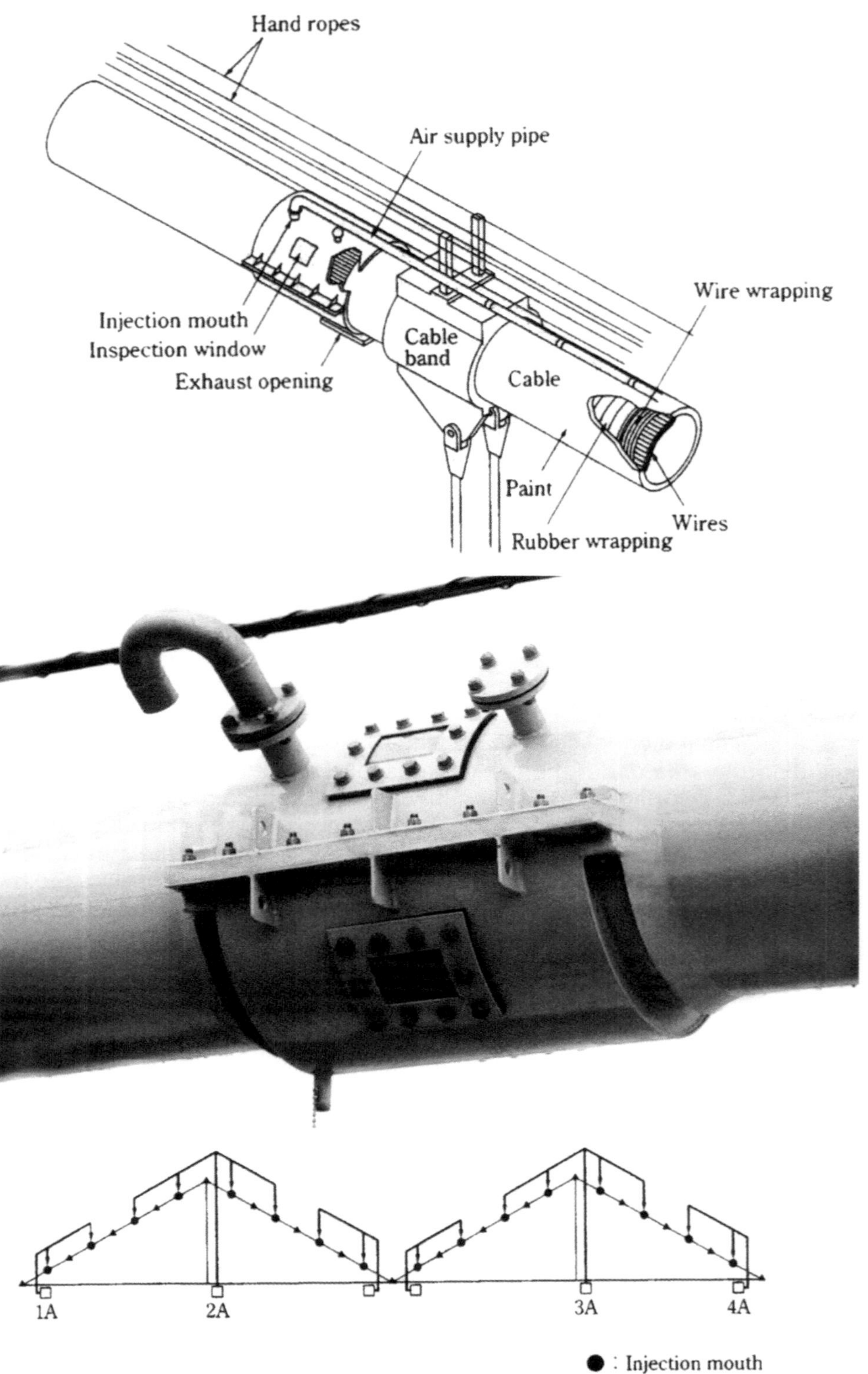

Figure A2.14 ► Details of the main cable dehumidification system on Akashi Kaikyo Bridge; schematic diagram (top), special cable clamp (centre) and location of dry air inlets and moist air outlets (bottom).

From the publication "The Akashi Kaikyo Bridge" – courtesy of Honshu-Shikoku Bridge Expressway Company Limited – Japan.

Different types of paste, including red lead paste, aluminum phosphate paste, zinc chromate paste, and polymerized organic lead paste were compared experimentally. None of them adequately sealed the main cables, especially in humid and hot weather. It should be noted that the relative humidity in some parts of Japan exceeds 80% in summer time.

Consequently, a new system for the corrosion protection of the main cables of the Akashi Kaikyo Bridge was ultimately developed, consisting of a water-tight Neoprene wrapping system, complemented by a dehumidification system. In order to ensure water- and air-tightness, a neoprene rubber sheet wrapping was applied over conventional wire wrapping, but the paste was omitted. It was recognized that the presence of existing paste on the perimeter of the cable would not affect airflow within the cable, but it was deemed unnecessary with the neoprene over-wrapping. The air-tightness at the cable bands is the most critical part of the system to maintain, and is ensured with sealants containing rubber and silicone. (Figure A2.14)

On the Akashi Kaikyo Bridge, dry air is injected from the periphery of the cables at intervals of about 140 m. The air pressure applied was determined by considering the durability of the sealing materials and the loss of air pressure at intakes and cable bands. These were determined from tests performed on model cables and on-site measurements. Since each set of bridge cables is unique in configuration, the required pressure (or more accurately, the required volume of air to maintain a pressure slightly higher than atmospheric pressure) is determined through the design process and the proper units are selected by consulting with the manufacturers.

The inside air pressure was designed to be less than 3 kPa. At each intake band, air is injected at a rate of 1.26 m^3 per minute. Fine filters remove salt particles with diameters of 0.1 to 0.2 μm from the air, prior to injecting it inside the cable. The relative humidity of the air at the outlet is set at 40%.

The operation of the dry air-injection system for the Akashi Kaikyo Bridge began on November 29, 1997 with an applied air pressure of 1 kPa for the first month, and then was increased to 2 kPa. After the first month, the relative humidity inside the cables dropped to approximately 10%. Subsequent inspections after ten years in service have confirmed that there is no evidence of corrosion in the cables. While the Akashi dry air system has been in service for just a decade, the "science" behind preventing corrosion by eliminating the moisture source is well proven.

It is not necessary to circulate large volumes of air inside the cable. If the cable is sufficiently sealed to permit positive air pressure to be maintained inside, then no outside air can enter. Only dry air supplied by the system will enter, thus preventing any corrosion. Positive air tightness over time must be maintained with proper routine maintenance. At the cable bands, damaged or deteriorated caulking will need to be repaired or replaced, just as with the existing cable protection system (although the type of material may differ).

At the tower saddles and cable anchorages where the forced air will vent a seal or enclosure is required. These are typically simple structures requiring minimal maintenance.

With the proper operation & maintenance service agreement, the chance of a malfunctioning unit should be all but eliminated. However, in such an event, if the dehumidifier stops working but the blower continues to operate, the effect of moist atmospheric air moving through the cable should not be significant, since humidity sensors and routine inspections of the physical plant should identify the situation and prompt immediate attention.

The dry-air-injection system is now in place on all of the Honshu-Shikoku suspension bridges and similar systems have been or will be installed on

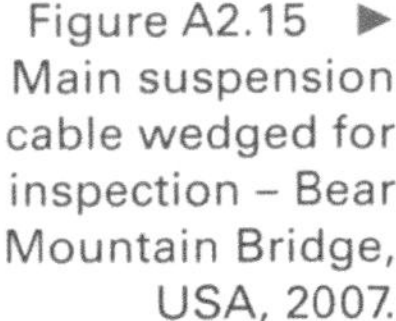
Figure A2.15 ► Main suspension cable wedged for inspection – Bear Mountain Bridge, USA, 2007.

bridges in Norway, Denmark and the United Kingdom. Several US bridge owners are considering installation of dry air injection systems in the near future. A dehumidification system has been specified for the main cables of the Messina Strait bridge.

A2.10 Inspection and evaluation of main suspension cables

There are over 100 major suspension bridges in the United States, and most are more than 40 years old. Based on the American experience, suspension bridge cables should be inspected in depth no later than 30 years after construction, and preferably sooner. This entails unwrapping portions of the cables and driving in soft wedges to spread the wires (or strands of a twisted strand cable) to allow visual inspection of the wires or strands. (Figure A2.15) If conditions warrant, wire samples may be removed for laboratory testing.

Based on the experience of bridge owners and consultants familiar with this special field of expertise, the Federal Highway Administration (FHWA) has developed guidelines for these special inspections that include some basic descriptions of how to open and wedge cables. Additional guidelines were developed with a concentration on the computational methods used to estimate remaining cable strength of a corroded cable.

The typical process of performing an in-depth cable inspection includes:

a. Visual inspection of the cables, bands, hand-strands, hangers, saddles, splay castings, cable strands and associated hardware in the anchorages.
b. Based on the visual inspection, suitable locations are selected where the cables will be unwrapped for inspection. The FHWA manual on Inspection of Fracture Critical Bridge Members recommends that four locations

be unwrapped at each cable. These usually include the low point on the main span, a low point on a side span, and part way up the main span on one cable and part way up a side span on the other cable. This should be considered an absolute minimum.

Each bridge is unique, and its particular features must be considered beyond the guidelines. For example, some bridges have long backstays, from a cable deviation saddle at deck level down to the anchorage, and these parts of the cables tend to be susceptible to damage. They are located below the roadway and often near to water. Sometimes, objects falling from the bridge hit the backstay and damage the wrapping, allowing water to penetrate. In such circumstances, it is prudent to allow for opening at least one such backstay section in addition to the general guideline locations, although access to these areas can be difficult. It is also often appropriate to remove the cover from a tower saddle to inspect the wires there if damage is suspected.

Depending on what is found, additional points may warrant inspection. Follow-up inspections are scheduled for 5, 10, 20 or 30 years depending on the bridge's age and condition. The inspected panels should be unwrapped from band to band to allow the wedges to be driven in to the centre of the cable.

The owner must weigh the importance of multiple inspection locations versus cost. The cost for a contractor's support can be significant depending on access, wrapping details, panel lengths and time spent waiting for the inspection to be performed. The local contracting environment is also an important factor. This becomes a significant issue when deciding how many locations should be opened. Some bridge owners elect to open fewer locations during the initial inspection and follow-up at five year intervals to look at additional locations. This is a valid way to proceed if the cable condition is good. A number of American suspension bridge cables have been inspected in stages like this.

c. Prepare the contract documents for bidding by qualified contractors to provide work platforms, labour, tools, equipment and materials necessary to remove the existing wrapping, assist in driving wedges for the inspection; cut and remove sample wires, splice in replacement wires, and finally re-compact and re-wrap the cables. Assuming the wrapping system has done its job well, the same material should be used to rewrap the inspected locations. Otherwise, an improved protection system might be used that should be extended throughout the length of the cables in the future.
d. Prepare the specifications for wire testing and solicit proposals from qualified testing laboratories. Even if the cable condition appears good, it is a good idea to remove some sample wires and test their chemical and physical properties. For example, assume some wires are pristine, some exhibit corrosion of the zinc, and some exhibit ferrous corrosion to varying degrees. At least one sample of each category can be cut out for testing. They can be replaced with new wires spliced in and tensioned close to the originals. This is a relatively minor expense, and after going through the trouble to open the cable, the opportunity to learn detailed information about the wire condition should not be missed.
e. Supervise the cable opening and wedging and perform the inspection of the internal wires. The conditions inside each wedge groove are noted and recorded in notes and photographs. From this data, the cable cross-section is divided into pie-slice-shaped sectors between the wedge lines and the conditions are extrapolated to create a "map" of the conditions throughout the cross-section at that panel. The data and

mapping diagrams are included in the inspection report. If the cables are in bad shape, this data is used along with the test data to compute the remaining cable strength at the sections investigated. If there are wire condition problems, the scope has to be revised to do a more extensive investigation.

It has become standard practice to classify wire corrosion grades as:

Stage 1 No Corrosion (spots of zinc oxidation).
Stage 2 White zinc corrosion product present (on entire surface).
Stage 3 Occasional spots of ferrous corrosion (up to 30% of surface).
Stage 4 Larger areas of ferrous corrosion (more than 30% of surface).

These basic definitions have been adopted by the FHWA and are illustrated in Figure A2.16. The additions in parentheses above are proposed clarifications included in the new guidelines of the National Cooperative Highway Research Program (NCHRP).

f. Supervise the testing program and prepare a report on the findings. The testing program will establish a baseline of wire properties for the various corrosion grades found. This will be available for comparison to wires removed during future inspections over the life of the bridge. If necessary, the data is also used to compute residual cable strength.
g. Prepare a written report that includes the foregoing information, an assessment of conditions and recommendations. If significant damage is found, analytical modeling and cable strength calculations may be carried out, with further investigation and/or rehabilitation and monitoring being recommended.

The recent practice in the United States has been to treat damaged cables by oiling and rewrapping them for their entire lengths. This entails major construction work and installation of work platforms below the cables to provide access. The existing wrapping is removed panel by panel, wedges are driven into the cable and oil (usually linseed oil with or without additives) is poured into the wedged grooves. The cables are then rewrapped, usually with wire and a sealing paste and sometimes with a neoprene overwrap. (Figure A2.17)

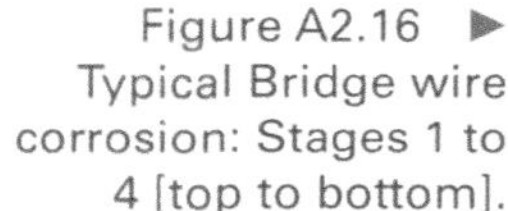
Figure A2.16 ▶ Typical Bridge wire corrosion: Stages 1 to 4 [top to bottom].

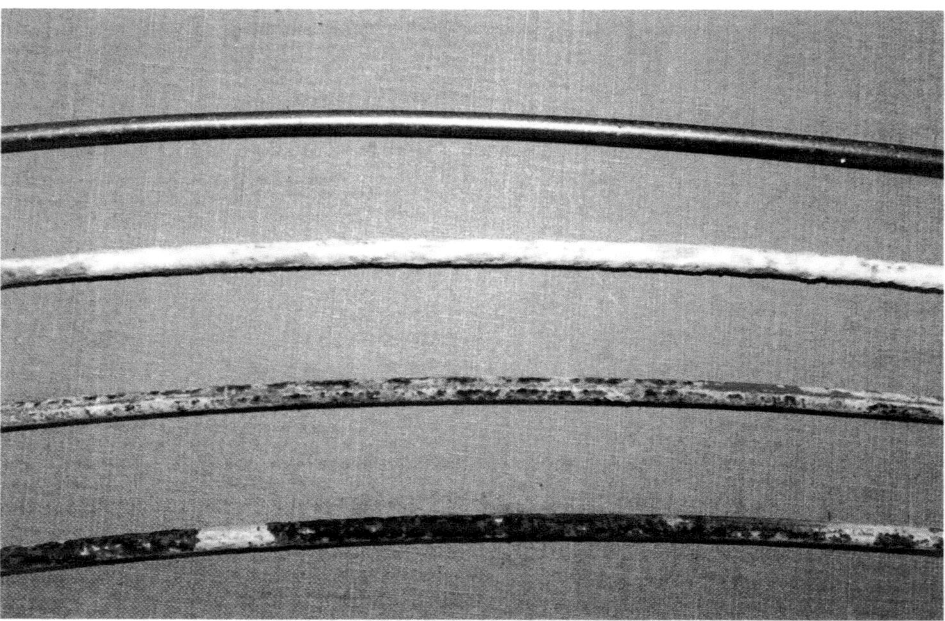

◄ Figure A2.17 Triborough Bridge, New York; main cable wedged for application of protective linseed oil.

Courtesy of Ammann & Whitney Consulting Engineers – New York.

Six major US bridges have been rehabilitated using this method. Currently, however, the alternative protection method of injecting dry-air into the sealed cable is being considered for several bridges.

A2.11 Continuous acoustic monitoring

Coupled with the rehabilitation methods which slow or stop corrosion, acoustic monitoring has been used to detect cable wires breaking due to corrosion and stress corrosion cracking on a number of existing bridges.

The principle of examining acoustic emissions to identify changes in the condition of structural elements is not new. However, until recently, continuous, unattended, remote monitoring of large structures was not practical or cost-effective. The availability of low-cost data acquisition and computing hardware, combined with powerful analytical and data management software, resulted in the development of a continuous acoustic monitoring system called SoundPrint® which has been successfully applied to un-bonded post-tensioned concrete structures in North America since 1994. The system was developed by Pure Technologies Limited of Canada.

The goal of continuous automated monitoring combined with low-cost, centralized data processing was central to the development of the technology. Original software consisted of a commercially available data acquisition package located at the site computer, and a proprietary data analysis and report generation package located at the processing facility. The data acquisition software was later replaced with more suitable proprietary software. As a result of tendon replacement work being undertaken on many of the early monitoring projects, it was possible to acquire data from many wire breaks. This information was used to train the data processing software to "recognize" wire breaks (Figure A2.18). When events possessed all the known properties of a wire break, they were classified as "probable wire breaks". Events possessing some of these properties were classified as "possible wire breaks". All other events were classified as "non-wire break events". By analyzing the time taken by the energy wave caused by the break as it travelled through the concrete to arrive at different sensors, the software was able to calculate the location of the wire break.

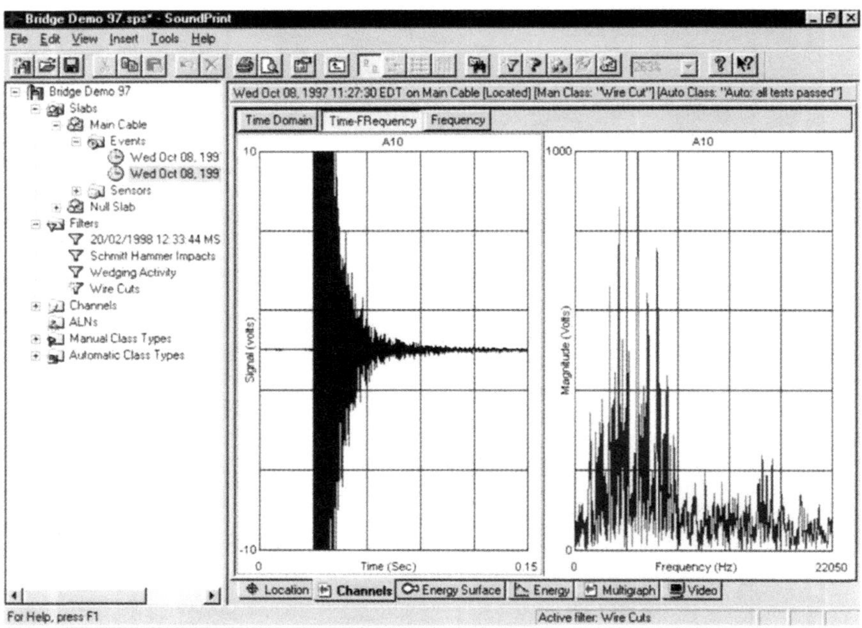

Figure A2.18 ► Time domain and frequency spectrum plots of wire breaks detected by sensor 5 m from event.

Courtesy of Pure Technologies US Inc. – USA.

Although a similar system, developed by Laboratoire Central des Ponts et Chausées, was used on the Tancarville suspension bridge in Normandy, France in the early 1990's, the absence of signal processing techniques limited its capability to differentiate between wire breaks and other acoustic events. In October 1997, the SoundPrint® system was tested on the Bronx Whitestone Bridge in New York City. This bridge, with a main span of 701 m, was opened to traffic in 1939 and is owned and operated by MTA Bridges and Tunnels, an agency of the Metropolitan Transportation Authority of New York. The monitoring system was installed during a rehabilitation of the main cables. This work involved removal of the circumferential wire wrapping, repair of broken wires and the application of corrosion-inhibiting oil to the wires. Consequently, it was possible to cut wires in the cable to test the system's recognition and location capabilities. Single sensors were attached to six cable bands, each 12.2 m apart. An array of three additional sensors was placed around two of the cable bands to evaluate radial location capabilities.

A portable acquisition system was set up at deck level and the testing was done while construction work was in progress. Six wires were cut within the monitored section in a blind test. The system correctly classified the events and located them longitudinally with errors ranging from 0.0 m to 0.7 m. Radial location using all four sensors on a cable band was accurate to within 7.5°. Acoustic events caused by steel chisels being driven between the wires were easily identified and filtered.

Analysis of the data generated during the test showed that single sensors mounted on alternate cable bands would be able to provide information of sufficient quality to permit reliable identification and location of wire breaks. Figure A2.19 shows the sensors installed on the Bronx-Whitestone Bridge in November 2000, and illustrates how the location of a break is identified by its distance from the sensors. The acquisition unit is located in an anchorage house, and data is transmitted from the sensors to the acquisition system through a coaxial trunk line attached to the existing messenger cable. Durability issues and ease of installation and maintenance were major factors in the design of the hardware. The sensor mounting brackets are designed to permit installation without modification to the cable band assembly and without damaging the paint system. (Figure A2.20)

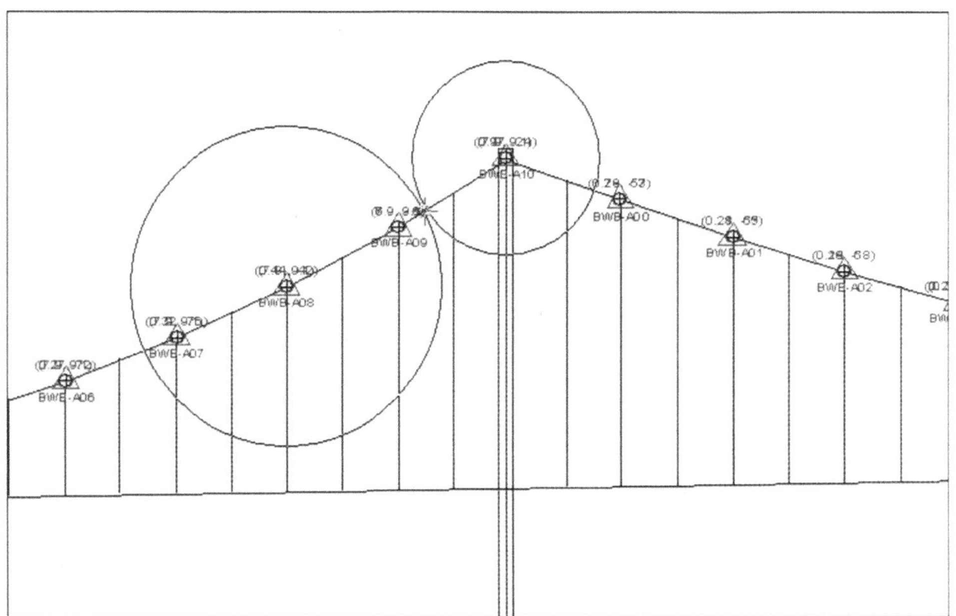

◄ Figure A2.19 Acoustic sensors located on Bronx-Whitestone Bridge (USA).

Courtesy of Pure Technologies US Inc. – USA.

◄ Figure A2.20 Acquisition system in bridge anchorage structure (left) and wired sensor (right).

The acquisition system includes a conventional telephone as well as a back-up cellular service. Data is transmitted over the Internet to a processing centre, where it is analyzed and archived. The data is also routed from this centre to a second data processing computer located at the bridge administration building. This permits bridge operation personnel to review the data and reports using the same proprietary processing software in use at the processing centre.

The installation of the coaxial trunk line along the main cables added significantly to the cost of the monitoring system. Also, the line is subject to damage by maintenance work being carried out on the cables. Hence, a wireless system was designed using analog wireless transmission from the sensors to the acquisition system. The transmitters can be powered either by a low-voltage hard-wired power supply or by solar panels mounted adjacent to the sensor. Because power is available at the sensor, it is possible to pre-amplify the sensor output, thereby increasing the range of each sensor, with a consequent reduction in sensor numbers from over a hundred, in the case of the Bronx Whitestone Bridge, to thirty, were this bridge to be monitored with a wireless system.

Figure A2.21 ► Wireless acoustic data transmitter on William Preston Lane Bridge, Maryland (USA).

Courtesy of Ammann & Whitney Consulting Engineers – New York.

Because of the sensitivity of the system, it is necessary to filter the data on site in order to reduce the amount of data being transmitted to the processing centre. The design of these filters is critical to the success of the monitoring program. If the filters are incorrectly calibrated, there is a danger that wire breaks will be missed. Conversely, if too broad a filter is applied, the amount of data being acquired and transmitted to the processing centre will be unmanageable.

Since the initial installation on the Bronx-Whitestone Bridge (Figure A2.21), the system has been installed in North America on the Bear Mountain Bridge in New York State, the Waldo Hancock Bridge in Maine, and in Europe on the Ancenis Bridge in France, the Forth Road Bridge in Scotland, and the Severn Bridge between England and Wales. A system based on similar principles is in use on the Benjamin Franklin Bridge in Philadelphia, Pennsylvania, USA.

A2.12 Hanger cables

Large amplitude oscillations of the longest hanger cables in the Akashi Kaikyo Bridge and the Storebælt East Bridge were observed from the opening of the two bridges where the longest hanger cables exceed 150 m.

The fact that the oscillations are most violent for the hangers adjacent to the towers may be partly due to the length and partly to the position in the wake of the tower legs.

In the Akashi Kaikyo Bridge dampers with high damping rubber were positioned between the two cables in each hanger pair. These dampers proved to be efficient in suppressing the vortex-induced oscillations but they broke under large amplitude oscillations due to wake-induced flutter.

In both the Akashi Kaikyo Bridge and the Storebælt East Bridge the strands forming the hanger cables had a polyethylene cover with a smooth surface. As the smoothness seemed to be a part of the problem the hanger cables of the Akashi Kaikyo Bridge were retrofitted with a 10 mm spiral rope wound around the critical hanger cables, and this measure seems to have been efficient in suppressing both the vortex-induced and the wake-induced oscillations.

At Storebælt it was initially tried to suppress the hanger oscillations by adding a horizontal stabilizing rope leading from the tower to the main cable halfway between the deck level and the tower top. The stabilizing rope proved to be efficient to suppress the in-plane oscillations but the effect was not convincing for the out-of-plane oscillations.

At present a system with a set of two hydraulic dampers positioned at a right angle 3 m above the deck is being tested and the experiences have so far been promising. (Figure A2.22) However, the most severe climatic conditions with strong winds combined with wet snow on the hangers have not yet been experienced.

The longest hanger cables of the Messina Strait Bridge will be considerably longer than those of the Akashi Kaikyo Bridge and the Storebælt

◄ Figure A2.22 Hanger damper on the longest hangers of the Storebaelt East Bridge (Denmark).

Courtesy of Storebælt A/S – Denmark.

East Bridge. Therefore, this behaviour has been thoroughly investigated theoretically and experimental test procedures are planned to determine the need to incorporate special measures for suppressing potential large amplitude oscillations in the longest hanger cables.

The approach for Messina is to avoid such hanger vibration problems and it is intended to design suitable measures from the start, such as by adding helical profiles to the smooth outer surface for example, rather than to deal with them later by a retrofit.

T - #0666 - 071024 - C334 - 297/210/15 - PB - 9780367577254 - Gloss Lamination